Comparative Biochemistry and Biophysics of Photosynthesis

COMPARATIVE BIOCHEMISTRY AND BIOPHYSICS OF PHOTOSYNTHESIS

Papers Presented at the Conference
"Comparative Biochemistry and Biophysics of Photosynthesis"
(Aug. 12th to 15th in 1967 at Hakone, Japan)
Sponsored by the Japan-U. S. Cooperative Science Program

Edited by

K. Shibata, A. Takamiya
A. T. Jagendorf and R. C. Fuller

UNIVERSITY OF TOKYO PRESS
Tokyo
UNIVERSITY PARK PRESS
State College, Pennsylvania

UTP No. 6717
Printed in JAPAN

Published jointly by
University of Tokyo Press
Tokyo
and
University Park Press
State College, Pennsylvania
Library of Congress Catalog Card Number 68-24152

INTRODUCTORY NOTE

HIROSHI TAMIYA

I have great pleasure in extending to you all a very hearty welcome to this symposium which, as one of the U.S.-Japan Cooperative Science Programs, we have been able to hold under the cosponsorship of the National Science Foundation of the U.S.A. and the Japan Society for the Promotion of Science. This is, indeed, the first photosynthesis symposium of an international character ever held in this country.

For the majority of our colleagues from abroad, I think this is their first visit to this country. On this occasion, therefore, I would like to speak briefly, by way of introduction, about the past history of photosynthesis studies and the "genealogy" of the researchers in Japan in comparison with those of the United States.

The pioneer in the study of the mechanism of photosynthesis in Japan was the late Dr. Keita Shibata (1877–1949), who was professor of plant physiology and biochemistry at the University of Tokyo until he retired in 1938. He was a student of the famous German plant physiologist W. Pfeffer and was an extraordinary scholar having a very wide scope ranging from taxonomy and morphology to physiology and biochemistry. It was he who, in 1901, erected a genus "Sasa" for the dwarf bamboos you find everywhere in Japan. And it was also he who wrote, in 1936, a now classical monograph entitled "Katalytische Wirkungen der Metallkomplexverbindungen." He also wrote several papers on protein chemistry in his later years. His first paper dealing with photosynthesis was published in NATURWISSENSCHAFTEN in 1933; it was entitled "Der Reaktionsmechanismus der Photosynthese." Although the so-called "catalase theory" he propounded in this paper has now become obsolete, it remains a monumental piece of work in having shown for the first time that hydroxylamine is a potent inhibitor of the dark reaction of photosynthesis.

With his enormously broad field of scientific activity Keita Shibata fostered a large number of disciples in different fields. Among them, to name only those who entered into the study of photosynthesis are, in order of age: A. Watanabe, myself, E. Yakushiji, K. Okunuki and H. Nakamura. Among these five disciples, Watanabe is well known for his work on the nitrogen-fixing blue-green algae. Yakushiji, who is now known as the discoverer of a chlorophyll-protein complex contained in *Chenopodium* leaves, is the man who in 1935 investigated cytochromes in green plants and isolated a *c*-type cytochrome from *Porphyra tenera*. No one anticipated at that time that cytochromes are involved in the mechanism of photosynthesis. Okunuki is also well known for his brilliant work on cytochromes. Although he himself has published no paper on photosynthesis, I mention here his name because he has trained a number of scientists who became later active photosynthesis researchers, mostly starting their work in the United States. Nakamura's name may be remembered for his classical papers on bacterial photosynthesis published in the late 1930's.

If we may call these five people the second generation in the study of photosynthesis in Japan, we now have a very active third generation. They are not necessarily students of the people of the second generation mentioned above. I call them the third generation because they were more or less influenced by the former in starting and pursuing their studies on photosynthesis. Typical of such non-student third generation people are A. Takamiya and Kazuo Shibata, organizers of this symposium on the Japanese side, besides whom we can classify as belonging to this generation—to name only those participating in this meeting—Y. Chiba, H. Huzisige, E. Hase, T. Sasa, S. Morita, T. Iwamura, A. Hattori, T. Horio, S. Miyachi and A. Mitsui.

Next comes a large body of the fourth generation which comprises most of the younger scientists you meet in this symposium. Some of these people are better classified as the 3.5th generation, because they once studied in the laboratories of the second generation, graduated from them, and then became coworkers of the third generation people. I should like to emphasize at this point that the Shibata school I mentioned above is by no means an academic faction. They, especially those belonging to the third and fourth generations, are quite multidisciplinary as you may notice from the papers they will present in this meeting.

The fifth generation of Shibata's line is not yet forthcoming, and I am very happy to tell you that we now have a number of photosynthesis researchers who do not belong to Shibata's pedigree. For instance, we have in this meeting Dr. G. Kikuchi of Tohoku University, Dr. K. Nishida of Kanazawa University, Dr. Y. Mukohata of Osaka University and Dr. S. Murakami of Tokyo University, all of whom started their researches independently of the Shibata school.

Perhaps our colleagues from abroad have been surprised to hear that photosynthesis studies in Japan started as late as in 1933. If I am correct, the first paper which appeared in the United States dealing with the mechanism of photosynthesis was that of the late Dr. W. J. V. Osterhout who published in 1918 a paper entitled "On the dynamics of photosynthesis". It was published as the very first paper in the first volume of the Journal of General Physiology. It must be said that, compared with the European countries (including Russia), the United States was a late-starter in the study of photosynthesis, but we Japanese were still far behind—as much as 15 years behind—the Americans in starting researches in the same field.

Another marked difference between the two countries is that the "genealogy" of photosynthesis researchers in the United States is, so to speak, quite "polygenetic". Shortly after Osterhout, Dr. H. A. Spoehr started to work on photosynthesis. His famous monograph on photosynthesis appeared in 1926. After these two pioneers there mushroomed numerous workers, most of whom, as you know well, developed their researches independently of the two senior scholars.

After World War II, many of our younger generation people have had the opportunity to visit your country and to study photosynthesis under scientists of different doctrines. Recent advances in photosynthesis study in Japan owe a great deal to the studies made by these younger people in the United States. In view of the high level of studies now attained by Japanese scientists, I have been harboring an ardent desire to have an international photosynthesis symposium in this country, which has finally come about thanks to the good offices of the U.S.-Japan Cooperative Science Committee.

In closing, I should like to thank, on behalf of the Japanese participants, Dr. E. L. Hess of NSF in Washington, D.C., Drs. R. R. Ries and W. H. Hodge and Mr. M. Miyahara of the NSF Tokyo Office, Dr. H. Miyayama of Ministry of Education, Japan, and also Dr. J. Ishida of the U.S.-Japan Cooperative Committee and Miss H. Nakayama of the Japan Society for the Promotion of Science for all they have done in realizing our long-cherished dream. Last but not least we wish to express our hearty thanks to Drs. R. C. Fuller and A. T. Jagendorf for their great efforts in making all the arrangements, in cooperation with Drs. A. Takamiya and Kazuo Shibata, for holding this meeting of historical significance in our study of photosynthesis.

CONTENTS

Part III. Developmental Aspects

Part IV. Regulatory Aspects

PART I

MOLECULAR AND STRUCTURAL ASPECTS

Properties of Subchromatophore Fragments from Photosynthetic Bacteria and Subchloroplast Fragments Prepared by the Action of Triton X-100*

LEO P. VERNON, AUGUSTO GARCIA, HILTON MOLLENHAUER AND BACON KE

Charles F. Kettering Research Laboratory, Yellow Springs, Ohio

INTRODUCTION

There have been many experiments over the past 25 years on the effects of detergents on chloroplast structure, photochemical properties, and solubilization of chloropast components. The latest era, however, began with Boardman and Anderson,[1,2] who showed that digitonin splits spinach chloroplasts into two fragments, each corresponding in general to one of the two photosystems which make up the plant photosynthetic apparatus. Similar fragmentation of the chloroplast is obtained with Triton X-100, another very effective nonionic detergent. Using the latter detergent, investigators at our laboratory have isolated and characterized the two particles so produced from chloroplasts,[3–5] which will be considered in more detail below.

The interaction of chromatophores from the photosynthetic bacteria with detergents has been studied by several workers, beginning with the observations by Komen[6] that sodium dodecyl sulfate produced changes in the location and intensities of the bacteriochlorophyll (Bchl) absorption bands of *Rhodospirillum rubrum* and *Chromatium* chromatophores. In 1958 Brill used the detergent Triton X-100 to convert *Rhodopseudomonas spheroides* chromatophores into two fractions, one enriched in the longwave Bchl form (B880) and the other enriched in the two shorter wavelength forms B850 and B800.[7] This investigator later studied the effects of deoxycholate upon *Chromatium* chromatophores,[8,9] and showed a bleaching of the B850 Bchl. In this case energy transfer between the Bchl forms was inhibited, as shown by a decreased fluorescence of the long wavelength Bchl. Clayton showed in 1962[10] that deoxycholate produced two fragments from *Chromatium* chromatophores, one of which was enriched with the B880 component.

* Contribution No. 296 from the Charles F. Kettering Research Laboratory. This investigation was supported in part by Research Grant GB-4797 (L.P.V.) from the National Science Foundation, and by Public Health Service Research Grant GM-12275 (B.K.) from the National Institute of General Medicine.

A systematic study of the effects of Triton X-100 upon the chromatophores of the purple photosynthetic bacteria was initiated in our laboratory in 1965. The results of studies on two bacteria, *Chromatium*[11] and *R. rubrum*,[12] have already been reported, and the data reported for those investigations may now be compared with recent results obtained with *Rhodopseudomonas palustris*[13] and *Rhodopseudomonas* species NHTC 133.[14] Chromatophores of all these bacteria are readily fragmented into two fractions which are separable by sucrose density gradient centrifugation. One of these fragments is a small particle, while the other is membranous in nature, which leads us to believe that the detergent removes a small particle from its position as a part of the membrane structure. It is the purpose of this communication to compare the properties of such particles from the four bacterial species, and further to draw attention to some basic similarities between the fragments so produced and the subchloroplast fragments produced by the same detergent when it acts on spinach chloroplasts.

Table I is a compilation of the various properties of the two fragments produced from bacterial chromatophores with this detergent. In each case two fractions are obtained, designated simply by heavy (H) and light (L) which indicates which one sediments ahead of the other during sucrose density gradient centrifugation. The small particle released from the chromatophore does not always appear in the same fraction from the different bacteria. Also, there may or may not be a separation of the Bchl forms between the two fragments. The small particle in all cases contains the reaction center Bchl, but in two cases the membranous fragment also shows reaction center Bchl. In general *Chromatium* and *Rhodopseudomonas palustris* resemble each other in their response to the detergent, while *R. rubrum* and *Rps. NHTC* are also quite similar. Let us now consider each bacterium separately.

METHODS

All cells were grown photosynthetically as previously described : *Chromatium* on malate-thiosulfate medium,[11] *R. rubrum* on malate,[12] *Rps. palustris* on malate medium[13] and *Rps. NHTC* on an enriched malate medium.[14] Chromatophores were prepared by sonication of the freshly harvested and washed cells for 3 minutes in the Raytheon 10 KC apparatus followed by treatment with 4 per cent Triton X-100 and centrifugation on discontinuous sucrose gradient consisting of three different layers of sucrose (57, 24 and 14 per cent) dissolved in 0.01 *M* Tris buffer pH 8.1. Centrifugation was for 15 hrs at $110{,}000 \times g$, following which the separated bands were removed, dialyzed overnight against Tris buffer and used for the experiments described below. Spectrophotometric measurements of the light-induced absorbance changes corresponding to reaction center Bchl photooxidation and ubiquinone photoreduction were made with the sensitive flash spectrophotometer built by one of us (BK) and used in the previous investigation.[4] Absorption spectra were obtained with a Cary 14 spectrophotometer.

RESULTS AND DISCUSSION

Bacteria showing a separation of the bacteriochlorophyll forms between the two fragments

(a) Chromatium

One salient difference between the two fragments obtained from *Chromatium* is the separation of the specific Bchl forms between the two fragments. The absorption spectra of the purified fractions have been given before,[11] and Table I gives only the ratio of absorbancies at the appropriate wavelengths. In the purple photosynthetic bacteria, Bchl *a* occurs in three forms, with absorption maxima at approximately 800, 850 and 880 mμ. Although the specific maxima will differ in different bacteria, Table I lists them as the forms corresponding to these three wavelengths, viz., B800, B850 and B880. It is apparent from the data that the heavy particle of *Chromatium* is enriched in the long wavelength B880 form, while the light membranous fragment contains the two shorter wavelength Bchl forms. Whereas ubiquinone is found in both fragments in appreciable amounts, the cytochromes, which are of the *c* type, are found only in the particulate fraction. Reduction of the particles with dithionite produces a peak at 553 mμ. There is some indication of cytochrome in the light, membranous fragment, but this could be due to some residual particles which were not removed by the Triton treatment.

Fig. 1 shows the fine structure of the two fragments, as shown by electron microscopy. The heavy fraction contains particles, which it is reasonable to assume are removed from the membrane structure by the detergent. Such particles are seen on the original chromatophore. These particles contain the photochemical apparatus of the chromatophore, as evidenced by a photooxidation of the reaction center Bchl and a coupled photoreduction of endogenous ubiquinone. Fig. 5 shows data representative of these reactions, carried out at room temperature and $-196°C$. The onset of these reactions is faster than the instrumentation used, and the rise times are less than 5×10^{-5} sec. Although the decay kinetics do not agree in their entirety, the general shapes of the curves are similar. The ratio of P890

TABLE I. Properties of the subchromatophore particles from four species of photosynthetic bacteria obtained through the action of the detergent Triton X-100.

Chromatophore fragment	Appearance	Bchl content[a]		Reaction center Bchl[b]	Endogenous ubiquinone Photo-reduction[c]	Fast ESR signal	Quinone content[d]
		A880/A800	A880/A850				
Chromatium		(0.58)	(0.60)				
Heavy	Particulate 70 Å	3.4	3.1	+	+	+	UQ
Light	Membranous	0.12	0.24	Absent	Absent	Absent	UQ
Rps. Palustris		(1.14)	(0.9)				
Heavy	Particulate, in strands 65–80 Å wide	2.4	1.3	+	+		UQ (6.7) Vit. K-like
Light	Membranous	0.63	0.56	Absent	Absent		UQ (5.4) Vit. K-like
R. rubrum		(4.7)					
Heavy	Membranous	3.5		+	+	+	UQ (2.8)
Light	Particulate 50 Å	3.6		+	+	+	UQ (3.4)
Rps. NHTC		(6.7)					
Heavy	Aggregated particles (130 Å spacing)	8.0		+	Absent		UQ (1.6) Vit. K-like
Light	Particles, some aggregated 135 Å	8.0		+	+		UQ (1.4) Vit. K-like

The procedures employed are reported in references 11 (*Chromatium*), 12 (*R. rubrum*), 13 (*Rps. palustris*) and 14 (*Rps. NHTC*), which also contain more information about the preparations listed here. The designation of heavy and light fractions refers to the relative position following a sucrose density gradient centrifugation. UQ stands for ubiquinone.

a. The absorbancies, 800, 850 and 880 $m\mu$, refer to the general locations of the three forms of Bchl. The exact locations in each case were as follows : *Chromatium,* 890, 850 and 800 $m\mu$; *Rps. palustris,* 873, 857 and 802 $m\mu$ (805 in chromatophores) ; *R. rubrum,* 877 and 800 $m\mu$; and *Rps.* NHTC, 1005 (1015 in chromatophores) and 830 $m\mu$. The values in parentheses are for chromatophores.

b. The reaction center Bchl was measured at the following wavelengths : *Chromatium,* 890 $m\mu$; *Rps. palustris,* 870 $m\mu$; *R. rubrum,* 890 $m\mu$; and *Rps. NHTC,* 940 $m\mu$.

c. Determined by absorbance change at 275 $m\mu$.

d. The numbers in parentheses are the ratios of Bchl / UQ on a molar basis. Similar data are not available for *Chromatium,* but ubiquinone is present in approximately equal amounts in the two fractions. The vit. K-like quinone migrates in the region of vitamin K during thin layer chromatography, but has not been positively identified.

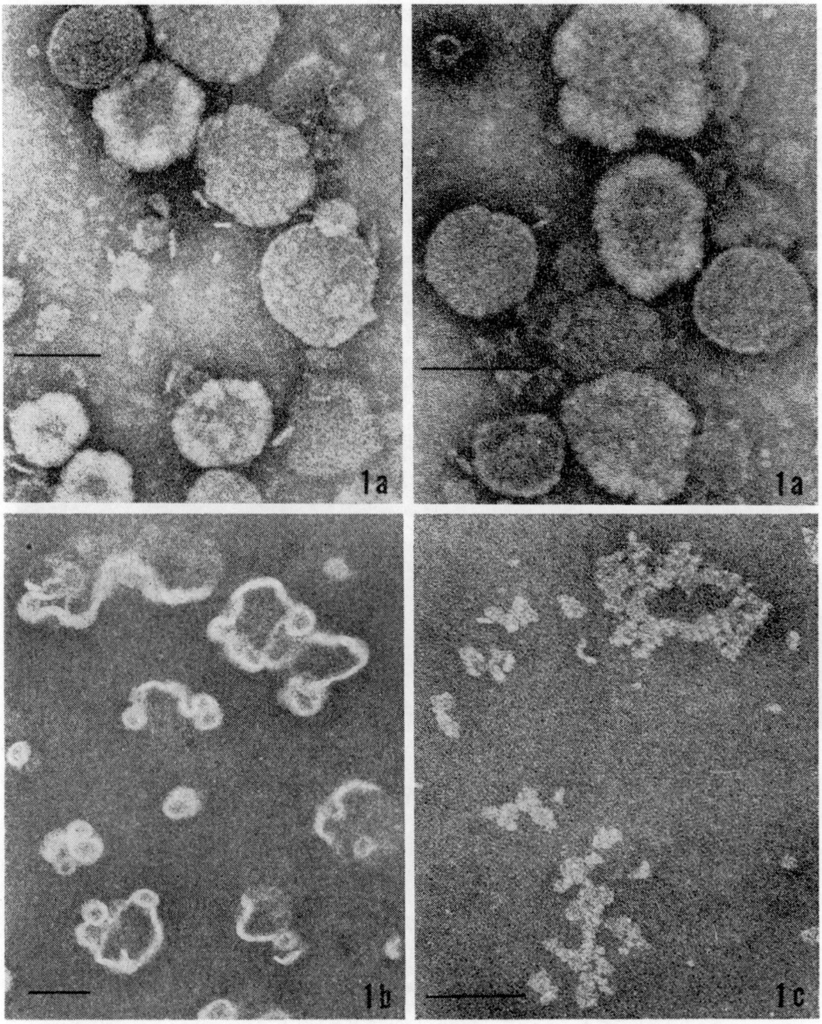

Fig. 1. Electron micrographs of *Chromatium* chromatophores and derived fragments, obtained by negative staining with phosphotungstic acid at pH 5.9 in the presence of 5×10^{-3} M $MgCl_2$. (a) Chromatophores prepared by sonication of whole cells. (b) Heavy and (c) light fragments produced by the action of Triton X-100. Subchromatophore particles which are approximately 70 Å in diameter are evident on the surface of the original chromatophore. The bar is equal to 0.1 μ.

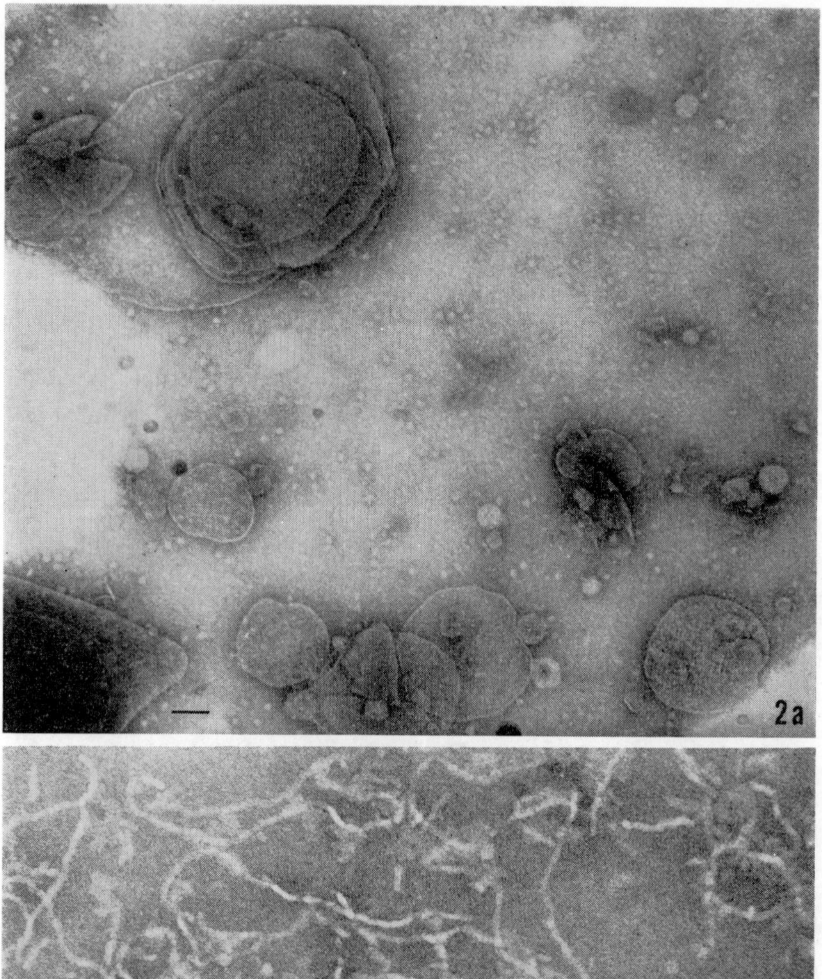

Fig. 2. Electron micrographs of *Rhodopseudomonas palustris* chromatophores and derived fragments, obtained by negative staining with phosphotungstic acid as in Fig. 1. (a) Chromatophore fraction which shows stacked thylakoids as well as some isolated discs which are more typical in appearance of chromatophores obtained by sonication. The stacked thylakoids represent the condition found in the cell. Some substructure is apparent in the chromatophore indicated by the arrow. (b) Heavy fraction showing the presence of particles which have joined together to form strands which are from 65–80 Å thick. In some cases alternate light and dark particles are seen. The bar is equal to 0.1 μ.

photooxidized to ubiquinone photoreduced is one, which is lower than the expected value of two since the reduction of ubiquinone is a two-electron change and P890 undergoes a one-electron change. This could be caused by an interaction of the semi-reduced quinone with other cellular components. Light-induced difference spectra show that the only observable changes reside in the Bchl and in the ubiquinone; the difference spectrum in the ubiquinone region shows a minimum at 275 mμ and isosbestic points at 247 and 290 mμ and closely corresponds to that of oxidized-minus-reduced ubiquinone in solution. Although the light fragment does contain ubiquinone, there is no photoreduction of this quinone upon illumination, which emphasizes the lack of significant photochemical reactions in this fragment.

(b) Rhodopseudomonas palustris

In most respects, chromatophores of this bacterium are similar to those of *Chromatium* in their response to Triton treatment, yielding a heavy particle which contains the photochemical apparatus as well as the long wavelength form of Bchl. The separation of the Bchl forms is not as complete as with *Chromatium*, however, but there is a clear separation of the primary photochemical activities, as shown in Fig. 6. In Fig. 6 are shown the absorbance changes corresponding to P870 and ubiquinone photoreactions, which are present only in the heavy particulate fraction. Again, ubiquinone is found in both of the fragments indicating that the lack of ubiquinone photoreduction by the light fragment is not due to the lack of quinone in the fragment.

One important difference between the fragments produced from *Chromatium and Rps. palustris* is the appearance of the heavy fraction. Fig. 2b shows that in contrast to the *Chromatium* heavy particle, this fraction from *Rps. palustris* contains small particles which join together to form long strands. The individual particles are visible in the strands, and in some cases there appears to be an alteration of light and dark particles. There are also other poorly defined small structures visible in these micrographs, some of which seem to have some substructure. The predominant feature found in preparations of this type, however, are the numerous strands of material composed of smaller subunits, which it is reasonable to believe are the photochemical units of this bacterium.

Fig. 2 also shows the structure of the isolated *Rps. palustris* chromatophores. Electron micrographs of the *Rps. palustris* cell have shown that the photosynthetic membrane system exists in closely associated lamallae.[15] The chromatophore preparation also shows that the chromatophores tend to associate to form stacks of individual units (thylakoids), which corres-

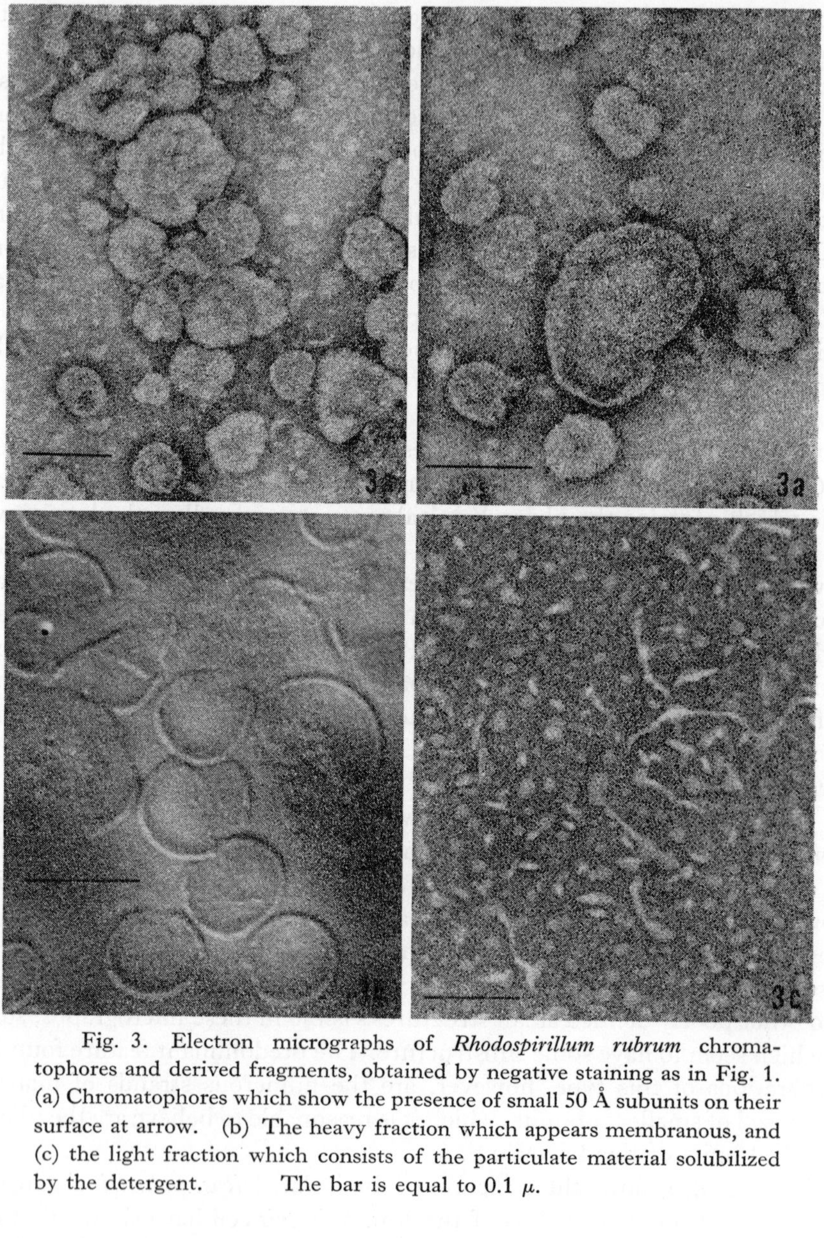

Fig. 3. Electron micrographs of *Rhodospirillum rubrum* chromatophores and derived fragments, obtained by negative staining as in Fig. 1. (a) Chromatophores which show the presence of small 50 Å subunits on their surface at arrow. (b) The heavy fraction which appears membranous, and (c) the light fraction which consists of the particulate material solubilized by the detergent. The bar is equal to 0.1 μ.

ponds to the situation *in vivo*. Occasionally it is possible to see some substructure within the chromatophore (arrow in Fig. 2a), in the form of small particles with some degree of regularity, but generally the surfaces of the chromatophores appear relatively flat. The light fraction, not shown, consists of small bodies of material with no detectable structure. Since this contains the 800 and 850 forms of Bchl, the Bchl in this fraction is not solubilized by the detergent, but is still attached to the carrier (protein) which is present in the original chromatophore. In all likelihood it is a lipoprotein which is complexed to the Bchl.

Bacteria showing no separation of bacteriochlorophyll forms

(a) R. rubrum

The details concerning the fragments prepared from this bacterium have been given previously.[12] The data given in Table I show there is no separation of the two Bchl forms between the two fragments produced with Triton X-100. As with the other bacteria, ubiquinone is found in both of the fragments. However, in this case, photochemical activity, measured as photooxidation of P890 and photoreduction of ubiquinone on the fragments, is present in both of the fragments. It is not known if this means incomplete removal of the 50 Å photoactive particle from the matrix of the membrane, or whether the photochemical unit can exist in more than one form.

In the case of this bacterium, the photochemical particle is lighter than the larger, membranous fragment.

(b) Rhodopseudomonas NHTC

Strains of this bacterium have been isolated and studied independently by Eimhjellen *et al.*,[16] who have designated it as Rhodopseudomonas sp. NHTC 133, and by Drews and Giesbrecht,[17] who have called it *Rhodopseudomonas viridis*. This bacterium has a new type of Bchl, named Bchl *b*,[18] which absorbs *in vivo* maximally at 1015 mμ, with a smaller band at 830 mμ. The fine structure of this bacterium has been studied in detail by Giesbrecht and Drews,[19] who showed that the photosynthetic apparatus is arranged in the form of stacked lamellae which, in many respects, resemble the thylakoids and the resultant grana of plant chloroplasts. Indeed, these authors call the bacterial membrane units thylakoids after the nomenclature used for plant chloroplasts. Subunits are clearly visible on the membrane surface of the individual thylakoid, and these are arranged in linear arrays to form a periodic pattern with an average spacing of 130 Å. In our experiments we have used the strain isolated by Eimhjellen, supplied by S. Holt, hence we will call the bacterium *Rps.* NHTC. It is the same or closely related to *Rps. viridis*.

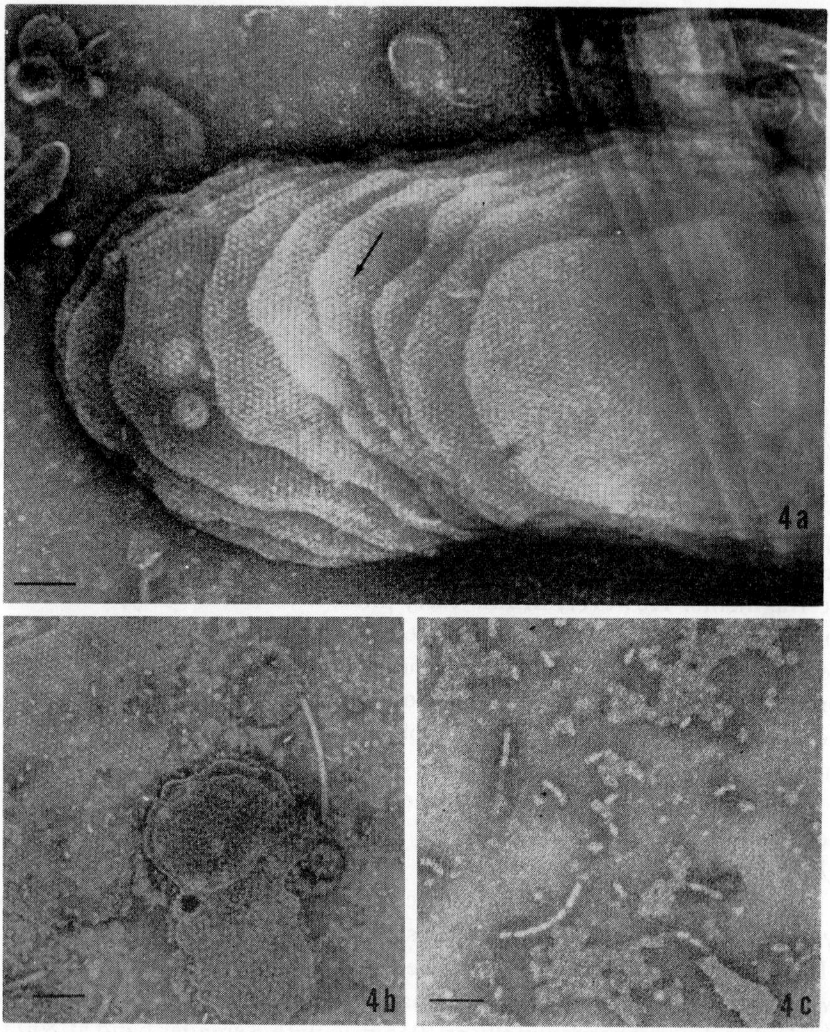

Fig. 4. Electron micrographs of *Rhodopseudomonas* NHTC, obtained by negative staining with phosphotungstic acid as in Fig. 1. (a) The chromatophore fraction, showing the presence of stacked lamellae chracteristic of this bacterium. Also visible are smaller fragments which look like the typical chromatophore obtained by sonication. Clearly visible are the ordered array of small subunits on the surface of this membrane which show a spacing of 130 Å. Also shown is one of the hexagonal arrangements (arrow) reported by Giesbrecht and Drews.[19] (b) Heavy fraction obtained by fragmentation with Triton X-100. This fraction contains some fragments which have the appearance of individual chromatophores, but in addition many planar aggregates of the subchromatophore particle are apparent, indicated by the arrow. Note the sharp boundaries of this aggregate, almost like a crystal face. (c) Light fraction, showing the presence of the more isolated subchromatophore particles. Small aggregates of the particles are visible, as well as some linear aggregates, indicated by the arrow. These particles are 135 Å in outer diameter with a 60 Å core. Note the change in shape of the particle when it exists as a spherical particle in the planar aggregates compared to its more sausage-like shape in the linear aggregate.
The bar is equal to 0.1 μ.

Fig. 4 shows some representative fractions prepared by the use of Triton X-100. The original chromatophore preparation consisted mostly of isolated chromatophores which resembled the usual chromatophore obtained from *R. rubrum*, except for the fact that the structural subunits of the membrane are plainly visible in rows with 130 Å spacing. Occasionally stacks of thylakoids were visible, as shown in Fig. 4a, which represent such structures as they exist in the original cell. On each thylakoid of the stack the subunit structure is clearly visible, as are the hexagonal arrangements of units seen by Giesbrecht and Drews.[19] The size of the individual particles making up the substructure in these thylakoids is approximately 130 Å in outer diameter with a 60 Å core. The various preparations show membrane structures of a wide range of sizes, with many discs of much smaller diameter than the thylakoids shown in Fig. 4a. Although we have not had extensive experience with whole cells, it would seem that the smaller typical chromatophores may be formed from the large thylakoid membranes through comminution via sonication. Only a few of the original thylakoid stacks come through the sonication procedure intact.

The heavy fraction from *Rps.* NHTC is shown in Fig. 4b. Two facets of the many samples taken of this fraction are shown in this micrograph. There are some fairly intact membranes left in this fraction, but none of the stacked thylakoids shown in the original chromatophore fraction. Along with the membranes which appear more or less like the usual chromatophore, there are also various sheets of material which appear to be made up of the structural subunit of the original membrane. Such sheets show a high degree of regularity, even to the extent of having sharp edges and faces reminiscent of crystals. In addition to these membranes and aggregated material, there are also a number of individual subunits. It is not possible to tell if the sheets of subunits are formed from the original membrane structure through the action of the detergent, or whether they are reformed from the basic subunit which has been released in the treatment. The rods seen in the micrograph could represent a virus.

The light fraction produced through the action of Triton X-100 is shown in Fig. 4c. By comparison with the heavy fraction, the light fraction contains a much higher proportion of the isolated and relatively free subunit. However, again there are aggregates of various size. The ones shown in Fig. 4c are quite small, but in other preparations much larger planar aggregates of the particle are observed. The plate shown in Fig. 4c was chosen to show both the isolated small particle, which is 135 Å in outer diameter and has a 60 Å core. Also shown are several small planar aggregates as well as a few linear aggregates of this particle. Again, it is not possible to determine if the aggregates are incomplete breakdown products of the original thylakoid membrane, or whether they are formed by reaggregation of the isolated particle.

In some of the preparations of the light particle from *Rps.* NHTC a unique arrangement of the derived particles is observed. Fig. 8 is a micrograph of such a preparation, in which several of the isolated particles appear to be elongated and flattened, as if they had formed miniature discs, and furthermore these particles are arranged in stacks. In this figure are shown both types of aggregates, the stacked-disc arrangement and the planar aggregation of the more spherical particle. To our knowledge, the method of preparation of this fraction was the same as with other chromatophore fragments, but this stacked disc arrangement is very prominent in this preparation. By inspection of the original chromatophore fraction it becomes apparent that such stacks of particles are not present originally, which means they must form spontaneously during or following the detergent treatment. The property of these particles which leads to this stacking phenomenon could be the one responsible for the packed lamellar structure of the membrane system of the original chromatophore.

The particles prepared from *Rps.* NHCT resemble those from *R. rubrum* in that there is no physical separation of the two Bchl forms between the two fragments; both show the same ratio of absorbancies at the two absorption peaks, 1005 and 830 mμ. Furthermore, ubiquinone is found in each of the two particles. A further similarity is seen in the presence of reaction center Bchl in both fragments, as shown by absorbance changes at 940 mμ in Fig. 7. This is not at the location of the absorption peak of the Bchl *b* form which undergoes light-induced bleaching, which is at 985 mμ. We were unable to follow the reaction at 985 mμ because of the lack of a suitable interference filter. One distinction between *R. rubrum* and *Rps.* NHTC, however, is that fragments from the former but not the latter show ubiquinone photoreduction in both the heavy and light fragment. With *Rps.* NHCT only the light particle shows ubiquinone photoreduction. This indicates that in the heavy fragments the photooxidation of the reaction center Bchl *b* must be coupled to the photoreduction of some other endogenous electron acceptor, the nature of which is unknown at present.

Comparison of the plant and bacterial membrane systems

In all of the photosynthetic bacteria examined to date, two fragments are obtained through the action of Triton X-100. One of these consists of small particles of varying sizes which exist free to a large extent, but with some aggregation. In the case of *Rps. palustris* these particles tend to form predominently linear aggregates. Some linear aggregates are also found with *Rps.* NHTC, but more often the aggregates are of a two-dimensional planar nature. In all cases, this fraction contains the reaction center Bchl, showing light-induced bleaching at appropriate wavelengths. Furthermore, all these particle preparations also show the photoreduction of endogenous ubiquinone, which appears to be coupled to the photooxidation of the reaction center Bchl. In *Chromatium*, where this reaction has been studied in more detail,[20] this is quite plainly the case. Therefore, this particle would appear to contain the photochemical unit which is designed to initiate the photosynthetic electron transfer reactions by transferring electron from the photoexcited Bchl at the reaction center to a neighboring ubiquinone molecule. Where tests have been made, the cytochromes tend to be found also in these fractions, although to date it has not been possible to show a photooxidation of the cytochromes. Perhaps the cyclic electron transfer system has been disrupted so that one sees primarily only the initial photochemical electron transfer act between the reaction center Bchl and ubiquinone.

The other fragment produced from each of the bacterial chromatophores with Triton is more membranous in nature, except in the case of *Rps.* NHTC in which case the original chromatophore membrane is made up of a network of highly ordered spherical subunits. In this case the frag-

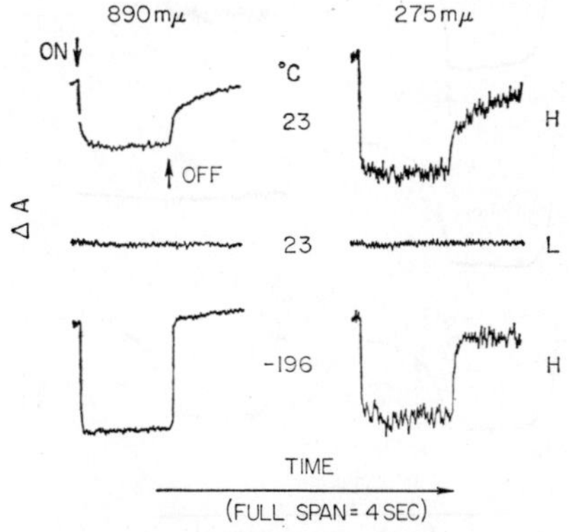

Fig. 5. Light-induced absorbance changes corresponding to photooxidation of reaction-center Bchl, P890, and reduction of ubiquinone at 275 mμ in *Chromatium* fragments. The actinic light used for measuring the 890 mμ reaction was 590 mμ with an intensity of 8×10^4 erg/cm^2 sec; that for the 275 mμ reaction was 875 mμ with an intensity of 10^5 erg/cm^2 sec. Both intensities are at the saturating level. The Bchl content of the suspensions of both fractions was 0.011 μmoles/3 ml.

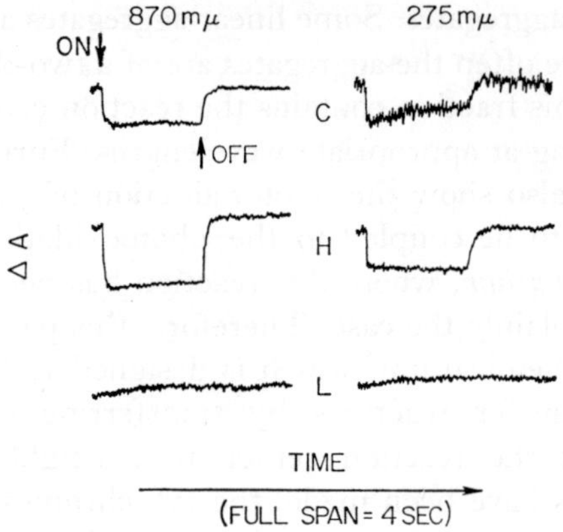

Fig. 6. Light-induced absorbance changes corresponding to photo-oxidation of reaction-center Bchl, P870, and reduction of ubiquinone in *Rps. palustris* chromatophores and derived fragments. The wavelength and intensity of the actinic light used for measuring the quinone changes were the same as in Fig. 5. A broad-band blue interference filter transmitting an intensity of $\sim 10^5$ erg/cm^2·sec was used for P870 measurements. The Bchl content was 0.013 μmole/3 ml for the chromatophore suspension.

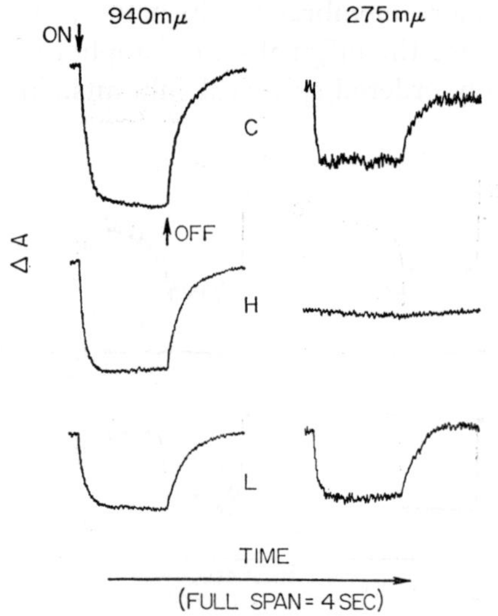

Fig. 7. Light-induced absorbance changes corresponding to the photo-oxidation of reaction-center Bchl *b* (P985) and photoreduction of ubiquinone at 275 mμ in *Rps.* NHTC chromatophores and derived fragments. The actinic light at 830 mμ (interference filter) with an intensity of $\sim 5 \times 10^4$ erg/cm^2 sec was used for measuring both the quinone changes and the changes at 940 mμ. The Bchl *b* content of the different fractions was 0.023 μmole/3 ml for the chromatophores (C), 0.014 μmole/3 ml for the heavy fraction (H) and 0.010 μmole/3 ml for the light fraction (L).

ment corresponding to the membranous fragment is composed of partially degraded membrane structures. Even so, the particulate subunits are still a part of these membrane fragments. There must be some distinction between the two fragments even with *Rps.* NHTC, however, since the two fragments can be separated by sucrose density gradient centrifugation, and they have one significant difference in terms of the photoreduction of endogenous ubiquinone. Both of the fragments produced through the action of Triton on all four bacterial species contain major quantities of ubiquinone. Even when present, it is not always photoreduced in a primary reaction, indicating that the ubiquinone photoreduced in the primary photoreaction is different in some way from the bulk of the ubiquinone. This is obvious, however, when one considers the extent of the primary photoreaction involving ubiquinone and the amount of ubiquinone present in the chromatophore and in the particles.

The bacteria examined to date fall into one of two groups. The first one, including *Chromatium* and *Rps. palustris*, have the photochemical particle in the heavy fraction and a distinct separation of the three Bchl forms between the two isolated fractions is obtained. Only the small particle carries out the primary photochemical reaction. The reverse is observed

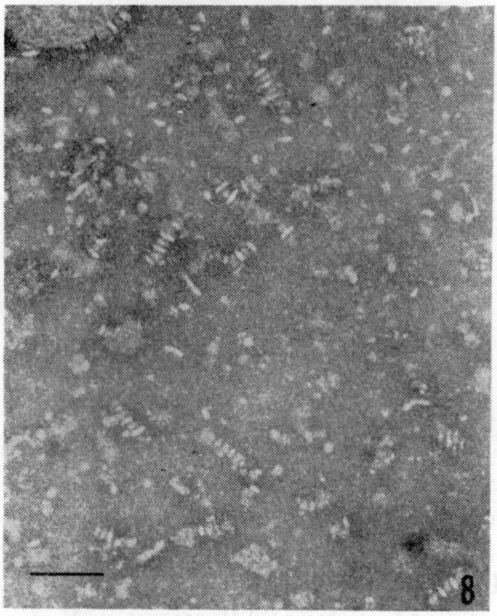

Fig. 8. Electron micrograph of the light fraction produced from *Rps.* NHTC through the action of Triton X-100. Note the occurrence of stacked particles, one of which is indicated by the arrow. Also present are small planar aggregates of the more typical spherical particles. The bar is equal to 0.1 μ.

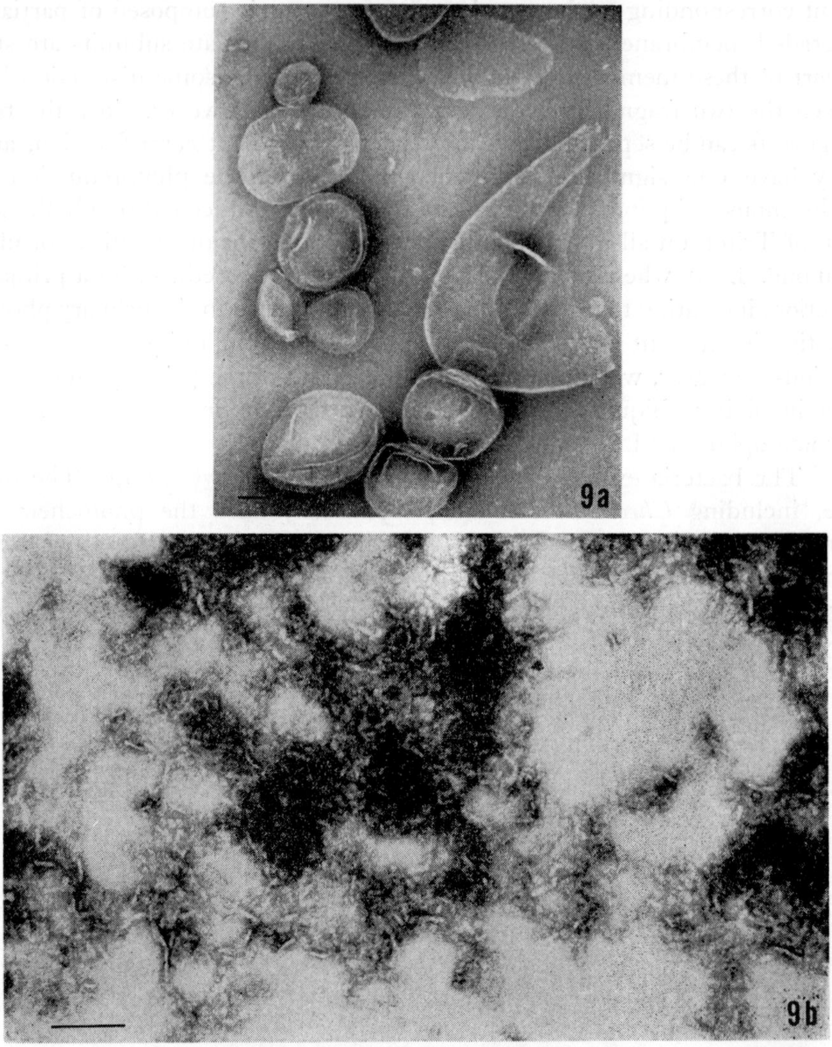

Fig. 9. Electron micrographs of fragments produced from spinach chloroplasts through the action of Triton X-100. (a) The heavy fraction which contains a low Chl *a*/Chl *b* ratio and a functional Photosystem 2, as evidenced by a DPIP Hill reation. (b) The light fraction which contains a high Chl *a*/Chl *b* ratio and a functional Photosystem 1, as evidenced by P700 photooxidation and the ability to photoreduce NADP with the addition of appropriate enzymes. Prior to the treatment with Triton X-100 the chloroplast lamellae were washed extensively and treated with EDTA according to the directions of Howell and Moudrianakis to remove the coupling factor.[21] The bar is equal to 0.1 μ.

for the other group, which includes *R. rubrum* and *Rps.* NHTC; the light fraction contains the photoactive particle, there is no separation of Bchl forms, and the primary photochemical oxidation of reaction center Bchl also takes place in the membranous fragment.

In all bacteria examined, it is shown that small, discrete particles are an integral part of the photosynthetic apparatus of the chromatophore. In the case of *Rps.* NHTC these particles are particularly obvious in orderly arrays across the face of the membrane. Obviously they form a major portion of the membrane system in this bacterium. In a general sense a similar situation is found with spinach chloroplasts which are fragmented by Triton X-100 into two fractions, one of which is a large, smooth membranous fraction which contains Photosystem 2. The other is a smaller particulate fraction which contains Photosystem 1. Fig. 9 shows electron micrographs of such fractions obtained through the action of Triton X-100. The predominant structure seen in the subchloroplast fragments which contain Photosystem 1 activity is a small strand composed of individual particles which are about 70–80 Å thick. Also seen are some planar aggregates of particles which are approximately 100 Å in diameter. Because of the selective concentration of the strands in the preparations with maximal Photosystem 1 activity, it appears these are the particles which contain the photochemical system responsible for this activity. The other aggregated material could be either another form of the same particle or some contaminating coupling factor. The preparations shown in Fig. 9 have been treated according to the directions of Howell and Moudrianakis[21] to remove both the carboxydismutase and coupling factor, which have been identified in the washings obtained by this procedure. This further substantiates our conclusion that the particles seen in Fig. 9b are concerned with Photosystem 1 activity.

Similar subchloroplast fractions can be prepared through the action of digitonin on spinach chloroplasts[1,2] and Henninger *et al.*[22] have published electron micrographs of such fractions. In their case, the fragment containing Photosystem 1 activity is of non-uniform size, but is clearly smaller than the large, smooth membranous fragment which constitutes the fraction which has Photosystem 2 activity. Whether Triton X-100 or digitonin is used to fragment the chloroplast, the small particulate fraction contains Photosystem 1, shows a photooxidation of P700 and carries out a photoreduction of NADP.[4] This is the plant counterpart of the bacterial particle which contains the reaction center Bchl. Investigation of the small particle from chloroplasts has shown that in the initial photoreaction there is a photochemical transfer of an electron from the P700 reaction center

Chl *a* to some acceptor molecule which is contained on the particle.[4] The analogy between this system and the bacterial system which carries out a photooxidation of P890 and a photoreduction of ubiquinone is obvious, and points out the similarity of the plant and bacterial systems in terms of their component parts and the primary photochemical processes carried out in these parts. In both cases, the complete photosynthetic apparatus includes distinct particles which are responsible for a primary photochemical act and are imbedded within and form part of the membrane system. Also, the membrane continuum contains other shorter wavelength chlorophyll forms, which, in the case of the bacteria, may be accessory Bchl molecules designed to harvest and transfer light energy to the reaction center. In the case of the plant chloroplast, the membrane continuum contains also the shorter wavelength chlorophyll form, but in this case, it is coupled to a separate and distinct photoact, that which we call Photosystem 2 of plants, and is responsible for oxygen evolution. One can speculate that the plant evolved a set of enzymes which carry out oxygen evolution coupled to the photochemical reactions of Photosystem 2, and this system could be derived by evolution from the membranous fragment we have observed in the photosynthetic bacteria.

Acknowledgement

The authors gratefully acknowledge the excellent assistance of Mrs. Ferne Lubbers for the growth of the bacteria used in this investigation and in the preparation of the chromatophores. They also express their thanks to Miss Elena Ngo for her help in the spectrophotometric assays involved.

REFERENCES

1. Boardman, N. K., and Anderson, J. M., *Nature,* **203**, 166 (1964).
2. Anderson, J. M., and Boardman, N. K., *Biochim. Biophys. Acta,* **112**, 403 (1966).
3. Vernon, L. P., Shaw, E. R., and Ke, B., *J. Biol. Chem.*, **241**, 4101 (1966).
4. Vernon, L. P., Ke, B., and Shaw, E. R., *Biochemistry,* **6**, 2210 (1967).
5. Ke, B., and Vernon, L. P., *Biochemistry,* **6**, 2221 (1967).
6. Komen, J. G., *Biochim. Biophys. Acta,* **22**, 9 (1956).
7. Brill, C., *Biochim. Biophys. Acta,* **29**, 458 (1958).
8. Brill, C., *Biochim. Biophys. Acta,* **39**, 296 (1960).
9. Brill, C., *Biochim. Biophys. Acta,* **66**, 50 (1963).
10. Clayton, R. K., *Photochem. Photobiol.*, **1**, 201 (1962).
11. Garcia, A., Vernon, L. P., and Mollenhauer, H., *Biochemistry,* **5**, 2399 (1966).
12. Garcia, A., Vernon, L. P., and Mollenhauer, H., *Biochemistry,* **5**, 2408 (1966).
13. Garcia, A., Vernon, L. P., Ke, B., and Mollenhauer, H., *Biochemistry,* **7**, 319 (1968).
14. Garcia, A., Vernon, L. P., Ke, B., and Mollenhauer, H., *Biochemistry,* **7**, 326 (1968).
15. Cohen-Bazire, G., and Sistrom, W. R., in *The Chlorophylls* (L. P. Vernon and

G. R. Seely, eds.), p. 313, Academic Press, New York, 1966.
16. Eimhjellen, K. E., Aasmundrud, O. and Jensen, A., *Biochem. Biophys. Res. Comm.*, **10**, 232 (1963).
17. Drews, G., and Giesbrecht, P., *Arch. Mikrobiol.*, **52**, 242 (1965).
18. Jensen, A., Aasmundrud, O., and Eimhjellen, K. E., *Biochim. Biophys. Acta*, **88**, 466 (1964).
19. Giesbrecht, P., and Drews, G., *Arch. Mikrobiol.*, **54**, 297 (1966).
20. Ke, B., Vernon, L. P., and Garcia, A., and Ngo, E., *manuscript in preparation.*
21. Howell, S. H., and Moudrianakis, E. N., *J. Mol. Biol.*, **27**, 323 (1967).
22. Henninger, M. D., Magree, L., and Crane, F. L., *Biochim. Biophys. Acta*, **131**, 110 (1967).

Distribution of Carotenoids in the Two Photochemical Systems of Higher Plants and Algae

Teruo Ogawa, Ryuzi Kanai and Kazuo Shibata

*The Tokugawa Institute for Biological Research, Mejiromachi, Tokyo,**
and *Tokyo Institute of Technology, Meguroku, Tokyo, Japan*

INTRODUCTION

Boardman and Anderson[1] have reported on the separation of photoactive pigment systems 1 and 2 by differential centrifugation of digitonin-treated spinach chloroplasts. A number of studies[2-10] have since followed dealing with physico-chemical and chemical properties of the two pigment proteins, including the distribution of pigments and other substances in these systems. The separation was greatly improved by the use of polyacrylamide electrophoresis and SDS-treated spinach chlorop lasts[11], and these were separated clearly into three colored components designated as components I, II and III according to their positions from the origin to the anode. Components I and II were pigmented proteins, and component III was a mixture of pigments released from these pigment-proteins by the action of SDS. The *a/b* ratios of the chlorophylls determined for components I and II were 7.0 and 1.9, and these pigment proteins were inferred from various properties to be those responsible for the photochemical systems 1 and 2, respectively. A study by the same electrophoretic technique was made by Thornber, Gregory, Smith and Bailey[12] with sodium dodecylbenzene sulfonate in place of SDS as the solubilizing agent. These authors obtained a similar distinct separation of components and a higher *a/b* ratio of 12 for component I (complex I according to their nomenclature) and a lower ratio of 1.2 for component II (complex II).

An interesting result obtained in the previous study[11] was that pertaining to the distribution of carotenoids in the two pigment proteins. Component II contained violaxanthin and neoxanthin with epoxide groups, whereas component I did not contain these xanthophylls. The content of lutein was higher in component II, while the content of β-carotene was

Abbreviations: SDS, sodium dodecyl sulfate; Chl, the total chlorophylls; X, xanthophyll, e.g. X-449 stands for the xanthophyll having its major absorption band at 449 mμ.

* Postal address.

higher in component I. The present paper deals with a similar line of investigation on *Anabaena variabilis*, *Porphyra yezoensis* and *Phaeodactylum tricornutum*. Green particles isolated from these organisms, when solubilized with SDS and subjected to polyacrylamide electrophoresis, were separated into three components, and two of them were pigment-containing proteins. The present paper describes the distribution of carotenoids in these pigment proteins as well as their spectral characteristics in comparison with the data obtained for spinach chloroplasts.

EXPERIMENTAL

Preparation of samples

Green particles were isolated from *Anabaena variabilis* and *Porphyra yezoensis* by the following procedure. *Anabaena* cells suspended in 0.04 *M* phosphate buffer at pH 7.5 containing 0.4 *M* sucrose were sonicated for 10 min at 20 kilocycles with a sonicator (Umeda Denki Co.). In the case of *Porphyra*, the thalli in the same medium were disintegrated by grinding with a mortar instead of by sonication. Cell debris was removed by a low speed centrifugation at $1{,}000\times g$, and the supernatant was centrifuged at $144{,}000\times g$ for 40 min. The sediment, which was green and will be referred to as "green particles," was suspended in 0.02 *M* borate buffer at pH 10.3 to be subjected to electrophoresis. The phycobilins in these organisms were dissolved out into the medium during the disintegration process, and remained in the supernatant after the high speed centrifugation. The green particles were free of phycobilins. The spectra of the sediments and the supernatants obtained by these low and high speed centrifugations showed that $90\pm5\%$ of the total pigments with the exception of phycobilins are contained in the green particles. *Phaeodactylum* cells were sonicated and green particles (chloroplast fragments) were obtained by differential centrifugation in the same manner as above. Practically no colored material was removed from the particles in this case.

A solution of SDS (Koso Kagaku Co.) in 0.02 *M* borate buffer at pH 10.3 was added to an equal volume of a suspension of green particles in the same buffer, and the mixture was shaken gently for 5 min in the dark at room temperature before being subjected to electrophoresis. Throughout the experiments in the present study, the concentration of green particles was maintained at 0.37 ± 0.02 m*M* in terms of total chlorophyll concentration, and the SDS concentration was varied in the range between SDS/Chl $=75$ and 500.

Electrophoresis

Electrophoresis of green particles solubilized with SDS was conducted by the same procedure as described previously,[11] using a synthetic polyacrylamide gel (Cyanogum 41) as the supporting medium or bed.

Spectrophotometry

The absorption spectra of the components *in situ* in the polyacrylamide gel were observed with a Shimadzu Multipurpose recording spectrophotometer model MPS-50. To see the distribution of chlorophylls, the gel after electrophoresis was scanned at 675 mμ with the same spectrophotometer, using the scanning attachment. In the case of *Anabaena*, the distribution of carotenoids was also measured at 480 mμ. Proteins in the gel were stained with 1% Amino Black 10B in methanol-water-acetic acid (4 : 5 : 1 by volume) for identification and the distribution was recorded at 600 mμ with the scanning attachment. Relative fluorescence intensities were measured with fluorescence attachment model I with a red filter which transmits the light above 600 mμ. The intensities were compared between the samples with the same height of the red band of chlorophyll *a*.

Assay of pigments

Pigments were extracted with methanol from a solution of each component prepared by Procedure B as previously described.[11] Pigments transferred into ethyl ether were analyzed by the following methods. Carotenoids in green particles of *Anabaena* and *Phaeodactylum* were separated by one dimensional thin layer chromatography with Avicel cellulose powder, and those in green particles of *Porphyra* with magnesium oxide powder. Light petroleum containing a trace of acetone was used as the developer for the separation of β-carotene, zeaxanthin, echinenone and myxoxanthophyll in *Anabaena*,[13] light petroleum containing 30% chloroform for the separation of β-carotene, diadinoxanthin+diatoxanthin, fucoxanthin and neofucoxanthin A+B in *Phaeodactylum*[14] and light petroleum containing 4% *n*-propanol for the separation of (α-+β-carotene), lutein and two other unknown xanthophylls in *Porphyra*. These xanthophylls had their major absorption maximum at 449 and 454 mμ, respectively. Light petroleum containing 2% acetone was used for further separation of α- and β-carotene in *Porphyra*.[15]

The contents of major carotenoids and chlorophylls were estimated from the absorbance values of their extracts from the spots on the chromatograms. Light petroleum, carbon disulfide or methanol was used as the solvent for extraction. The contents of minor carotenoids were estimated from the chromatographic patterns recorded at 460 mμ with the scanning attachment of the spectrophotometer. The following values of molar extinction coefficient (ε_M) calculated from $E^{1\%}_{1\,cm}$ values were used in these determinations: 1.35, 1.21 and 1.34×10^5 mole^{-1}cm^{-1} for β-carotene, echinenone and zeaxanthin in light petroleum, 1.28 and 1.23×10^5 mole^{-1} cm^{-1} for fucoxanthin and lutein in carbon disulfide, respectively[16]. The ε_M value of β-carotene was assumed for other carotenoids, of which the $E^{1\%}_{1\,cm}$ value are unknown. The ε_M value 6.58×10^4 mole^{-1}cm^{-1} was used for chlorophyll *a* in methanol.[17] The contents of chlorophylls *a* and *c* in the components of *Phaeodactylum* were determined by the method of Humphrey and Jeffrey.[18,19]

RESULTS

Electrophoretic patterns of SDS-solubilized green particles of *An-*

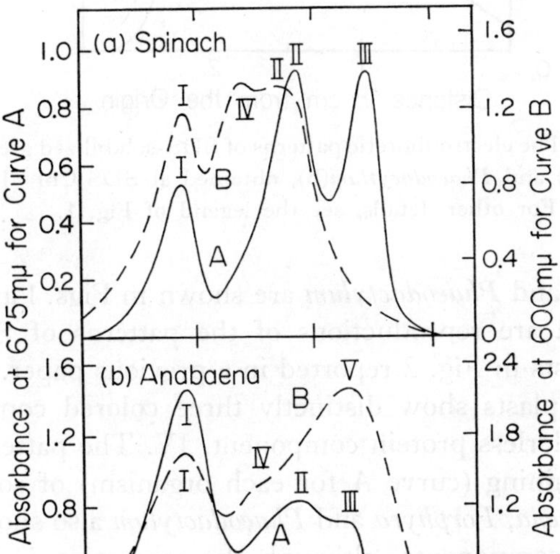

Fig. 1. The electrophoretic patterns of SDS-solubilized spinach chloroplasts(a) and green particles of *Anabaena* (b), obtained at SDS/Chl=190 and 75, respectively. Curve A was recorded at 675 mμ before staining, and curve B at 600 mμ after staining the proteins with Amino Black 10 B. The electrophoresis in polyacrylamide gel was conducted at pH 10.3 for 2 hrs.

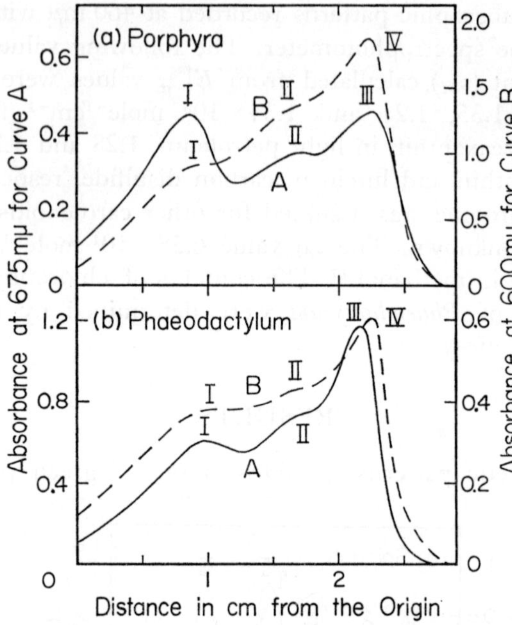

Fig. 2. The electrophoretic patterns of SDS-solubilized green particles of *Porphyra*(a) and *Phaeodacytlum*(b), obtained at SDS/Chl=100 and 75, respectively. For other details, see the legend of Fig. 1.

abaena, *Porphyra* and *Phaeodactylum* are shown in Figs. 1 and 2, where the curves in Fig. 1a are reproductions of the patterns of SDS-solubilized spinach chloroplasts in Fig. 2 reported in a previous paper.[11] The patterns of spinach chloroplasts show distinctly three colored components, I, II and III, and a colorless protein component, IV. The patterns recorded at 675 mμ before staining (curve A for each organism) of solubilized green particles of *Anabaena*, *Porphyra* and *Phaeodactylum* also show the presence of three colored components, although the separation is less marked as compared with that in the pattern obtained for spinach chloroplasts. These three components will be referred to as components I, II and III for each organism according to their positions from the origin to the anode as denoted previously. Electrophoresis in these experiments was performed for green particles solubilized at the critical concentration of SDS above which the solubilization was complete. Incomplete solubilization below the critical concentration was easily noticed from the colored materials remaining at the origin upon electrophoresis. As described previously, the critical concentration was dependent on the concentration of green particles to be solubilized, so that the molar ratio, SDS/Chl, was taken as a measure of SDS concentration relative to the material to be solubilized. The critical SDS/Chl ratio for complete solubilization of spinach chloroplasts was 125 (ref. 11) and those estimated in the present experiments for green particles of *Anabaena*, *Porphyra* and *Phaeodactylum* were 75, 100 and 75, respectively.

The patterns recorded at 600 mμ after staining are shown by curve B in each figure. The curve for *Anabaena* shows a peak at the same position as that of component I on curve A, which indicates that component I is a pigment protein. There is another peak (component V) and a shoulder (component IV) on curve B. The protein band of component II, which is considered to be a pigment protein, judging from other properties, may have been masked by this peak and shoulder. It is, therefore, not certain from this result alone, because of the overlapping of these other bands, that component II is a pigment protein and that component III is a mixture of free pigments. Curve B in Fig. 2a for *Porphyra* shows three bands at the same positions as components I, II and III, respectively, although the agree-

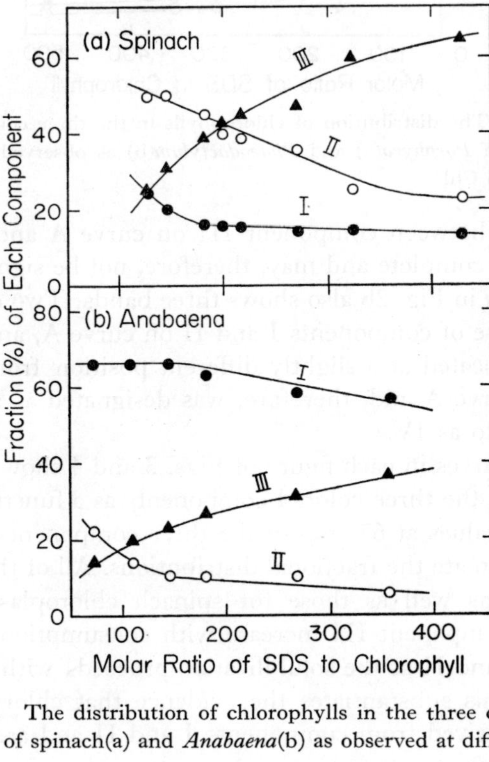

Fig. 3. The distribution of chlorophylls in the three components, I, II and III, of spinach(a) and *Anabaena*(b) as observed at different ratios of SDS/Chl.

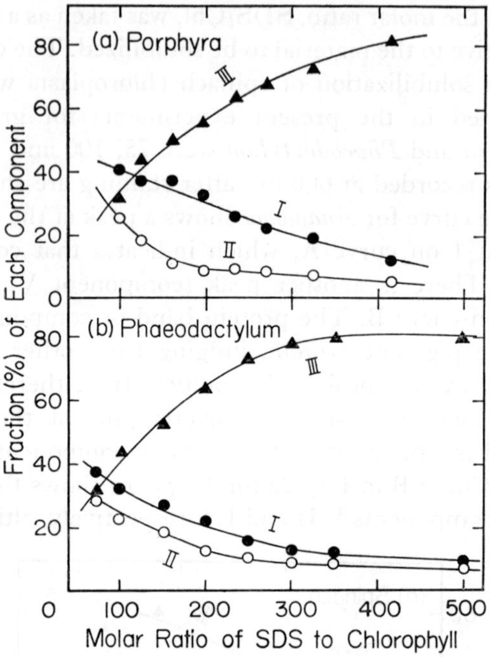

Fig. 4. The distribution of chlorophylls in the three components, I, II and III, of *Porphyra*(a) and *Phaeodactylum*(b) as observed at different ratios of SDS/Chl.

ment in position between component III on curve A and component IV on curve B is not complete and may, therefore, not be significant. Curve B for *Phaeodactylum* in Fig. 2b also shows three bands. Two of them agree in position with those of components I and II on curve A, and the remaining protein band is located at a slightly different position from that of component III on curve A and, therefore, was designated as a different component, referred to as IV.

The three curves in each figure of Figs. 3 and 4 show the distribution of chlorophylls in the three colored components as a function of SDS/Chl. The absorbance values at 675 mμ of the three components in the gel were normalized to estimate the fractional distributions. All of the curves for the three organisms as well as those for spinach chloroplasts indicate that chlorophylls in component III increase with consumption of chlorophylls in components I and II as the solubilization proceeds with increasing SDS concentration. This substantiates the evidence that chlorophylls in component III are derived from components I and II and are free from protein.

This interpretation is supported by a much higher fluorescence intensity of component III of each organism as compared with those of components I and II of the same organism. As summarized in Table I, the fluorescence from component III was 3 to 20 times higher than that from the native green particles before solubilization. The fluorescence from component II was 1.5 to 6.4 times higher, and the fluorescence from component I was nearly equal to or at most twice as much as that from the native chloroplasts or green particles.

TABLE I. Relative fluorescence intensities were compared for the samples having the same height of the red absorption band, and the intensities of the three components of each organism relative to the intensity of its native green particles are listed in the table.

Organism	I	II	III
Spinacia oleracea	1.3	6.4	20.3
Anabaena variabilis	0.9	7.5	16.3
Porphyra yezoensis	1.9	2.6	6.2
Phaeodactylum tricornutum	1.2	1.5	3.0

The position of the red band of chlorophyll *a* is also different for the three components. The maxima for component I of the different organisms were 675 mμ except for a slightly shorter wavelength, 673 mμ, obtained for the component of *Phaeodactylum*, and the maximum wavelengths for component II were shorter, ranging from 670 to 672 mμ (Table II). This may indicate a difference in state of chlorophyll *a* in these pigment proteins. The higher content of P700 in component I (ref. 4, 7) and the higher content of chlorophyll *b* in component II (ref. 1, 11, 12) may also be responsible for the difference in these maxima; overlapping of the P700 band will shift the red band of chlorophyll *a* toward longer wavelengths, and overlapping of the chlorophyll *b* band toward shorter wavelengths. The maxima for component III of different origins were 670±1 mμ, being the shortest in wavelength of the maxima shown by the three components.

TABLE II. Red absorption maxima in mμ of the three components.

Organism	I	II	III
Spinacia oleracea	675	672	671
Anabaena variabilis	675	670	669
Porphyra yezoensis	675	671	670
Phaeodactylum tricornutum	673	672	671

Pigment compositions and contents in the three components of these organisms, which were determined at the critical SDS/Chl ratio for each organism, are listed in Table III, where the relative content of chlorophyll *a* in each component is taken as 100, and carotenoids are listed in decreasing order of the R_f value in thin layer chromatography. The number of hydroxyl, keto and epoxide groups per molecule of carotenoids are also listed for reference, except in the case of *Phaeodactylum*. In the case of *Anabaena*, the content of β-carotene is higher in component I while the xanthophyll content is higher in component II. In particular, zeaxanthin with two hydroxyl groups and myxoxanthophyll with an unknown structure are con-

TABLE III. Distribution of carotenoids and chlorophylls in the three components; the relative content of each component in each organism was expressed taking the content of chlorophyll *a* as 100.

Organism and Pigment	(-OH)	(-CO-)	(-O-)	I	II	III
Spinacia oleracea (SDS/Chl=190)						
Chlorophyll *b*	—	—	—	14	53	29
(*a*/*b* Ratio	—	—	—	7.0	1.9	3.5)
β-Carotene	0	0	0	16	8	22
Lutein	2	0	0	6	25	14
Violaxanthin	2	0	2	0	13	12
Neoxanthin	3	0	1	0	8	6
Anabaena variabilis (SDS/Chl=75)						
β-Carotene	0	0	0	13	9	8
Echinenone	0	1	0	2	7	8
Zeaxanthin	2	0	0	0	2	2
Myxoxanthophyll	—	—	—	0	27	24
Porphyra yezoensis (SDS/Chl=100)						
α-Carotene	0	0	0	3	3	2
β-Carotene	0	0	0	12	13	12
X-449	—	—	—	2	2	2
Lutein	2	0	0	12	24	23
X-454	—	—	—	3	7	6
Phaeodactylum tricornutum (SDS/Chl=75)						
Chlorophyll *c*	—	—	—	20	25	25
(*a*/*c* Ratio	—	—	—	5.0	4.0	4.0)
β-Carotene	0	0	0	13	8	14
Diato-X+Diadino-X*	—	—	—	16	24	16
Fucoxanthin (C=C=C)	—	—	—	38	48	96
Neofucoxanthin A+B	—	—	—	6	8	13

* Diato-X and Diadino-X stand for diatoxanthin and diadinoxanthin, respectively. Diadinoxanthin has an epoxide group.

tained only in component II, as in the case of violaxanthin and neoxanthin in spinach chloroplasts. Myxoxanthophyll moved most slowly on the chromatography, which suggests that this xanthophyll molecule contains a greater number of oxygen-containing side chains than the other two xanthophylls.

This uneven distribution of carotenoids is certainly interesting, but it raises the question as to whether this is a true reflection of the distribution in native chloroplasts or is merely due to the effect of SDS which may have dissolved the carotenoids unevenly out of the two pigment proteins. To answer this question the absorbance value at 480 mμ of each component, which may be regarded as a measure of the total carotenoid content, was measured as a function of the concentration of SDS relative to chlorophyll. The result shown in Fig. 5 is similar to the result

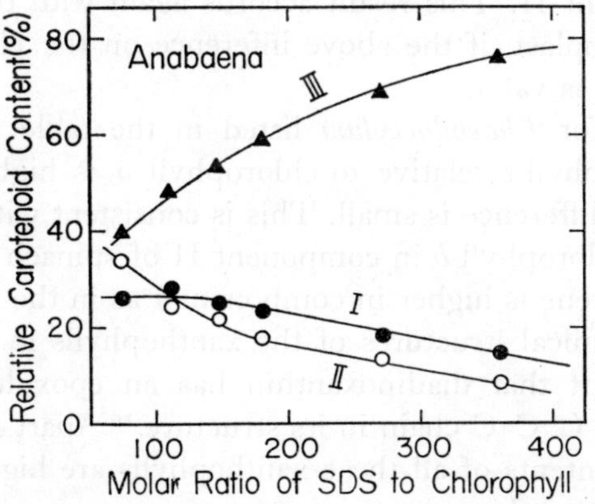

Fig. 5. The distribution of carotenoids in the three components of *Anabaena* as observed at different ratios of SDS/Chl; relative carotenoid contents of the three components, I, II and III, were estimated from the absorbance value at 480 mμ. They were normalized and plotted against SDS/Chl.

obtained at 675 mμ for chlorophylls (Fig. 3) in that the pigment content in component III increases and the contents in components I and II decrease as solubilization proceeds at higher SDS/Chl ratios. However, it must be noted concerning the distribution of carotenoids that the carotenoid content in component I was nearly constant and the content in component II was changing at lower SDS concentrations below SDS/Chl=150. This implies that the carotenoids in component III at the lower SDS concentrations were derived from component II and not from component I. In fact, the relative contents of the four carotenoids in component II and those in component III, both measured at SDS/Chl=75, are very similar (Table III). It may, therefore, be concluded that the uneven distribution of carotenoids in the two pigment proteins determined at the low SDS concentration is not due to an artifact caused by the action of SDS.

Two unknown xanthophylls were found in the chromatograms of the components of *Porphyra*. These xanthophylls showed a major absorption maximum at 449 and 454 mμ, respectively, so that they were designated as X-449 and X-454. X-449 moved on chromatography between carotenes and lutein, and X-454 moved most slowly. This strongly suggests that X-449 is more hydrophobic than lutein and that X-454 is most hydrophilic, probably due to a high content of oxygen-containing side chains in the molecule. The two isomeric carotenes and X-449 are almost evenly distributed in components I and II, while lutein and X-454 are distributed more in component II. This result accords again with the result obtained for spinach chloroplast, if the above inference on the chemical nature of X-449 and X-454 is valid.

The result for *Phaeodactylum* listed in the table indicates that the content of chlorophyll *c* relative to chlorophyll *a* is higher in component II, although the difference is small. This is consistent with the higher relative content of chlorophyll *b* in component II of spinach chloroplasts. The content of β-carotene is higher in component I as in the cases of *Anabaena* and spinach. Chemical structures of the xanthophylls in this organism are not known, except that diadinoxanthin has an epoxide group and that fucoxanthin has a C=C=C chain in its structure.[16] Apart from the chemical structures, the contents of all these xanthophylls are higher in component II than in component I.

DISCUSSION

Green particles obtained from the three different organisms examined in the present study, when solubilized with SDS and subjected to electrophoresis, were all separated into three colored components as has been found previously[11] for spinach chloroplasts. The three components from different organisms were numbered in the same manner according to their positions in electrophoresis from the origin to the anode, and corresponding components with an identical number showed similar spectroscopic properties and pigment compositions. Component I showed a red band near 675 mμ, and the fluorescence from this component was nearly equal in intensity to that from native chloroplasts or green particles, while the red band of component II was at a shorter wavelength of 671$\pm$1 mμ and the fluorescence was stronger. The red maximum of component III was located at an even shorter wavelength of 670±1 mμ, and the fluorescence was much stronger. As the solubilization proceeds at higher SDS concentrations, component III as estimated from the absorbance value at 675 mμ increased and components I and II decreased. It is certain from these and other data presented in this and the previous papers, that components I and II are pigment proteins and component III is a mixture of free pigments released from these pigment proteins. The similarity of components I and II to the respective component in spinach chloroplasts strongly suggests that components I and II in these organisms are those responsible for photochemical reactions 1 and 2, respectively.

The pigment compositions of these two components are most interesting in connection with this deduction. Most xanthophylls, especially those with a greater number of oxygen atoms in the side chains, are distributed more in component II and, in the case of *Anabaena*, zeaxanthin and myxoxanthophyll are contained only in component II, as is the case of violaxanthin and neoxanthin with epoxide groups localized in component II of spinach chloroplasts. This general trend of distribution found for the four species of organisms may be closely related to the fact that the photochemical reaction of system 2 is coupled with the oxygen-evolution system.

The possibility that epoxide carotenoids might be involved in photosynthetic oxygen evolution was first proposed by Dorough and Calvin.[20] Several years later, Sapozhnikov *et al.*[21,22] presented a result of a light-induced conversion of violaxanthin to lutein and the reverse reaction in the dark, and Blass, Anderson and Calvin[23] found a decrease in the light of the violaxanthin content of both *Chlorella* and *Scenedesmus*. Experiments carried out recently with oxygen isotope by Saakov[24] suggested that exchange may occur between the oxygen from water and the epoxide groups of violaxanthin and neoxanthin during photosynthesis, and Yamamoto, Nakayama and Chichester,[25] using spinach and bean leaves as the samples, demonstrated the transformation by light under nitrogen pressure from violaxanthin to zeaxanthin through antheraxanthin and the reverse transformation in the dark with oxygen. The same transformation was found by Krinsky[26] for *Euglena* cells. Yamamoto *et al.*[27] also found that O_2 is utilized for the hydroxyl groups of xanthophylls while the oxygen of H_2O is utilized for the epoxide groups. These previous observations more or less suggest the participation of xanthophylls in the oxygen evolution in photosynthesis, although the mechanism of their participation is not clear.

Most recently, Lundegårdh[28,29] by measuring the effect of blue and red light on the ratio of carotene to xanthophylls in spinach chloroplasts proposed a mechanism involving a reduction of NADP in the light by carotene through ferredoxin, and demonstrated the reducing power of illuminated carotene *in vitro*. The fact found in the present study that carotenes are distributed more in component I is a strong support for this mechanism. Regardless of the mechanism, the fact that the more reduced forms of carotenoids such as carotenes are distributed more in component I and that the more oxidized forms such as violaxanthin and neoxanthin are localized in component II may well be accounted for by the fact that system 1 is on the more reduced side of the electron transport system and system 2 on the more oxidized side. Elucidation of the mechanism of participation of various carotenoids in each photochemical system is awaited.

Acknowledgement

The authors wish to express their heartfelt gratitude to Dr. T. Oohusa at Yamamoto Nori Research Institute for his ample supply of *Porphyra* thalli and to Dr. Y. Morimura and Mr. T. Nishimura for their supply of *Phaeodactylum* cells.

SUMMARY

Green particles isolated from *Anabaena variabilis*, *Porphyra yezoensis*, and *Phaeodactylum tricornutum*, when solubilized with sodium dodecyl sulfate and subjected to polyacrylamide gel electrophoresis, were all separated into three colored components as observed previously for spinach chloroplasts. Components I and II, numbered according to their position in electrophoresis from the origin to the anode, were pigment proteins, which are inferred to be responsible for the photochemical reactions of system 1 and 2, respectively, and component III was a mixture of free pigments released by SDS from these pigment proteins. In all the three organisms investigated component I showed a red absorption maximum near 675 mμ, while component II showed the red band at 671 mμ. The fluorescence from component I was nearly equal in intensity to that from native chloroplasts or green particles, while that from component II was 1.5 to 6.4 times higher. The more reduced forms of carotenoids such as carotenes were distributed more in component I and the more oxidized forms with a greater number of oxygen atoms in the side chains were distributed in higher ratios or exclusively in component II. This general trend of carotenoid distribution was discussed in connection with the two photochemical reactions in the electron transport system.

REFERENCES

1. Boardman, N. K., and Anderson, J. M., *Nature*, **203**, 166 (1964).
2. Anderson, J. M., Boardman, N. K., and David, D. J., *Biochem. Biophys. Res. Comm.*, **17**, 685 (1964).
3. Anderson, J. M., and Boardman, N. K., *Biochim. Biophys. Acta*, **112**, 403 (1966).
4. Anderson, J. M., Fork, D. C., and Amesz, J., *Biochem. Biophys. Res. Comm.*, **23**, 874 (1966).
5. Vernon, L. P., Shaw, E. R., and Ke, B., *J. Biol. Chem.*, **241**, 4101 (1966).
6. Wessels, J. S. C., *Biochim. Biophys. Acta*, **126**, 581 (1966).
7. Kok, B., and Rurainski, H. J., *Biochim. Biophys. Acta*, 126, 584 (1966).
8. Cederstrand, C. N., and Govindjee, *Biochim. Biophys. Acta*, **120**, 177 (1966).
9. Sironval, C., Clyster, H., Michel, J. M., Bronchart, R., and Michel-Wolwertz, R. M., in *Currents in Photosynthesis* (J. B. Thomas and J. H. C. Goedheer, eds.), p. 111, Donker, Amsterdam, 1965.
10. Wessels, J. S. C., in *Currents in Photosynthesis* (J. S. Thomas and J. H. C. Goedheer, eds.), p. 129, Donker, Amsterdam, 1965.
11. Ogawa, T., Obata, F., and Shibata, K., *Biochim. Biophys. Acta*, **112**, 223 (1966).
12. Thornber, J. P., Gregory, R. P. F., Smith, C. A., and Bailey, J. L., *Biochemistry*, **6**, 391 (1967).
13. Goodwin, J. W., *J. Gen. Microbiol.*, **17**, 467 (1957).
14. Jeffrey, S. W., *Biochem. J.*, **80**, 336 (1961).
15. Strain, H. H., *J. Am. Chem. Soc.*, **70**, 588 (1948).
16. Davis, B. H., in *Chemistry and Biochemistry of Plant Pigments* (T. W. Goodwin, ed.), p. 530, Academic Press, London, 1965.
17. Ogawa, T., and Shibata, K., *Photochem. Photobiol.*, **4**, 193 (1965).
18. Humphrey, G. F., and Jeffrey, S. W., in preparation.
19. Jeffrey, S. W., and Allen, M. B., *J. Gen. Microbiol.*, **36**, 277 (1964).
20. Dorough, G. D., and Calvin, M., *J. Am. Chem. Soc.*, **73**, 2362 (1962).
21. Sapozhnikov, D. I., Krasovskaya, T. A., and Maevskaya, A. N., *Dokl. Akad. Nauk, SSSR*, **113**, 465 (1957).
22. Sapozhnikov, D. I., Maevskaya, A. N., Kravoskaya-Antropola, T. A., Prialgauskaita, L. L., and Turchina, V. S., *Biokhimia*, **24**, 34 (1959).
23. Blass, U., Anderson, J. M., and Calvin, M., *Plant Physiol.*, **34**, 329 (1959).
24. Saakov, V. A., *Fiziol. Rastenii*, **12**, 377 (1965).
25. Yamamoto, H. Y., Nakayama, T. O. M., and Chichester, C. O., *Arch. Biochem. Biophys.*, **97**, 168 (1962).
26. Krinsky, N. I., *Biochim. Biophys. Acta*, **88**, 487 (1964).
27. Yamamoto, H. Y., Chichester, C. O., and Nakayama, T. O. M., *Arch. Biochem. Biophys.*, **96**, 645 (1962).
28. Lundegårdh, H., *Proc. Natl. Acad. Sci.*, **55**, 1062 (1966).
29. Lundegårdh, H., *Physiol. Plantarum*, **19**, 754 (1966).

The Nature of β-Carotene Association in Chloroplast Lamellae*

Tae H. Ji, John L. Hess[†] and A. A. Benson

Scripps Institution of Oceanography, University of California, San Diego, La Jolla, California

INTRODUCTION

The nature of the chloroplast lamellar membrane has been extensively investigated. As a predominant intracellular membrane of plant tissue, its molecular organization is directly related to physiological functions of the chloroplast. Several models of this lamellar membrane structure have been proposed, based upon electron microscopic studies, X-ray scattering analyses, and chemical composition.

Although pigment-protein complexes have been reported,[1–11] only recently have several laboratories successfully isolated pigment-protein complexes from chloroplasts which function in particular photosynthetic reactions.[1, 3, 9, 10] The ability to isolate physiologically active complexes with rather defined pigment compositions implicates specific association between pigments and lamellar protein. This concept is also supported by electrophoresis[7, 8] and optical rotatory dispersion[12, 13] studies which demonstrate strong interaction between pigment and lipid molecules and/or between these molecules and the lipoprotein molecules.

A β-carotene-protein complex has been isolated by several workers.[‡4, 6] This complex absorbs light maximally at 538 nm, 498 nm, and 460 nm. Platt[14] considered that polarization effects altered $\pi \rightarrow \pi^*$ transitions of the β-carotene, and such a β-carotene-protein complex may lead to a red shift in light absorption properties of the pigment. This polarization could result from the close proximity of β-carotene to charged moieties within the complex, these being most probable in the isolated β-carotene-protein complex.

Carotenoid-protein complexes are found in bacteria[15] and marine

---

* We acknowledge the support of this work by Grant GM-12310, U.S. Public Health Service.

† Present address: Department of Biochemistry and Nutrition, Virginia Polytechnic Institute, Blacksburg, Virginia.

‡ Personal communication, N. I. Bishop, Department of Botany, Oregon State University, Corvallis, Oregons.

organisms.[16,17] Using the astaxanthin-protein complexes, crustacyanin and ovorubin, Cheesman and coworkers demonstrated that carotenoid absorption properties are greatly altered by the protein associated with it.[16,17] The data of Jencks and Buten[18] clearly demonstrate the influence of protein conformation of the carotenoid spectral properties in crustacyanin. Such studies support the concept of specific protein-carotenoid association.

METHODS

Lamellae Preparation

Chloroplasts were prepared from commercial spinach obtained from a local market. The de-ribbed leaf tissue was macerated in a Waring Blendor in a volume of buffer, 0.8 *M* sucrose, 0.01 *M* NaCl, and 0.02 *M* Tricine at pH 7.9, equal to the weight of tissue. After 15 sec at maximum grinding speed, the homogenate was filtered twice through four layers of cheese cloth. The chloroplasts, Fig. 1, which sedimented between $250 \times g$ and $650 \times g$ for 10 minutes were resuspended in 0.01 *M* NaCl, 0.02 *M* Tricine at pH 7.9 and sonicated at 20 KHz for 60 sec. After removal of whole chloroplasts, the lamellae were centrifuged for 40 minutes at $35{,}000 \times g$. The chloroplast fragments were then treated as outlined in Fig. 2 in order to obtain the red β-carotene-protein complex. Bean leaves from 10 to 14-day-old seedlings also serve as a source for the carotene-protein complex, but 0.5 *M* cysteine or other sulfhydryl reagent was required throughout the preparation. Alternatively, good yields of the complex could be obtained if the chloroplasts were fragmented in a tight fitting Teflon-glass homogenizer.

Spectrophotometry

For spectral measurements, lamellar particles, approximately 0.5 mg dry wt/ml buffer, were suspended by sonication for 60 sec or less at 5°–10° in 0.01 *M* NaCl, 0.02 *M* Tricine at pH 7.9. Particles that were sequentially extracted with different solvents were always lyophilized prior to the addition of solvent. Many of these protein suspensions had scattering properties which interfered with spectral measurements. Hence, we measured all spectra with the Shimadzu Multipurpose Recording Spectrophotometer Model MPS-50, the optics of which minimize scattering effects. All spectra were measured at room temperature using a tungsten lamp source between 340 nm and 800 nm.

Analysis of Pigments and Lipids

Extracted pigments were assayed spectrophotometrically using the ab-

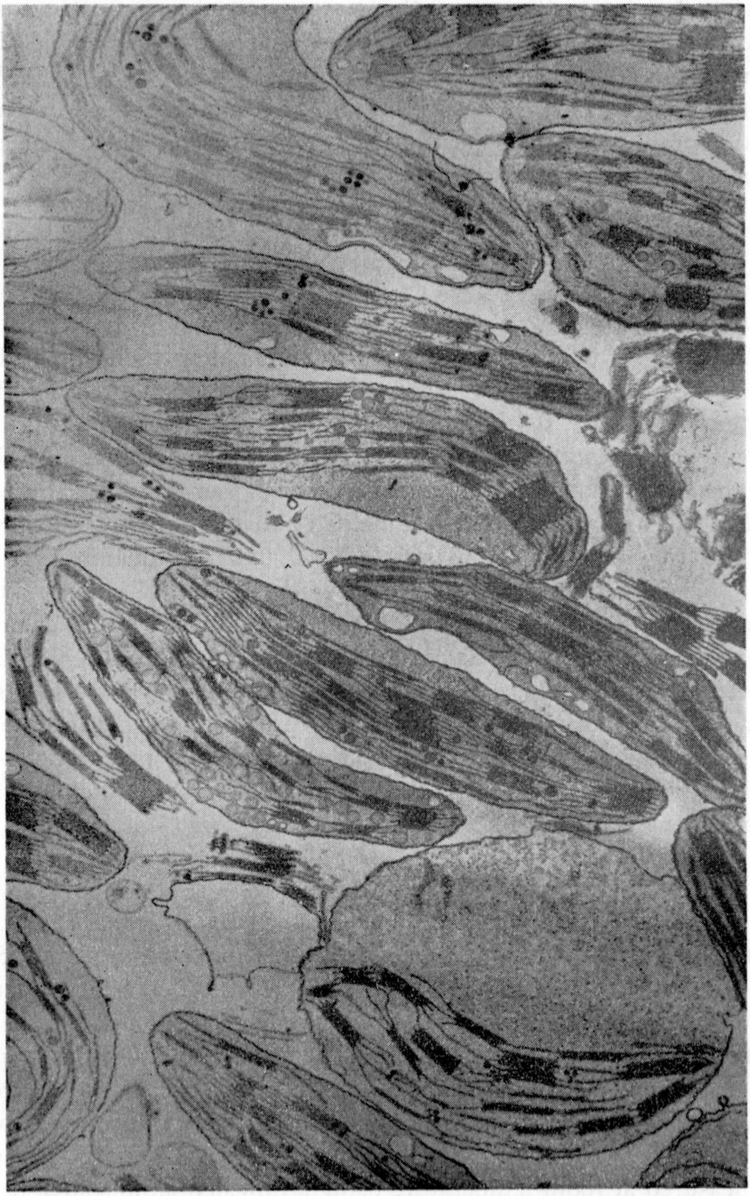

Fig. 1. Electronmicrograph of a representative sample of chloroplasts prepared as described in *Methods*, magnification ×10,000. Electronmicroscopy performed by R. Sjolund in the laboratory of E. T. Weier, University of California, Davis.

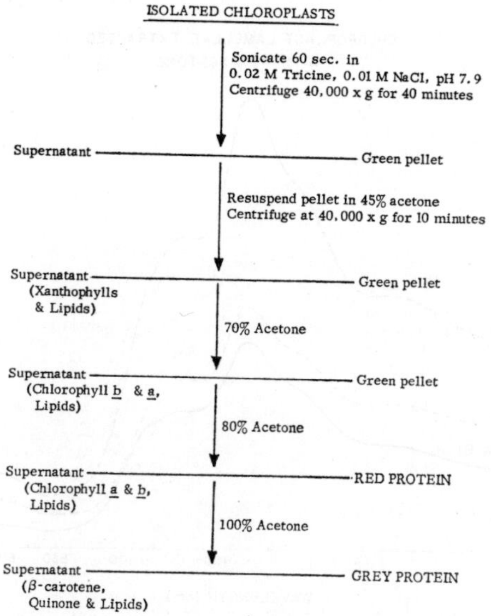

Fig. 2. Sequential extraction procedure for the preparation of the β-carotene-protein complex (red protein) and lamellar protein (grey protein).

sorption coefficients of MacKinney[19] for chlorophyll and Davies[20] for carotenoids. Pigments were purified on thin layer plates of silica gel G using a solvent system of 80 parts hexane and 30 parts acetone.

Thin layer chromatography of silica gel G in a solvent of chloroform : methanol : water, 75 : 25 : 2, provided good separation of the lipids from chloroplast extracts. After development of thin layer plates, phospholipids were initially detected with the molybdate spray reagent of Dittmer and Lester.[21] The plates were then sprayed with a 5% dichromate sulfuric acid solution and heated for 15 minutes at 100°. In this manner, the principal chloroplast lipids were detected. The positions of mono- and digalactosyldiglycerides were determined independently with diphenylamine spray.[22] The R_f values are : monogalactolipid 0.72, digalactolipid 0.54, phosphatidylglycerol 0.49, sulfolipid 0.32, and chlorophyll 0.94. Quantitative assays for these glycolipids were based on the determination of sugar contents in lipids.[23] Phosphorous assay according to the method of Yang *et al.*[24] was used for phospholipid determination.

Reassociation of Lipoprotein

The reassociation of protein with lipophilic components demands an

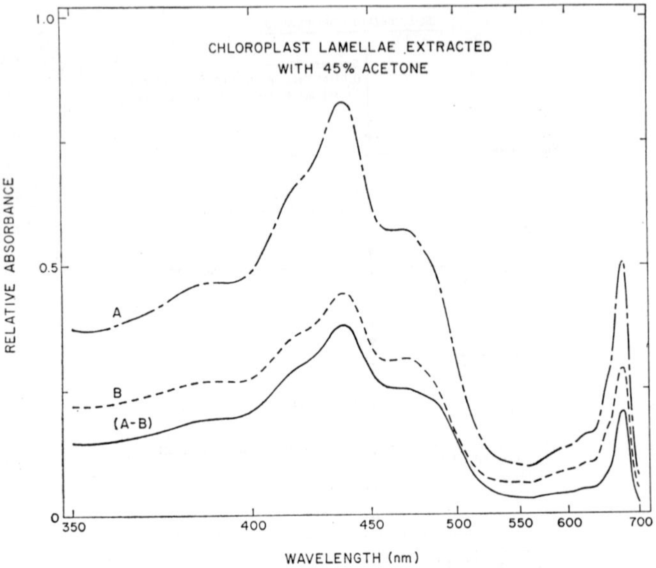

Fig. 3. Chloroplast lamellae were extracted with 45% acetone, divided into two equal portions and lyophilized. Half of the lamellae were further extracted with heptane. Then both portions were suspended in buffer and spectra measured. A: lamellae extracted with 45% acetone; B: lamellae extracted with 45% acetone and heptane; (A–B): difference spectrum—A was sample and B was reference.

experimental process which permits sufficient solubilization of both lipid and protein. We approached this requirement by using a double dialysis system. A uniform suspension of lamellar protein in methanol (0.1 mg/ml) was put into a small diameter dialysis bag. The membrane material for this bag was acetylated (anhydrous pyridine: acetic anhydride, 2:1) for 1 hour in order to enhance lipid permeability. This bag was sealed and put into a larger dialysis bag containing 5 μg of β-carotene per ml of methanol. After 9–10 hours, equilibrium between the two solutions was interrupted by the addition of aqueous methanol to the exterior of the large dialysis bag. When the system approached 60% methanol in water, the protein was recovered and analyzed for β-carotene contents as were the supernatant solutions. From this technique we have been able to evaluate the association of β-carotene with the lamellar protein and the properties of the reassociated complex.

RESULTS AND DISCUSSION

Selective extraction of lipids and difference spectroscopy

The difference spectrum (A-B) in Fig. 3 revealed no spectral characteristics of the red β-carotene-protein complex (Fig. 5). From Table I and the spectra in Fig. 3, we learn that extraction of approximately 50% of the lipids and all xanthophylls by 45% acetone apparently does not alter the β-carotene absorption spectrum. Significantly, no maximum was observed at 538 nm.

Fig. 4 presents spectra from particles initially extracted with 70% acetone (curve C), and extracted with heptane after lyophilization (curve D). Absorption maxima in the difference spectrum (C-D) occur at 680 nm, 538 nm, 498 nm, 460 nm, and 438 nm. From the quantitative data in Table I, we conclude that maxima at 680 nm, 460 nm, and 438 nm result from the additional extraction of chlorophylls *a* and *b* by heptane. The 498 nm and, in particular, the 538 nm peaks are related to the absorbance of β-carotene.

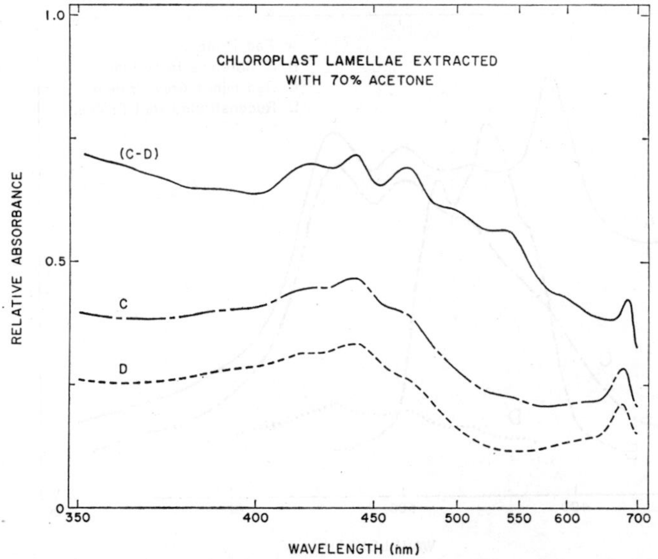

Fig. 4. Chloroplast lamellae were extracted with 70% acetone, divided into two equal portions and lyophilized. Half of the lamellae were further extracted with heptane. Both portions were suspended in buffer and spectra measured. C: lamellae extracted with 70% acetone; D: lamellae extracted with 70% acetone and heptane; (C–D): difference spectrum —C was sample, D was reference—scale expanded five times.

TABLE I. Lipids Obtained by Sequential Extraction of Chloroplast Lamellae.

Lipid	Analyses of Lamellae Extracts (in %)					
	Initial extraction with 45% acetone			Initial extraction with 70% acetone		
	45% Acetone	Heptane	100% Acetone	70% Acetone	Heptane	100% Acetone
Monogalactolipid	59.5	15.8	24.7	61.7	16.1	22.2
Digalactolipid	48.9	19.3	31.8	53.0	19.5	27.5
Phosphatidylglycerol	37.2	20.8	42.0	39.5	23.1	37.4
Chlorophyll *a*	1.1	9.2	89.7	15.1	8.9	76.0
Chlorophyll *b*	14.6	4.7	80.7	73.0	8.7	18.3
Chlorophyll *a*+*b*	4.5	8.0	87.5	32.1	8.8	59.1
Xanthophylls	100	—	—	100	—	—
β-Carotene	—	99.4	0.6	—	97.2	2.8

Lamellae were prepared as described in *Methods*, and initially extracted with either 45% or 75% aqueous acetone. After centrifugation, the particles were lyophilized and then extracted with heptane. The particles were again lyophilized and finally extracted with 100% acetone. Each of the extracts were analyzed for pigment and lipid composition as described in the text. Extractions were performed at 0°–5°.

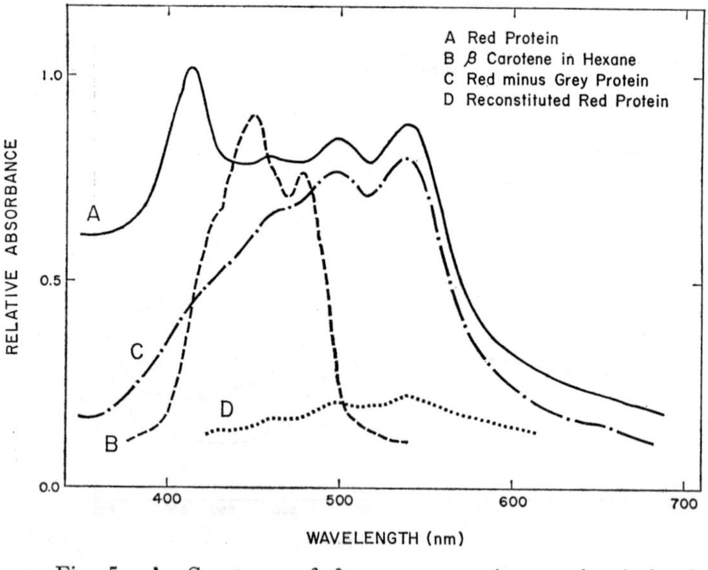

Fig. 5. A: Spectrum of β-carotene-protein complex isolated from lamellae. B: Spectrum of β-carotene in hexane solution. C: Difference spectrum, red protein was sample and an equal quantity of grey protein was the reference —scale expanded five times. D: Difference spectrum, reassociated red protein (see *Methods*) was the sample and an equal quantity of grey protein was the reference—scale expanded five times.

The percentage of lipids extracted by either 45% acetone or 70% acetone is similar (Table I). Significantly, however, 70% acetone extracted almost 10 times more chlorophyll than did 45% acetone. Possibly the absorbance at 538 nm resulted from removal of chlorophyll from the lipoprotein matrix. Alternatively, the extensive extraction of lipid and pigments could result in a conformational change in the lamellar protein.

Treatment of lamellae with 80% acetone removes most lipids and all pigments except β-carotene (and quinones). The spectrum of this residual carotene-protein complex, Fig. 5, is caused by β-carotene which can be

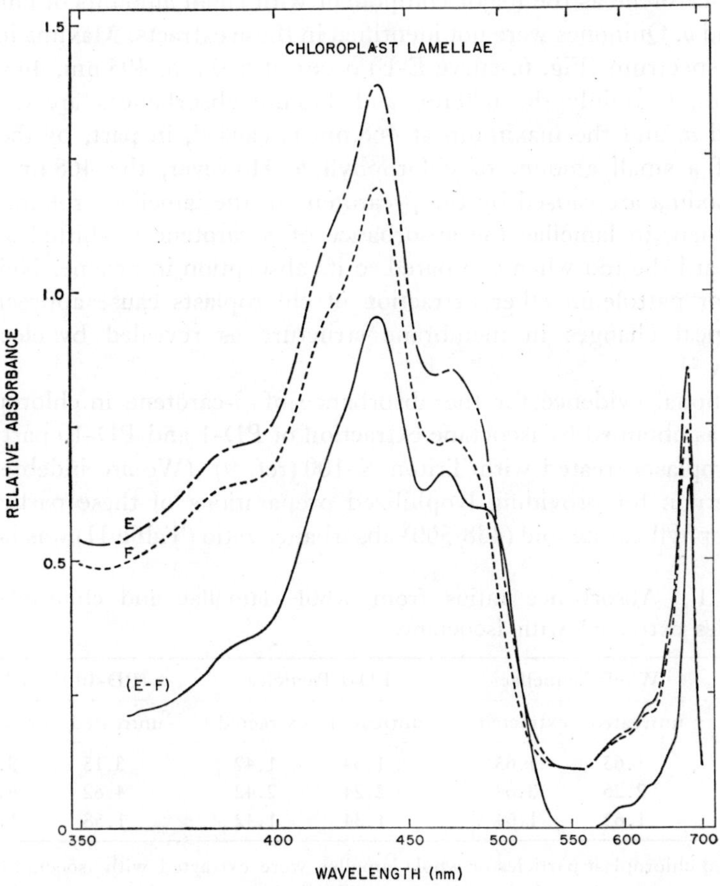

Fig. 6. Lamellae were lyophilized and extracted with either heptane or isooctane. After suspension of particles in buffer, spectra were measured. E: untreated lamellae; F: lamellae extracted with heptane or isooctane; (E–F): difference spectrum—E was sample, F was reference—scale expanded five times.

extracted with 100% acetone, hexane or heptane to yield "grey" lamellar protein. The absorbance at 418 nm is due to cytochromes; difference spectroscopy at −196° also supports the presence of cytochromes (low temperature spectra measured by W. Cramer in W. L. Butler's laboratory). As revealed by the difference spectrum, red protein minus grey protein (Fig. 5), the remaining peaks at 538 nm, 498 nm, and 460 nm are caused by a shifted β-carotene absorption.

Lyophilized chloroplast lamellae were extracted with isooctane or heptane directly. Analyses of these extracts by thin layer chromatography revealed β-carotene as the major component with small amounts of chlorophylls *a* and *b*. Quinones were not identified in these extracts. Maxima in the difference spectrum (Fig. 6, curve E-F) occur at 680 nm, 493 nm, 468 nm, and 438 nm. Certainly the 680 nm and 438 nm absorbances are due to chlorophyll *a*, and the maximum at 468 nm is caused, in part, by the extraction of a small amount of chlorophyll *b*. However, the 468 nm and 493 nm maxima are caused by the β-carotene in the lamellar protein. Apparently, then, in lamellae the absorbance of β-carotene is shifted about 15 nm toward the red when compared to its absorption in hexane. Neither heptane nor petroleum ether extraction of chloroplasts cause appreciable morphological changes in membrane structure as revealed by electron microscopy.[25]

Additional evidence for the absorbance of β-carotene in chloroplast lamellae was obtained by isooctane extraction of PD-1 and PD-10 particles from chloroplasts treated with Triton X-100 (ref. 9). (We are indebted to Dr. L. Vernon for providing lyophilized preparations of these particles.) The chlorophyll/carotenoid (438/500) absorbance ratio (Table II) was larger

TABLE II. Absorbance ratios from whole lamellae and chloroplast particles extracted with isooctane.

Wavelength nm	Whole Lamellae		PD-1 Particles		PD-10 Particles	
	untreated	extracted	untreated	extracted	untreated	extracted
438/475	1.63	1.65	1.34	1.42	3.15	3.30
438/500	2.26	3.04	2.24	2.42	4.82	6.05
438/675	1.62	1.64	1.34	1.42	1.58	1.62

Lyophilized chloroplast particles or whole lamellae were extracted with isooctane at 0°–5°. The extracted particles and untreated particles were suspended in pH 7.9 buffer as described in *Methods*. The ratios were obtained from the absorbance at the designated wavelengths. PD-1 and PD-10 particles were obtained from chloroplasts treated with Triton X-100 (see text). The 438 nm is the chlorophyll *a* maximum chosen as a reference wavelength.

for PD-10 particles than for PD-1 or whole lamellae, since PD-10 particles contain mainly chlorophylls *a*, *b*, and β-carotene while PD-1 particles and whole lamellae contain a larger quantity of the other carotenoids. Difference spectra of isooctane-extracted particles appeared similar to the spectra for whole lamellae in Fig. 6. The absorbance change, revealed by the 438 nm/500 nm absorbance ratio (Table II) was greatest for the PD-10 particles. The 490–500 nm absorbance was selected since it is least affected by chlorophyll absorption. Thus, in these PD-10 particles the β-carotene absorbance appears shifted in a manner similar to that discussed for whole lamellae.

Lipid extraction experiments (Table II) demonstrated that lipid solubility did not correlate with lipid extractability. Extractability may depend upon two parameters: 1) The lipoprotein may assume different conformations related to the solvent system. 2) Various hydrophobic associations exerted by lipids and lamellar protein may minimize the solubilizing effects of polar groups. Apparently phosphatidyl glycerol maintains stronger hydrophobic associations with lipoprotein than do the galactolipids, since it is extracted poorly by aqueous acetone and more easily with heptane. Also, the fatty acids of galactolipids are more polar, because of unsaturation, than fatty acids of phosphatidyl glycerol.[26] Thus extractability of lipids from lamellar systems probably depends upon polarity of fatty acid moieties and

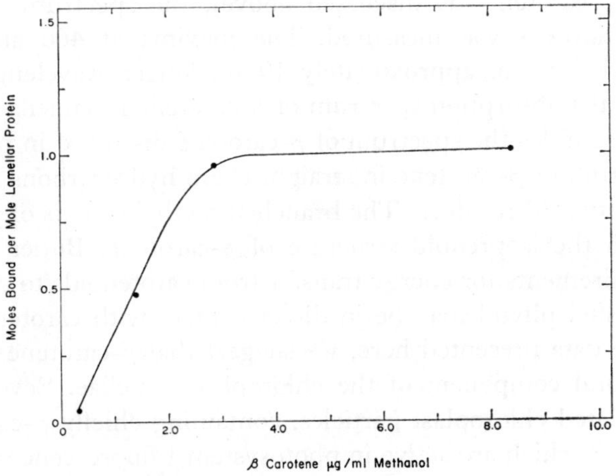

Fig. 7. The association of β-carotene with lamellar protein was measured as a function of the initial concentration of β-carotene in the methanol solution. The double dialysis system for the reassociation of β-carotene with protein was used as described in the Methods.

the hydrophobic associations between lipids and lipoprotein. For example, Benson[27] suggested that hydrophobic affinity may result from the association of the π-orbital systems of the lamellar protein chain with the π-electron system of the three olefinic groups near the outer end of α-linolenic acid, the principal component of the galactolipids.

The reassociated complex (see *Methods*) had spectral properties similar to the red protein isolated directly from lamellae (Fig. 5). β-carotene was absorbed to the extent of one mole β-carotene per mole of lamellar protein, assuming a molecular weight of 23,000 (Fig. 7). Formation of this complex appears to be specific; similar association was not observed with bovine serum albumin.

Our spectra of the β-carotene-protein complex (Fig. 5) differ from those reported by Shibata for β-carotene crystal suspensions in water.[28] Spectral properties of the red protein remain unchanged even after solubilization of the complex of 5% SDBS. However, 5% Triton X-100 or SDBS solutions of free β-carotene reveal absorption spectra markedly different from the spectra of the red protein (*cf.* Figs. 5 and 8). The spectrum of a detergent solution of lamellar protein added to a detergent solution of free β-carotene maintained light absorption properties of the free β-carotene solution. Hence, the data indicate directly that the red protein is not a product of coprecipitating β-carotene with protein in an aqueous system.

Since the removal of phytol from the lamellar system alters the spectral properties of β-carotene (as discussed above), the spectrum of a phytol solution of β-carotene was measured. The maxima at 460 and 487 nm (Fig. 8), absorb light at approximately 10 nm longer wavelength than in hexane. Thus, the absorption spectrum of β-carotene in lamellae at 468 nm and 493 nm resembles the spectrum of β-carotene dissolved in phytol more than the spectrum of β-carotene in straight chain hydrocarbons like hexane or saturated fatty acid residues. The branched methyl groups of phytol may fit closely with the isoprenoid structure of β-carotene. Butler has shown structural requirements for energy transfer from carotenoids to chlorophyll, and has suggested phytol may be in direct contact with carotene.[29]

From the data presented here, we suggest that β-carotene serves as a specific structural component of the chloroplast lamellae. Several laboratories have isolated chloroplast particles, containing chiefly β-carotene and chlorophylls,[1,8,9] which are active in photosystem I fluorescence and photoreduction. Specific protein-carotenoid interactions are readily demonstrated in crustacyanin[17] and ovorubin[16] in which astaxanthin apparently stabilizes protein structure. β-carotene's resistance to extraction may be related to its stabilizing influence in chloroplast protein structure.[30]

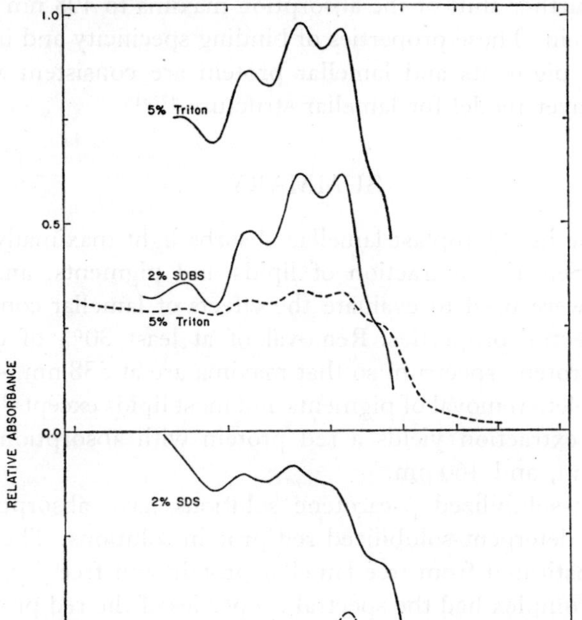

Fig. 8. Absorption spectra of various β-carotene solutions. β-Carotene was dissolved in 5% Triton X-100, 2% SDS, 2% SDBS, and phytol. Crystalline β-carotene was shaken with the detergent solution for at least 4 hours, and then the solution was centrifuged for 20 minutes at $30{,}000 \times g$. Absolute quantities of β-carotene were not determined. Solid lines indicate five time scale expansion.

Thus, the shift in β-carotene absorption maxima from 450 nm and 478 nm in hexane to 468 nm and 493 nm in lamellar protein may result from the polarizing influence of the lamellar protein, and the solvating effects of the phytol moiety of chlorophyll. The phytol, perhaps, insulates β-carotene from the effects of polar groups in the lamellar protein. We conjecture, then, that acetone extraction of chlorophyll leads to stronger polarization of β-carotene and further shift of the absorption maxima to 498 nm and 538 nm in the red protein. These properties of binding specificity and interrelationships between pigments and lamellar protein are consistent with a lipoprotein monolayer model for lamellar structure.[27,31]

SUMMARY

β-carotene in chloroplast lamellae absorbs light maximally at 468 nm, and 493 nm. Selective extraction of lipids and pigments, and difference spectroscopy were used to evaluate the effects of lamellar constituents on β-carotene spectral properties. Removal of at least 30% of chlorophylls alters the β-carotene spectrum so that maxima are at 538 nm, 498 nm, and 460 nm. Complete removal of pigments and most lipids except β-carotene by 80% acetone extraction yields a red protein with absorption maxima at 538 nm, 498 nm, and 460 nm.

Detergent-solubilized β-carotene solutions have absorption maxima different from detergent-solubilized red protein solutions. The red protein could be reconstituted from free lamellar protein and free β-carotene. The reconstituted complex had the spectral properties of the red protein isolated from chloroplast lamellae. These data indicate that formation of the red β-carotene complex does not result from a simple coprecipitation of β-carotene micells or aggregates with lamellar protein.

We discuss the requirement for the removal of chlorophyll prior to red protein formation as it applies to the structural properties of lamellar membranes. The intimate association of the phytol of chlorophyll with β-carotene presents an environment which reduces the polarizing effects of the lamellar protein. The data are consistent with the lipoprotein monolayer model for lamellar molecular structure, but not with the lipid bilayer model.

REFERENCES

1. Boardman, N. K., Thorne, S. W., and Anderson, J. M., *Proc. Natl. Acad. Sci.*, **56**, 586 (1966).
2. Harton, A. A., and Whatley, F. R., *Plant Physiology Supp.*, **37**, v (1962).
3. Kahn, J.S., *Biochim. Biophys. Acta*, **79**, 234 (1964).
4. Menke, W., *Naturwiss.*, **28**, 31 (1940).
5. Nieman, R. H., Nakamura, H., and Vennesland, B., *Plant Physiol.*, **34**, 262 (1959).
6. Nishimura, M., and Takamatsu, K., *Nature*, **180**, 699 (1957).
7. Ogawa, T., Obata, F., and Shibata, K., *Biochim. Biophys. Acta*, **122**, 223 (1966).
8. Thornber, J. P., Gregory, R., Smith, C. A. and Bailey, J. L., *Biochemistry*, **6**, 391 (1967).

9. Vernon, L. P., Shaw, E. R., and Ke, B., *J. Biol. Chem.*, **241**, 4101 (1966).
10. Wessels, J. S. C., *Biochim. Biophys. Acta*, **126**, 581 (1966).
11. Smith, E. L., *J. Gen. Physiol.*, **24**, 565 (1941).
12. Ke, B., *Arch. Biochem. Biophys.*, **112**, 554 (1965).
13. Urry, D. W., Mednieks, M., and Bejnarowicz, E., *Proc. Natl. Acad. Sci.*, **57**, 1043 (1967).
14. Platt, J. R., *Sci.*, **129**, 372 (1959).
15. Salton, M. R. J., and Elishan-Ud-Din, A. F. M., *Aust. J. Exptl. Biol. Med. Sci.*, **43**, 255 (1965).
16. Cheesman, D. F., *Proc. Roy. Soc., Lond.*, **149**, B, 571 (1957).
17. Cheesman, D. F., Zagalsky, P. F., and Ceccaldi, H. J., *Proc. Roy. Soc., Lond.*, **164** B, 130 (1966).
18. Jencks, W. P., and Buten, B., *Arch. Biochem. Biophys.*, **107**, 511 (1964).
19. MacKinney, G., *J. Biol. Chem.*, **140**, 315 (1941).
20. Davies, B., in *Chemistry and Biochemistry of the Plant Pigments* (T. W. Goodwin, ed.), Academic Press, New York, 1965.
21. Dittmer, J. D., and Lester, R. C., *J. Lipid. Res.*, **5**, 126 (1964).
22. Wagner, H., Horhammer, L., and Wolf, P., *Biochem. Zeit.*, **334**, 175 (1961).
23. Dubois, M., Giles, K. A., Hamilton, J. K., Rebers, P. A., and Smith, F., *Anal. Chem.*, **28**, 350 (1956).
24. Yang, S. F., Freer, S., and Benson, A. A., *J. Biol. Chem.*, **242**, 477 (1967).
25. Margree, L., Henninger, M. K., and Crane, F. L., *J. Biol. Chem.*, **241**, 5197 (1966).
26. Allen, C. F., Hirayama, O., and Good, P., in *Biochemistry of Chloroplasts*, Vol. I (T. W. Goodwin, ed.), Academic Press, New York, 1965.
27. Benson, A. A., *J. Am. Oil Chem. Soc.*, **43**, 265 (1966).
28. Shibata, K., *Biochim. Biophys. Acta*, **22**, 398 (1956).
29. Butler, W.L., *Arch. Biochem. Biophys.*, **92**, 287 (1965).
30. Walles, B., *Heriditas*, **53**, 247 (1965).
31. Weier, T.E., and Benson, A.A., *Am. J. Bot.*, **54**, 389 (1967).

Structure and Photochemistry in Tobacco Chloroplasts*

Peter H. Homann, Georg H. Schmid and Hans Gaffron

Department of Biological Science and Institute of Molecular Biophysics, Florida State University, Tallahassee, Florida

During the last two years Schmid and Gaffron[1,2] were able to establish a correlation between photosynthetic activity and the general microstructure of the chloroplast. We have extended these studies to include isolated chloroplasts. The following is a summary of our results. We have come to accept purely on metabolic and photochemical evidence that there must be a combination of two, or, as we have reasons to believe, three distinct pigment systems to obtain complete and maximal photosynthesis in a plant cell. Only a few of the non-photochemical enzymatic reactions belonging to this or that partial system can be made to proceed in homogeneous solutions free of particles. The success of living photochemistry to achieve the synthesis of new organic matter depends on the arrangement of the participating cofactors and enzyme complexes on the structure visible in electron microscopes. Our latest finding is that the need for a cooperation between two pigment systems in order to obtain complete photosynthesis is apparently reflected in the requirement that at least two lamellar layers touch one another. The structural dimensions which determine how absorbed light energy is made use of are much larger than those of the semi-crystalline structural subunits (quantasomes) which Dr. Park's systematic studies have brought to light.[3]

In vigorously growing plants we have found the following types of differently organized chloroplasts (see Fig. 1a–d). (1) Chloroplasts which are full of normal grana (Fig. 1a). These chloroplasts occur generally in plants which contain plenty of chlorophyll and are easily light-saturated. (2) Chloroplasts with long, widely separated lamellae, which here and there either touch or are folded at the ends (Fig. 1b and c). Such chloroplasts usually contain less chlorophyll, about one-fifth to one-tenth of the chlorophyll of normal leaves. The leaves, therefore, look yellowish. (3) Chloro-

* These studies were supported by contract No. AT-(40-1)-2687 from the United States Atomic Energy Commission, and by grants No. AF-AFOSR-662-65 from the Air Force Office of Scientific Research and GB-345 from the National Science Foundation.

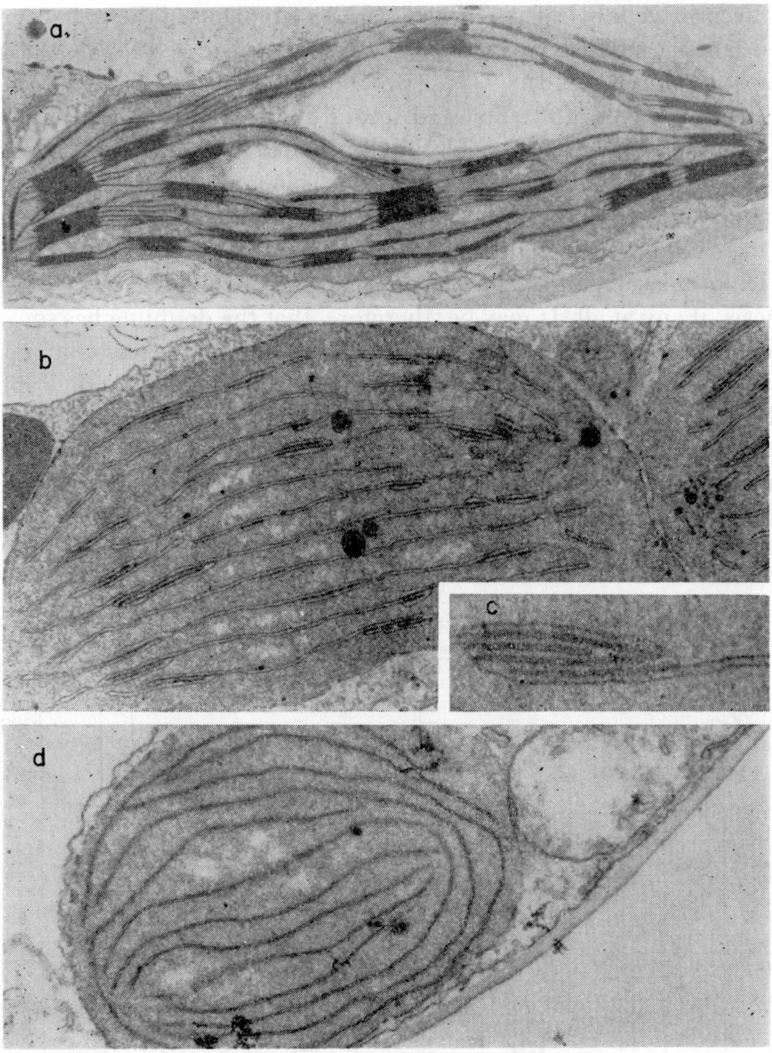

Fig. 1. Chloroplasts from tobacco, glutaraldehyde fixation: post-fixed with OsO_4.

a: Chloroplast from normal green *Nic. tabacum* John Williams Broadleaf × 13,000.

b: Chloroplast of *Nic. tabacum* aurea mutant Su/su 6 weeks after germination × 16,000.

c: Lamellar stack of Su/su 6 weeks after germination × 77,000.

d: Chloroplast of the yellow-green section of the variegated *Nic. tabacum* NC 95 × 13,300.

plasts seen in yellow leaf patches of variegated plants. These have widely spaced lamellae that stay separated along their entire length, do not touch, overlap or fold (Fig. 1d).

Healthy leaves of the first and second types photosynthesize normally, except that the capacity per chlorophyll is better in the yellow leaves, sometimes so much better that the chlorophyll-poor plants fix more CO_2 per square centimeter of leaf area, provided the flux of quanta is adequately high to maintain such high rates.

Chloroplasts of the third type give only negligible rates of photosynthesis, if any at all. It may be, therefore, that the doubling has something to do with the proper function of the lamellae. Type 3 leaves contain about as much chlorophyll as leaves with chloroplasts of type 2. To find out what is behind the difference in photosynthetic activity, it was necessary to know something about the capacities not only for normal photosynthesis, but also for its partial reactions. This required experiments with isolated chloroplasts. The results are summarized in Fig. 2. The grana with overlapping lamellae in the structural types 1 and 2 gave rates of the Hill reaction which

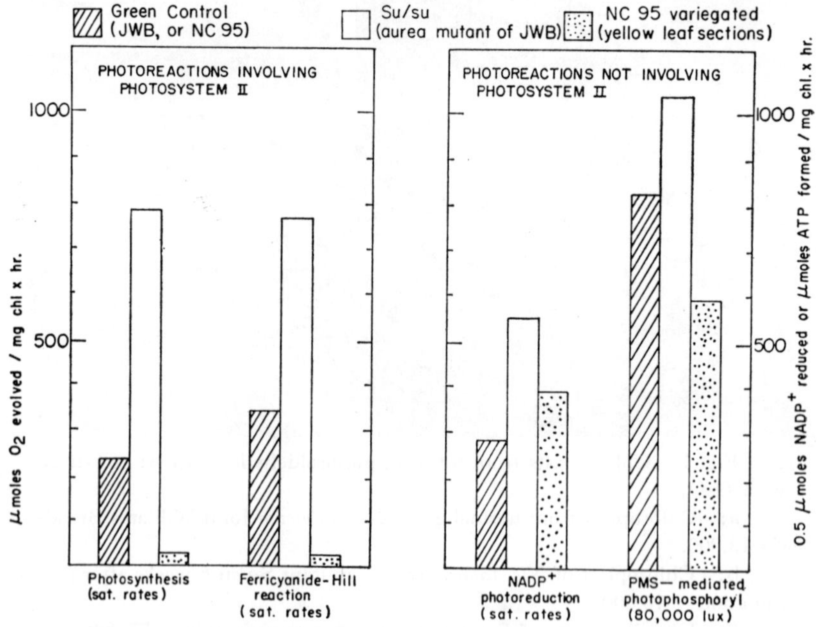

Fig. 2. Maximal rates of photosynthesis and various chloroplast reactions obtained repeatedly with three different tobacco varieties (for the average values of all our data and their standard deviations see ref. 5).

paralleled those of the rates of photosynthesis in corresponding intact leaf sections. As Fig. 2 shows, chloroplasts from the non-photosynthetic type 3 were inactive also in the Hill reaction. The obvious question was: Are these type 3 chloroplasts inactive under all circumstances or do they retain some useful photochemistry?

The data on the right side of Fig. 2 compare the activity of partial reactions due to system I in our three structures. Here chloroplasts of type 3 are at least as good as those of the normal type 1. This novel and telling result requires that we refer to the detailed model for chloroplast lamellae developed by Weier and Benson.[4] Our data clearly show that active chlorophyll does occur outside the grana in separate, single or "fret" lamellae. Although the isolated lamellae in our chloroplast type 3 performed only photoreactions not dependent on photosystem II, they carried a normal amount of manganese.[5] Very likely, they contain the binding site for the manganese needed in photosystem II. The quantum requirement for cyclic photophosphorylation in type 1 was 6 times higher in the red than in the far-red. The same was true with chloroplasts of type 3. This odd ratio in the red/far-red efficiencies indicates that the single lamellae contain chlorophyll which is not part of system I. Their normal manganese content, and recent results from fluorescence studies, speak for the presence of photosystem II. Yet, oxygen is not evolved. Therefore it may be the process of water oxidation which demands a more complex lamellar arrangement than just a single lamella. The additional requirement might be a protected, elaborately designed niche in the interlamellar space of either lamellar foldings or grana. The inefficiency of the cyclic photophosphorylation in red light, finally, raises again the question whether there exists a special far-red absorbing pigment system in the lamellae of green plants as proposed by Gaffron *et al.*,[6] and by Schmid and Gaffron,[7] to explain three related observations concerning far-red light: the excellent photoreduction in adapted algae, the assimilation of acetate and the extremely strong Emerson effect on the low efficiency of photosynthesis in blue light.

With all this variability of the saturation rate for the entire process or for some of its partial reactions, it should be emphasized that the one parameter which is not governed by structure but plainly by thermodynamics, is the minimum number of quanta needed for the complete fixation of one molecule of CO_2. This remains the same as it has been found in hundreds of earlier measurements. It oscillates little around the value of ten.[7] We have good evidence that a quantum number of ten is obtained only when all pigment systems are fully cooperating—that is, are working in balance with each other.[7] This is the secret of the Emerson effect.

Another basic question concerns the nature of the photosynthetic unit. Discovered by Emerson and Arnold in 1932 (ref. 8) it gave rise to a very extended literature. In recent years, it was believed that super-magnification of micrographs would permit us to see the units in the form of clusters of enzyme molecules. These have even been given a name, quantasomes.[3] The metabolic photosynthetic unit is defined by the number of chlorophyll molecules involved in the fixation of one molecule of carbon dioxide as the consequence of a very strong light flash which activates all available reaction centers at once, but is so short as to make a repetition of the ensuing normal photosynthetic sequence of reactions during its lifetime unlikely.

Fig. 3 shows our results. We found that the size of the unit varies with the type of structure studied. The correlation is not rigid, but shows a reg-

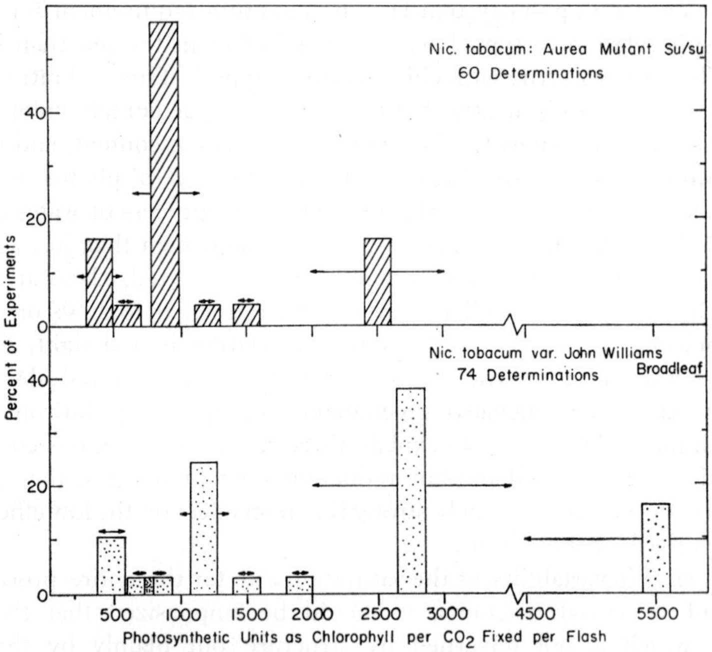

Fig. 3. Distribution of photosynthetic units as chlorophyll per CO_2 fixed per flash. Flash duration 25 μsec. Upper graph: Aurea mutant Su/su from *Nic. tabacum* (leaves with a maximal photosynthetic capacity of 230 ± 55 μl O_2/mg chl/min). Lower graph: Normal green variety John Williams Broadleaf from *Nic. tabacum* (leaves with a maximal photosynthetic capacity of 47 ± 17 μl O_2/mg chl/min). Background 900 ergs/sec cm^2 red light 575 mμ $< \lambda <$ 700 mμ. Absolute maximal deviations are indicated by arrows.

ular pattern. First of all, units vary considerably within normal green leaf tissue. There are units as small as 400 and some as large as 5,000. But the smaller units occur more often with the chlorophyll-poor chloroplasts of type 2. Yellow leaves tend to fix more carbon dioxide per chlorophyll per flash than the dark green ones. "Units" giving values twice as large as those of Emerson and Arnold[8] were found only in those leaves which otherwise showed a preponderance of results which checked with the Emerson and Arnold number of 2,500. Two points must be made. First, the separation of the values into groups that do not overlap is justified because any general scattering of the results between 400 and 5,000 should not give us empty spaces in this graph.

Second: the idea of the unit as it emerges here does not fit well with the rigid physical energy transfer model first proposed by Gaffron and Wohl thirty years ago,[9] but more with a kinetic model alluded to but not further developed by Rabinowitch.[10]

This is true, because we found the critical time between flashes as well as the shape of the light intensity curve for the flash yield also subject to variations. Space does not permit us to go into the details of these further developments. To us the unit looks like a cluster A of pigment molecules to which a variable but small number of reaction centers B can be attached. It is the ratio of centers to chlorophyll molecules which roughly determines "unit" size. The centers react further with a variable concentration of enzyme C. This gives the time for the "waiting period" between flashes. If each center B needs not one but two quanta to be stably changed, different shapes of light curves for the flash yield are easier to understand. What presents a puzzle is why each of the structural systems we have been talking about should have a preference for a particular unit type at any one time instead of giving us some average value all the way from 300 to 5,000. Since there is no continuous spectrum of possible units, one may postulate that a large unit consists of several, well-defined latent subunits. Which subunit predominates may be governed by those structural characteristics which regulate the contact between units, electron carriers and enzymes.

REFERENCES

1. Schmid, G. H., and Gaffron, H., *Brookhaven Symp. Biol.*, **19**, 380 (1966).
2. Schmid, G. H., and Gaffron, H., *J. Gen. Physiol.*, **50**, 563 (1967).
3. Park, R. B., and Biggins, J., *Science*, **144**, 1009 (1964).
4. Weier, E. T., and Benson, A. A., *Amer. J. Bot.*, **54**, 389 (1967).
5. Homann, P. H., and Schmid, G. H., *Plant Physiol.*, **42,** 1619 (1967).
6. Gaffron, H., Wiessner, W., and Homann, P.H, in *Photosynthetic Mechanisms of Green*

Plants (B. Kok and A. T. Jagendorf, eds.), p. 436, Nat. Acad. Sci-Nat. Res. Council, 1963.

7. Schmid, G. H., and Gaffron, H., *J. Gen. Physiol.*, **50,** 2131 (1967).
8. Emerson, R., and Arnold, A., *J. Gen. Physiol.*, **16**, 191 (1932).
9. Gaffron, H., and Wohl, K., *Naturwiss.*, **24**, 81, 103 (1936).
10. Rabinowitch, E., *Photosynthesis* **II**, 2, pp. 1280–1299, Interscience, New York, 1956.

The Ultrastructure of Fracture and Deep Etch Faces of Spinach Thylakoids*

RODERIC B. PARK AND L. K. SHUMWAY†

Botany Department, University of California, Berkeley, California

INTRODUCTION

Fifteen years ago Steinmann first observed the particulate character of shadowed chloroplast thylakoids in the electron microscope.[1] Though these observations were not expected on the basis of myelin studies, they were consistent with the observed low birefringence of chloroplasts[2] and suggested that thylakoid structure was more complex than that of myelin.

Subsequent investigations of thylakoid structure used probing radiation varying in wavelength from X-ray through visible light. These studies confirmed the existence of various sized particles and subunits associated with the chloroplast thylakoid. Though there is general agreement on the existence of such subunits within the membrane, there is considerable disagreement both as to their size and location. The source of these disagreements lies in the interpretation of experimental observations both at the ultrastructural and X-ray level. The interpretations of ultrastructural observations based on thin sectioning,[3] negative staining,[4] heavy metal shadowing[5] and freeze-etching[6,7] often disagree with each other and sometimes with interpretations of X-ray diffraction.[8] This paper is concerned with a major discrepancy in the interpretation of ultrastructural observations, namely, do the membrane associated particles lie principally within the membrane or are they situated on the thylakoid surface? In 1961 Park and Pon proposed a thylakoid model in which the largest particles (quantasomes) were buried within the membrane.[5] This model was subsequently challenged by Oda and Huzisige,[4] who claimed that the membrane-associated quantasomes could be removed from isolated thylakoids by centrifugal washing and that they therefore existed on the outside of the membrane in a manner similar to the elementary particle of mitochondrial cristae. The Park and Pon model was also challenged by Moor and Mühlethaler[6] who proposed the structure

* This work was supported by the following grants: NSF GB-4245 and USPHS, Institute of General Medicine GM 13943–02.

† Present address. Botany Department, Washington State University, Pullman, Washington.

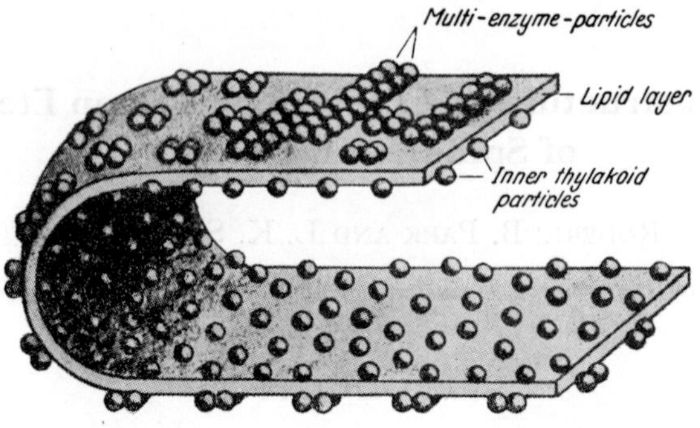

Fig. 1. Model of thylakoid membrane proposed by Mühlethaler, Moor and Sarkowski[6] based on freeze-etch observations.

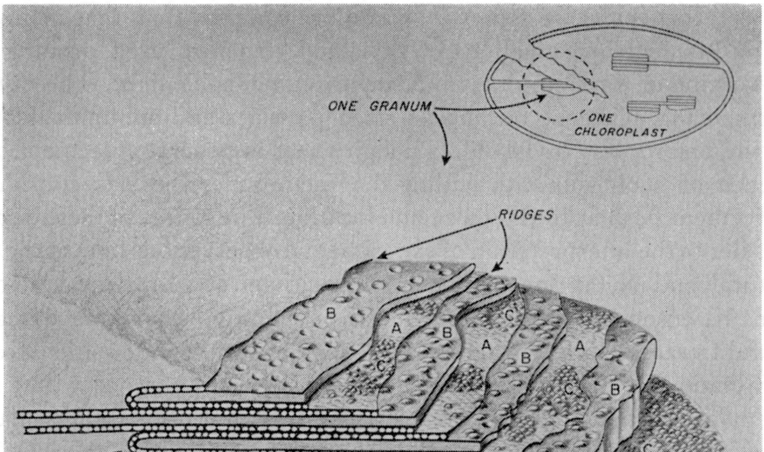

Fig. 2. Model of thylakoid membranes proposed by Branton and Park[9] based on freeze-etch observations.

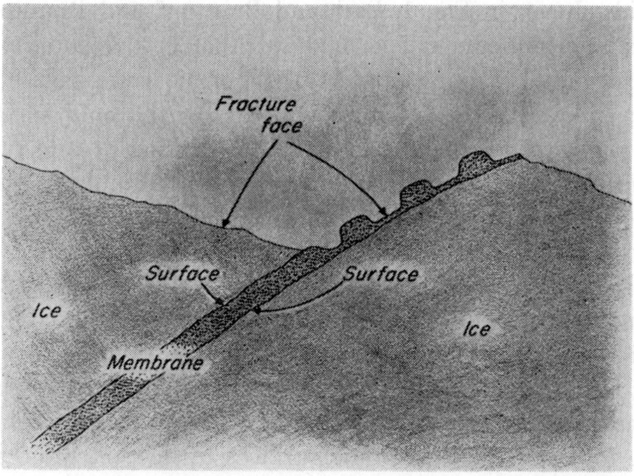

Fig. 3. Cross section of a fracture plane intercepting a single thylakoid membrane. The relation of the fracture plane to the membrane surface follows the interpretation of Branton and Park.[9]

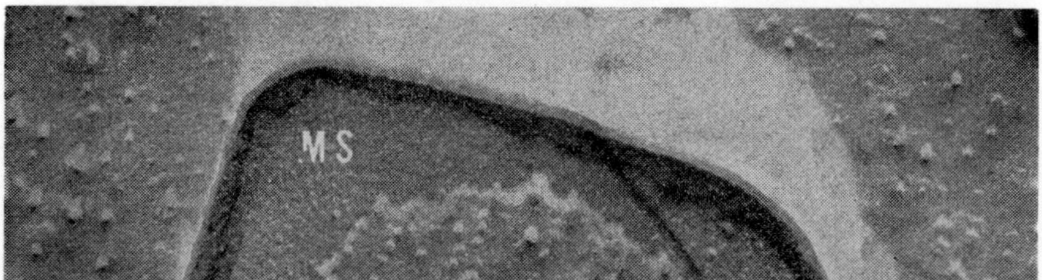

Fig. 4. A deep-etched thylakoid membrane. MS—membrane surface, I—ice. ×88,000.

for thylakoids shown in Fig. 1. Park and Branton[7] and Branton and Park[9] answered these arguments and have shown that 1) thylakoids which appear smooth by negative staining contain the normal complement of quantasomes when split by freeze-etching and that 2) the bold surface relief of membrane particles demonstrated by freeze-etching is due to membrane splitting and removal of membrane matrix from around the particles. The Park-Branton model is shown in Fig. 2.

This paper presents further evidence that the thylakoid fracture faces exposed during freeze-etching differ from the actual membrane surface and that the quantasome subunits actually lie buried within the membrane. The technique we have used here is to freeze thylakoids in water for subsequent fracturing. We then deep-etch the ice to expose the thylakoid surface. The generally used technique of freezing in glycerol solutions is not suitable for exposing membrane surfaces by deep-etching, since a non-etchable eutectic mixture of glycerol and water forms adjacent to the membrane surface during freezing. The question may be asked: Is the deep-

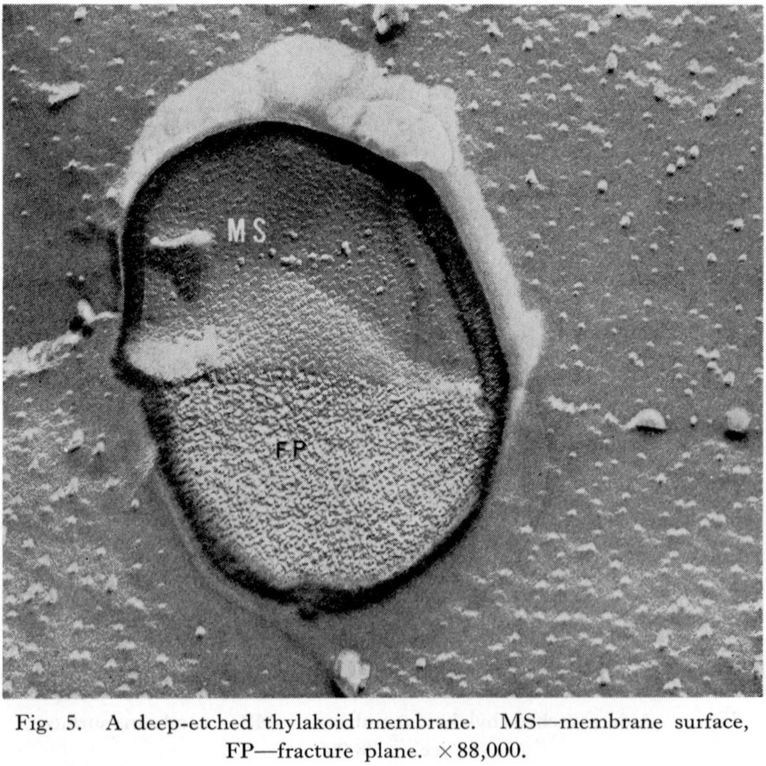

Fig. 5. A deep-etched thylakoid membrane. MS—membrane surface, FP—fracture plane. ×88,000.

etched thylakoid surface the same as the membrane face exposed by fracturing? If the faces are the same, the Moor and Mühlethaler model is supported. The Branton-Park model predicts, on the other hand, that the deep-etched face and the fracture face will differ (see Fig. 3), and that only the deep-etch face should resemble the thylakoid surface while the fracture face will show particles in bold relief.

METHODS

Spinacia oleracea chloroplasts were isolated from commercially grown mature plants in 0.5 *M* sucrose, 0.02 *M* tricine, pH 7.4. The thylakoids were separated from the chloroplasts by washing the isolated chloroplasts in 0.01 *M* tricine (pH 7.4). The washed thylakoids were then studied by air drying and heavy metal shadowing,[10] by freeze-etching,[7] or by deep-etching. In deep-etching the sample is frozen in pure water rather than in a glycerol water solution. When the sample is fractured it is then possible to sublime enough water to expose thylakoid surfaces and to compare these surfaces exposed by sublimation with the fracture faces.

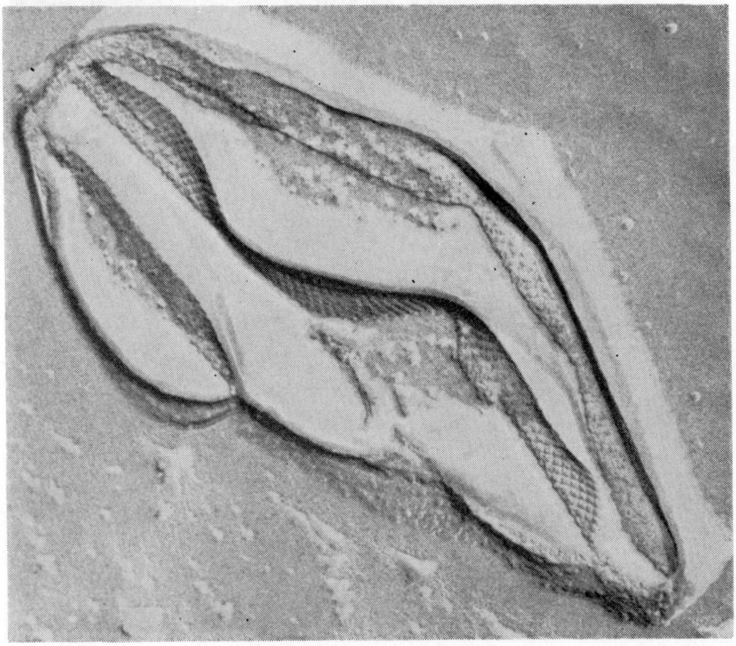

Fig. 6. Deep-etched thylakoids showing a paracrystalline quantasome array on the inside surface of a thylakoid. ×88,000.

In those experiments using tetrazolium dyes the reaction mixtures usually contained chlorophyll (as chloroplasts) $2 \times 10^{-5} M$, tetrazolium dye $10^{-4} M$, methylamine 0.1 M, HEPES buffer (pH 7.6) 0.05 M, Sorbitol 0.25 M. The inhibitor 3-(3,4-dichlorophenyl)-1,1-dimethyl-urea (DCMU) $10^{-5} M$ was present in control reaction mixtures and completely inhibited tetrazolium reduction. The reaction mixtures were placed in a water bath at room temperature and were illuminated with a 500 watt tungsten bulb at a distance of 20 cm. The samples were centrifuged and the ultraviolet absorption spectra of the resulting supernatants were used to calculate the amount of tetrazolium dye reduced during the course of the experiments. The ratio of molecules of dye reduced to molecules chlorophyll present in the reaction mixture varied from 1 to 6.

RESULTS AND DISCUSSION

Figs. 4 and 5 are a comparison of two similar thylakoid membranes.

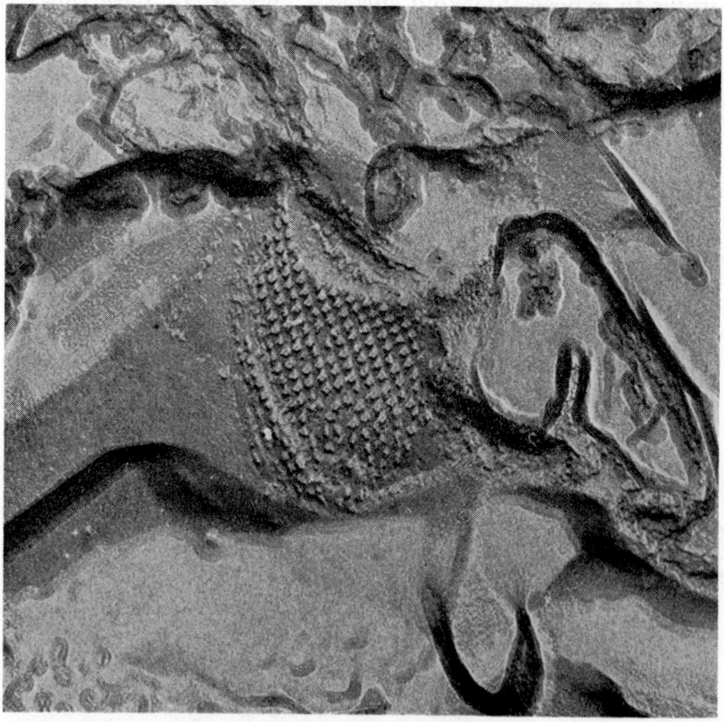

Fig. 7. A paracrystalline quantasome array revealed by fracturing during freeze-etching. $\times 88{,}000$.

In Fig. 4 not all the water has been removed from the cup-shaped membrane. It can be seen that the remaining water is highly structured. The particulate regions are characteristic of frozen water. Since the particles are totally removed by extensive etching (Fig. 5), they are apparently ice regions with slightly lower vapor pressures than the surrounding matrix. Fig. 5 is similar to Fig. 4 in that it also shows a cup-shaped thylakoid membrane. Here the water within the membrane cup has been totally removed. The discontinuity in thylakoid texture at A is the result of fracturing. The region below A was exposed by fracturing during the freeze-etch process while that above A was exposed by deep-etching. This difference would be expected on the basis of the Park-Branton model in which the fracture plane follows an inner membrane face and not the boundary between the membrane and the external medium. Occasionally deep-etching exposes a paracrystalline array as shown in Fig. 6. Since only the interior of the loculus is exposed by deep-etching, the array appears to lie on the interior thylakoid surface. The difference between fracture faces and deep-etch faces is shown even more dramatically

Fig. 8. A paracrystalline quantasome array exposed by deep-etching. ×88,000.

Fig. 9. A paracrystalline quantasome array as seen in an air-dried heavy metal shadowed preparation. ×88,000.

in Figs. 7, 8 and 9 where paracrystalline quantasome arrays are seen in fractured, deep-etched and air-dried shadowed preparations. The similarity between the deep-etched and heavy metal shadowed preparations is immediately evident as contrasted with the fractured preparations. This result could not occur if the Moor and Mühlethaler model was correct. In their model the membrane surface and fracture face should be identical. As shown in Figs. 7 and 8 this is not the case. Rather the deep-etched surface and shadowed preparation are similar as is predicted by the Branton-Park model.

Fig. 10 shows a deep-etched thylakoid preparation which had previously been treated with nitro blue tetrazolium (see *Methods*) in light in the absence of DCMU. The formazan deposits occur only on faces exposed by deep-etching and never on the fracture faces. Since histological experiments have shown that tetrazolium dyes are deposited on the exterior of the thylakoid,[11] this experiment again shows that the surfaces exposed by deep-

Fig. 10. A deep-etched thylakoid preparation previously exposed to a tetrazolium dye (NBT) in light. MS, membrane surface; FP, fracture plane. ×40,800.

etching are the true surfaces of the membrane, while those exposed by fracturing are generally on internal membrane faces. When two membranes are appressed it is possible to expose a membrane surface by fracturing.

CONCLUSIONS

The observations reported here show that the fracture face exposed by freeze-etching is generally not the membrane surface. Deep-etching, on the other hand, exposes the real membrane surface which closely resembles thylakoid surfaces as seen by shadowing air-dried or frozen-dried material. These results, combined with the obserbations that light dependent reduction of tetrazolium dyes by the thylakoids modifies the deep-etched faces and not the fracture faces, gives additional support to the Branton-Park thylakoid model in preference to that proposed by the Zürich group. The particles which Moor and Mühlethaler place on the membrane surface are probably largely buried within the thylakoids.

REFERENCES

1. Steinmann, E., *Experientia*, **8**, 300 (1952).
2. Frey-Wyssling, A., *Protoplasma*, **29**, 279 (1937).
3. Weier, T. E., Bisalputra, T., and Harrison, A., *Ultrastructure Research*, **15**, 38 (1966).
4. Oda, T. and Huzisige, H., *Exptl. Cell Res.*, **37**, 481 (1965).
5. Park, R. B., and Pon, N. G., *J. Mol. Biol.*, **3**, 1 (1961).
6. Mühlethaler, K., Moor, H., and Sarkowski, J. W., *Planta*, **67**, 305 (1965).
7. Park, R., and Branton, D., *Brookhaven Symposia in Biology*, No. 19, p. 341 (1966).
8. Kreutz, W., *Ber. deut. bot. Ges.*, **79**, 34 (1966).
9. Branton, D., and Park, R., *J. Ultrastructure Research*, in Press.
10. Park, R., *J. Cell Biol.*, **27**, 151 (1965).
11. Weier, T. E., Stocking, C. R., and Shumway, L. K., *Brookhaven Symposium in Biology*, No. 19, p. 353 (1966).

Characterization of the " Quantasome " and Its Role in Photosynthesis*†

EVANGELOS N. MOUDRIANAKIS, STEPHEN H. HOWELL‡ AND ALEXANDER E. KARU

Department of Biology, The Johns Hopkins University, Baltimore, Maryland

Grana lamellae isolated from chloroplasts of several green plants and examined in the electron microscope by a variety of preparation techniques exhibit a number of particulate units on their surfaces. Park has named one of these particles "quantasome"[1] and postulated it to be a " . . . morphological expression of the physiological photosynthetic unit."[2,3] The concept of photosynthetic unit (PSU) was introduced by Emerson and Arnold[4] as a unit of chlorophyll molecules which collectively could reduce one CO_2 molecule. Subsequent work by Park and Biggins[5] led to the morphological and chemical "characterization" of the quantasome. The material used in that study consisted of the lamellar pellet of a $145,000 \times g$ centrifugation (60 min) of sonically disrupted chloroplasts that had been previously washed three times in distilled water.[6] This preparation technique yielded membranes that retained reasonable Hill reaction rates and exhibited, among other morphological details, rectangular-shaped particles distributed at random on the surface of the membranes.[5] A less frequent distribution (below 2%) of these particles was found to be the linear or paracrystalline array. From the latter array the dimensions of the single quantasome were determined to be 150 Å × 180 Å × 100 Å. For these calculations the assumption was made that the particles were in contact with each other and no corrections were made for the increase in size contributed by the coat of metal deposited on the particles during the shadowing of the preparation.

Our analysis of chloroplast lamellae by both shadowing and negative staining[7,8] has revealed membrane-bound particles that exhibit morphological characteristics identical to those mentioned above for quantasomes,

---

* This investigation was supported by grants from the National Science Foundation (GB-3238) and the National Institutes of Health (GM-13518).

† This paper summarizes results and arguments presented in references 8 and 14.

‡ Present address: Department of Biology, Revelle College, University of California at San Diego, LaJolla, California 92307.

i.e. the particles are rectangular in shape with a central depression and appear to contain 4 subunits on their surface. They are found most frequently in a random array (Fig. 1) but occasionally appear in linear or paracrystalline array. When in paracrystalline array (Fig. 2) they exhibit a center-to-center separation of 150 Å × 180 Å, exactly the dimensions given for the quantasome. It can be seen, though, that individual particles are

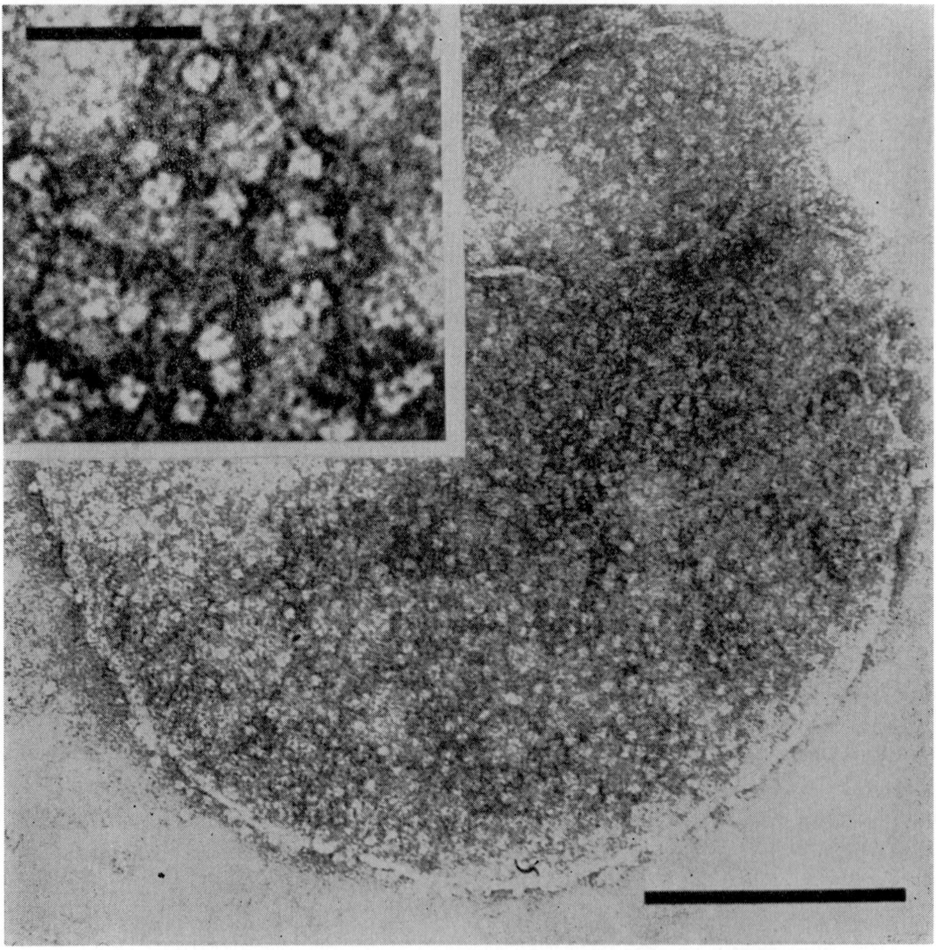

Fig. 1. Chloroplast membrane from a preparation of water-washed lamellae negatively stained with ammonium molybdate. 100 Å particles are arranged on the membrane in random array. Bar = 2,000 Å.

Inset shows detail of membrane-bound particles negatively stained with ammonium molybdate. Bar = 500 Å in inset.

Fig. 2. Chloroplast membrane-bound particles in the paracrystalline array. Individual particles measure 100 Å × 120Å. The center-to-center separation of the ordered particles is 150 Å × 180 Å. Note that the particles are not in contact with each other.

Preparation negatively stained with ammonium molybdate.[14] (A) Bar = 2,000 Å. (B) Detail of the same area. Bar = 500 Å.

not in contact with each other but are separated by definite clefts. The size of the particles in these preparations is about 100 Å, the axial ratio being 1 : 1.2. The dimensions of the particles in the shadowed preparations were larger and depended on the extent of shadowing. With excessive shadowing they could be made to approach 150 Å × 180 Å, the clefts disappeared and the particles appeared to be in contact with each other.

From these results we concluded that the particle observed by us was identical to the "quantasome" of Park and Biggins. The difference in the reported size was attributed to different methods of specimen treatment prior to microscopy. As mentioned earlier, in our hands shadowing was found to increase greatly the dimensions of the particle. Park reported a similar effect[9] (a metal coat of 20 Å thickness surrounded the shadowed particle) although no attempt was made to correct the size of the particles accordingly. If the correction is made the dimensions of the particle should be *ca.* 110 Å × 140 Å. We believe that dimensions obtained from negatively stained preparations can approximate more closely the true particle dimensions. Our measurements from such preparations gave a quantasome size of *ca.* 100 Å × 120 Å. The dimensions of 150 Å × 180 Å represent the periodicity of the quantasomes when in paracrystalline array rather than the size of the particles.

The theory that the quantasome represents the Emerson and Arnold photosynthetic unit[3,4] and the later modified hypothesis[9] that the particle is involved in electron transport associated with photosynthetic system I are both based on circumstantial evidence. The preparation that exhibited the morphological entities called quantasomes was also active in photoreduction.[5] No *quantitative* analysis was reported for other biochemical activities that possibly could be associated with the particle, i.e. CO_2-fixation or photophosphorylation.[2,6] The chemical composition of the quantasome[5] used for the estimation of its M.W. was that of the rather *heterogeneous* lamellar pellet obtained after centrifugation of chloroplast membranes at 145,000 × *g*. The same was true for the values of partial specific volume. It should be pointed out that even in this earlier paper it was realized by the authors that from the membrane preparation used in chemical analysis only a minute percentage exhibited the paracrystalline array of quantasomes, yet the above-mentioned calculations of M.W. and density were based on the assumption of a homogeneous, pure preparation of quantasomes.

Thus it appeared that most of the properties assigned to the quantasome needed further investigation and documentation. Our efforts then were turned toward the direct characterization of the function of the quantasome.

In a preliminary effort we attempted to localize the site of Hill reaction on chloroplast membranes by histochemical staining. Salts of several electron-dense tetrazolium compounds had been used previously[10–12, 23] to demonstrate the site of reduction reactions in tissue sections and on membranes. The principle of this technique is based on the difference in solubility properties between the oxidized tetrazolium salt (readily soluble in water) and the product of its reduction, the water-insoluble formazan. The precipitated formazan provides sufficient electron density to be visible in the electron microscope. A crucial factor determining the success of this method, especially in the case of membranous material, is the extent of solubility of the formazan into the lipid phase of the membranes. By the use of TNBT as an electron acceptor, it has been possible to demonstrate the location of succinic dehydrogenase on mitochondrial cristae; TNBT granules of 30–40 Å were resolved in that case.[12] From a variety of tetrazolium salts tested,[8, 12, 13] it was found that INT (2-*p*-iodophenyl-3-*p*-nitrophenyl-5-phenyl-monotetrazolium chloride) was the most suitable for the present study.

We have demonstrated that INT can be used very efficiently as a Hill oxidant by purified lamellar preparations of spinach chloroplasts.[8] It is reduced only in the presence of chloroplast lamellae active in Hill reaction, with a concomitant evolution of oxygen. There is no reduction of INT by the same chloroplast lamellae if they are kept in the dark. Evidence has also been presented concerning the specificity of the deposition of the reduced formazan on the chloroplast lamellae. Only formazan, produced during the action of light on the lamellar preparations, appeared bound to the membranes.[8] Formazan produced by added soluble reductants such as ascorbate, in the presence of non-illuminated chloroplast lamellae, did not associate with the membranes. A time course of Hill reaction with INT as hydrogen acceptor resulted in membranes having increased amounts of formazan associated with them. With long photoreduction periods aggregates of INT-formazan of rather large dimensions were produced; this result would make any correlation between the reducing site and the site of deposition impossible. To overcome this the process of the reaction was carefully controlled and microscopy was performed on preparations that had been illuminated from a few seconds to several minutes. Only those preparations that exhibited more limited formazan deposition were examined in the microscope. Aliquots of the *same* preparations used for microscopy were analyzed for the amount of formazan deposited and the chlorophyll content of the membranes by extracting the preparation with toluene or methanol, respectively, as described previously.[8]

It was observed that by increasing the time of illumination of the INT-lamellar suspension there was a progressive increase in contrast over the lamellae, corresponding to an increased amount of formazan deposited on the membranes; this increase was always uniform throughout the membrane surface and it never appeared in the form of distinct granules as would be expected if the quantasome were the site of photoreduction.[8] Previous tests had indicated that this technique was capable of resolving granular deposits of formazan in the size range of *ca.* 100 Å. Their absence from the chloroplast preparations was taken to indicate either that the photoreduction by chloroplast membranes was not localized on distinct sites on the membranes but was rather a total membrane phenomenon, or that the dimensions of such sites have to be much smaller than the dimensions of quantasomes so as to remain unresolvable by present techniques of electron microscopy.

A more direct analysis of the determination of the function of the quantasome could be achieved if the particle could be purified from membranes and if the functions of the two separate fractions were tested. We employed a variety of chemical and physical treatments on the purified chloroplast lamellae in our effort to separate the particles from the membranes.[8] Although with some treatments the number of particles per unit area of the membrane was *drastically reduced*, the rate of Hill reaction of such preparations remained *unchanged*. Complete removal of the particles from the membranes was achieved through the combination of an extraction of prewashed lamellae with a solution of 1 m*M* EDTA (pH 8.0), followed by an isopycnic centrifugation of the extracted membranes in sucrose density gradient.[8] After the lamellae had formed a rather wide band as they were approaching their equilibrium position in the gradient, they were collected as narrow fractions and were analyzed for their Hill reaction rates and their morphology in the electron microscope. It was observed[8] that the denser fractions of the gradient contained chloroplast lamellae that were completely devoid of particles. These lamellae exhibited undiminished rates of Hill reaction when compared to control lamellae. The term "control" is applied to membranes that had not been subjected to this fractionation procedure but had been kept in sucrose for the same period of time. These control membranes were completely covered with quantasome-like particles. This evidence excludes the particle from participation in any of the photochemical events that were ascribed to it by earlier investigators. Moreover, pelleting the EDTA-extracted lamellae left in the supernatant solution a unique protein species entirely devoid of any pigment, and when examined in the electron microscope this protein exhibited the characteristic structure of the quantasome.[14] Fig. 3 is an electron micrograph of a membrane purified through the sucrose density gradient centrifugation and active in Hill reac-

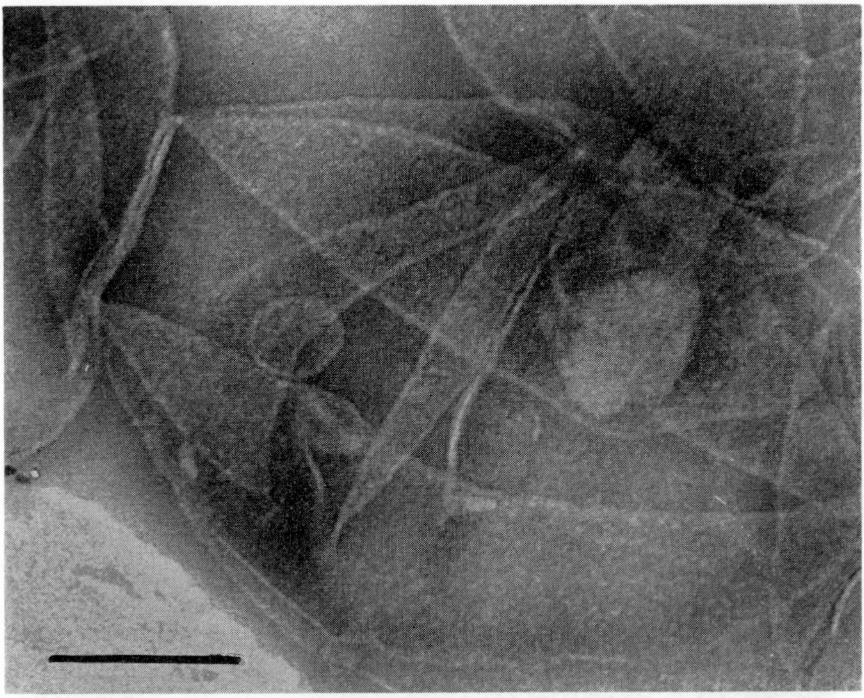

Fig. 3. Chloroplast lamellae purified through sucrose density equilibrium centrifugation. These lamellae, free of quantasomes, are highly active in Hill reaction.

Preparation negatively stained with ammonium molybdate. Bar= 1,000 Å.

tion. It can be seen that the membrane is completely devoid of any particles. Fig. 4 is from a purified preparation that contained the main protein species released from the 1 m*M* EDTA treatment of chloroplast lamellae. The isolated, non-chlorophyllous particles exhibit the morphology of quantasomes.

At this point a few details of the purification-fractionation scheme should be emphasized.

Chloroplasts prepared in the STN medium[15] are intact chloroplasts highly active in CO_2-fixation and photophosphorylation. From a sequence of water and EDTA washes it was found that *ca.* 90% of the total soluble protein was released from the membranes by three water washes. The remaining dissociable protein was released by two subsequent washes with 1 m*M* EDTA (pH 8.0) (Fig. 5).

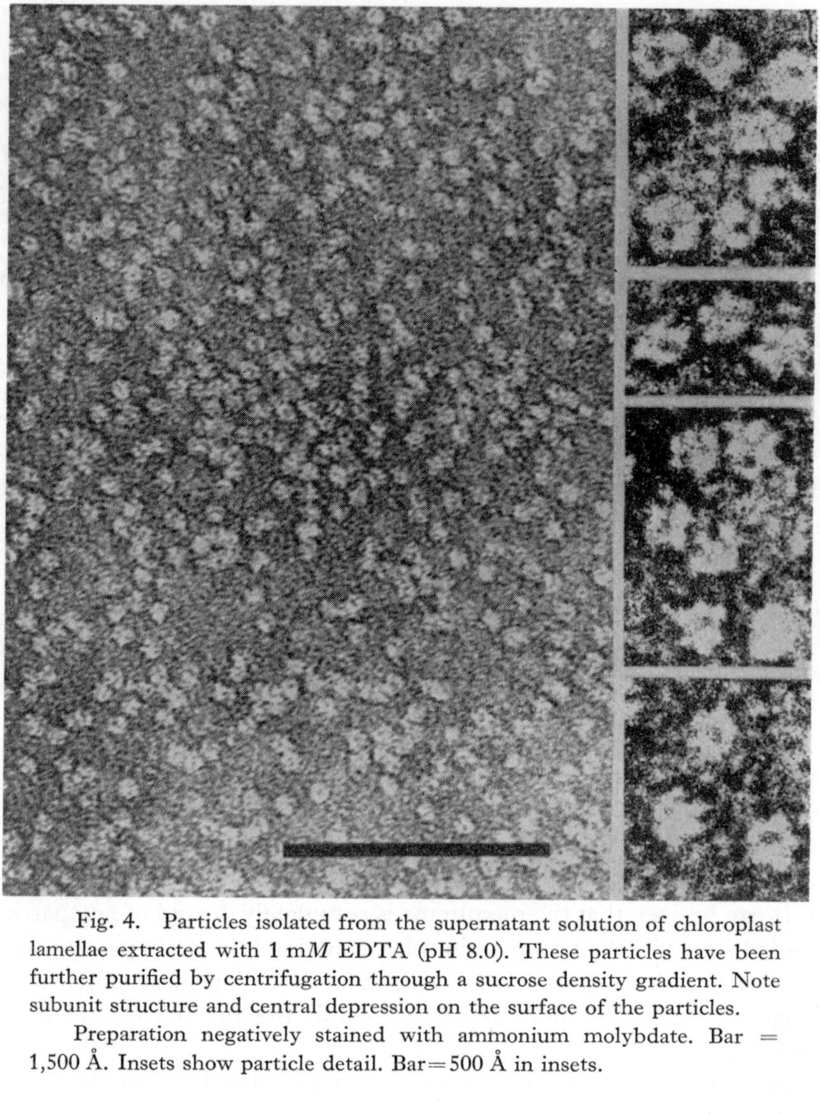

Fig. 4. Particles isolated from the supernatant solution of chloroplast lamellae extracted with 1 m*M* EDTA (pH 8.0). These particles have been further purified by centrifugation through a sucrose density gradient. Note subunit structure and central depression on the surface of the particles.

Preparation negatively stained with ammonium molybdate. Bar = 1,500 Å. Insets show particle detail. Bar = 500 Å in insets.

Disrupting the chloroplasts by exposure to distilled water causes a release of *ca.* 80% of their carboxydismutase activity in the supernatant solution. The remaining 20% (16 μmoles CO_2/min/mg Chl) is still associated with the membranous pellet. After a total of three repeated washes of the chloroplast pellet in water, there was still a significant amount of CO_2-fixing activity (less than 5%) associated with the membranes. With two fur-

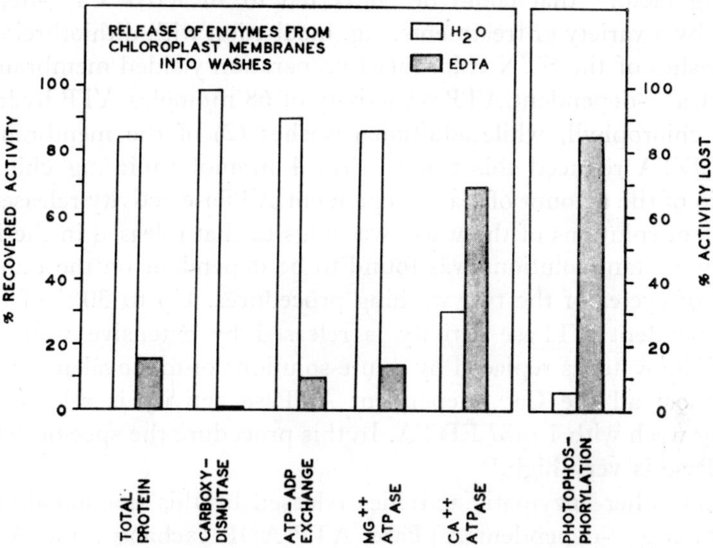

Fig. 5. Release of enzymatic activities from chloroplast membranes into supernatant solutions of membrane washes. Chloroplast membranes were treated with a sequence of three water washes followed by two 1 m*M* EDTA washes as described elsewhere.[14] Activities are reported as % recovered activity of combined supernatant solutions from either water or subsequent 1 m*M* EDTA washes, or as % activity lost from membrane pellets. Total activities of carboxydismutase, ATP-ADP exchange, Mg^{++}-dependent ATPase and Ca^{++}-dependent ATPase were determined from supernatant solutions of washes. Photophosphorylation activity was determined using the pelleted membrane fractions.

ther washes of the pellet in 1 m*M* EDTA solution, only 0.4% of the initial carboxydismutase activity remained associated with the membrane fractions. These results suggest the likelihood of earlier chloroplast membrane preparations[2,5,6] used by Park and co-workers being contaminated with significant amounts of carboxydismutase.

The three water washes of STN chloroplasts resulted in a small decrease of their photophosphorylation activity,[14] i.e. only 5%. When these membranes were washed one more time in 1 m*M* EDTA, a drastic decrease (85%) of their photophosphorylating activity was observed. Further washes in EDTA completely abolished photophosphorylation.[16] It has been reported[16,17,18] that EDTA extraction of chloroplast membranes causes uncoupling of photophosphorylation from electron transport with a concomitant release in the supernatant solutions of these extractions of a "coupling factor" that could be converted to an active Ca^{++}-dependent ATPase by a variety of treatments, e.g. incubation with dithiothreitol. Two water washes of the STN chloroplast preparation yielded membranes with specific Ca^{++}-dependent ATPase activity of 68 mμmoles ATP hydrolyzed/min/mg chlorophyll, while additional washes (2) of the membranes with 1 m*M* EDTA reduced this value to *ca.* 4 mμmoles/min/mg chlorophyll. The ratio of the amount of Ca^{++}-dependent ATPase activity released in the supernatant solutions of the water washings to that released in the EDTA-wash supernatant solutions was found to be dependent on the ratio of the number of cycles of the two washing procedures. Up to 30% of the total Ca^{++}-dependent ATPase activity is released by extensive water washes while, if the water is replaced by dilute solutions of monovalent or divalent salts, almost all the Ca^{++}-dependent ATPase activity is released in the following wash with 1 m*M* EDTA. In this procedure the specific activity of the ATPase is very high.[19]

Some other enzymatic activities released by this fractionation procedure are: Mg^{++}-dependent ATPase, ATP-ADP exchange, and Adenylate kinase. They have been discussed elsewhere[14, 19] and we will not be concerned with them here since their size and morphology are distinctly different from those of the quantasome.

It is evident, then, that the major protein species recovered in the water washes is that of carboxydismutase, while the predominant or almost exclusive component released by EDTA treatment is the coupling factor or Ca^{++}-dependent ATPase. The size and morphology of these two particles are of great interest in respect to the "quantasome" issue. The sedimentation coefficient and molecular weights have been computed for these particles and have been found to be $S_{20,w}=17.5$ (M.W. 515,000)[14, 20] for carboxydismutase and $S_{20,w}=12.7$ (350,000)[14] for the Ca^{++}-dependent ATPase.

These two protein species have been purified by sucrose density gradient centrifugation (Fig. 6) and when subjected to acrylamide disc electrophoresis each was found to migrate as a single band. The Ca^{++}-dependent ATPase was moving to the anode faster than the carboxydismutase. When examined with the electron microscope (compare Figs. 4 and 7) the two particles exhibited a remarkable similarity in shape and substructure. Both particles approximated squares and both appeared with a central depression on one surface. They differed in that the carboxydismutase particle was markedly cuboid and was larger (*ca.* 120 Å on edge)[14, 21] than the Ca^{++}-dependent ATPase (*ca.* 100 Å on edge).

The Ca^{++}-dependent ATPase in negatively stained preparations appeared identical to the negatively stained quantasome on membranes in

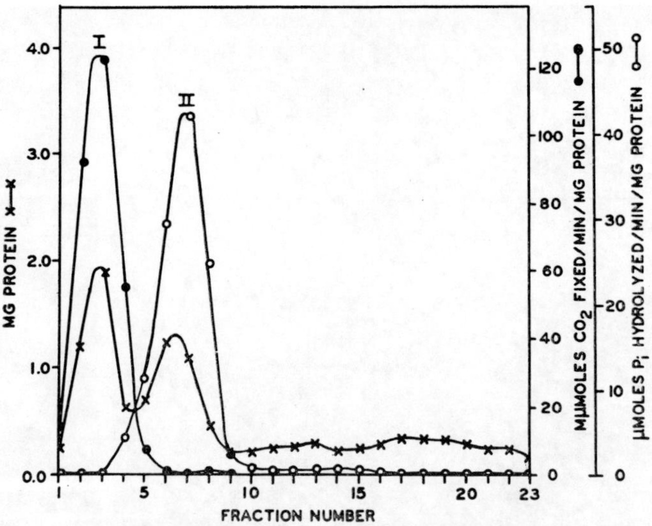

Fig. 6. Sucrose density gradient profile of 1 m*M* EDTA extract. Fractions were assayed for carboxydismutase and Ca^{++}-dependent ATPase activities. (I) denotes maximum in specific carboxydismutase activity (123 mμmoles CO_2 fixed/min/mg protein), and (II) denotes maximum in specific Ca^{++}-dependent ATPase activity at 37°C (42.1 μmoles ATP hydrolyzed/min/mg protein).

The chloroplast preparation used in this extraction was not thoroughly washed in water, in order to recover carboxydismutase activity in the EDTA extract for the purpose of comparison with the Ca^{++}-dependent ATPase.

the random array or paracrystalline array. When the particles were viewed from the surface which exhibited a central depression, the axial ratio of the particle approached unity. On this surface the particle appeared as a 4to6 sided polygon with subunits comprising the vertices (Fig. 4). In the negatively stained paracrystalline arrays the particle surface with the central depression was seen most frequently. It appeared in the ordered arrays that two sides of the particle were parallel to the long axis of the paracrystalline formation.

Observations of the isolated Ca^{++}-dependent ATPase particle on the surfaces that lacked the central depression indicated that the axial ratio increased to *ca.* 1.2.[14] The subunits generally seen on this face appeared to be parallel to the long axis of the particle. The dimensions and the appearance of subunits on this face were consistent with the structure of many quantasomes found in the random array.

Further investigations on the function of the Ca^{++}-dependent

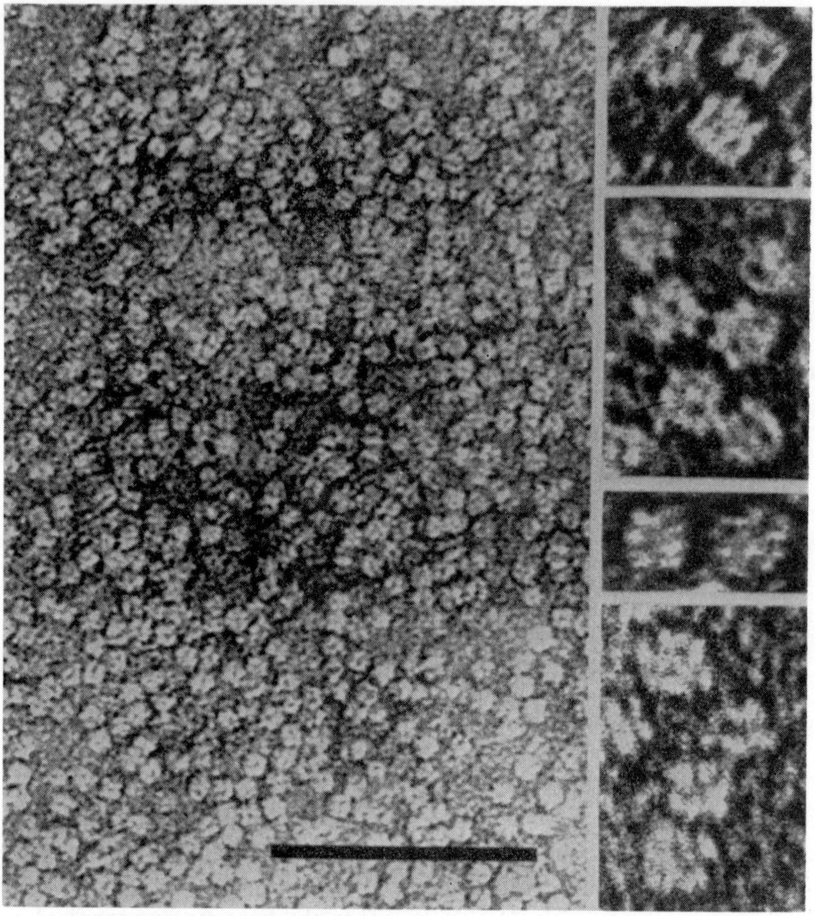

Fig. 7. Carboxydismutase particles isolated from the supernatant solution of chloroplast lamellae extracted with water. These particles have been further purified by centrifugation through a sucrose density gradient.

Preparation negatively stained with ammonium molybdate. Bar= 1,500 Å. Insets show particle detail. Bar=500 Å in insets.

Particles are larger in size than the Ca^{++}-dependent ATPase, and the number and arrangement of subunits is distinctly different in the two particles.

ATPase particle offered evidence[14] that the particle can restore photophosphorylating activity to a chloroplast lamellar preparation that has been uncoupled by means of extraction with 1 mM EDTA (pH 8.0). In these reconstitution experiments, photophosphorylation was carried out in two steps as separated light and dark reactions. The light step is thought to result in the high energy intermediate state known as X_E,[15] and the dark step supports the final phosphorylation of ADP to ATP. It was found that the purified particle derived from the supernatant of the 1 mM EDTA extraction was operating in the dark step of the phosphorylation reaction[14] in this reconstituted system. This phosphorylation reaction is simply visualized as the reversal of the ATP hydrolysis reaction and is considered to be driven by X_E: $ADP + P_i \xrightarrow{X_E} ATP$. It is therefore concluded that the particle does not participate in the light-mediated formation of the high energy state of the chloroplast membranes, X_E.

Because of the above-mentioned observations, it is most likely that the particle described as the quantasome[5] by Park and Biggins is identical to the particle exhibiting both "coupling factor" activity in photophosphorylation, and latent Ca^{++}-dependent ATPase. Hence we have suggested the term "13S photophosphorylase" in lieu of the term quantasome.[14]

It should be pointed out, though, that due to preparation techniques employed by the previous investigators and to the change in dimensions and limited resolution found associated with shadowed preparations, the possibility exists that the *larger* of the quantasomes described in the literature[2, 9] referred to carboxydismutase particles. This latter view is strengthened by the fact that both the size and the location of the quantasome as reported by Park and co-workers varied in several consecutive publications.[5, 22, 9]

The quantasome particles were originally described as ". . . 200 Å entities . . . " found " . . . attached . . . " on the surface of chloroplast lamellae.[2] The technique employed for specimen preparation in these electron microscopic studies was that of metal shadow-casting. This technique reveals only the morphological characteristics of the membrane surface. Application of the negative-staining technique to the study of quantasome morphology was found unsuccessful by Park and Biggins,[5] but it gave very satisfactory results in our hands.[8, 14] This disagreement can be explained by differences in the methodology employed. A careful selection of the metal salt solution used and of the conditions of staining must be made for every different type of preparation if maximum resolution of information is to be achieved. If these requirements are not met critically, negative staining results in preparations with either too high or too low contrast, usually very smooth-looking surfaces, *unsuitable* for high resolution analysis. We have examined chloroplast lamellae, with attached quantasomes or after complete removal of quantasomes, by shadowing and negative staining,[8, 14] and the results of both techniques are in perfect agreement with each other.

The location of the quantasome has been altered for the third time,[9] as a result of the application of the freeze-etching technique to the study of chloroplast lamellae. It has now been suggested that the particle lies within the lamella matrix. This hypothesis is in direct disagreement with several experimental observations presented for the formulation of the quantasome concept and during its subsequent characterization.[2,5,6] Our experiments[8,14] have offered evidence in support of the earlier observations of Park and co-workers that the morphological entity known as quantasome is an independent structural unit that can associate strongly with the chloroplast lamellae, and not a chemical and/or physical subunit of the lamella itself. The quantasome particle can be dissociated from the lamellae and both the purified particle and the quantasome-free lamellae are biologically active and their morphology remains unchanged. Their association can be restored *in vitro*. Whether or not freeze-etching splits the lamellae along inner surfaces, as well as what is the ultrastructure of these faces, are interesting questions but are entirely unrelated to the question of the nature and function of the originally described quantasome.

We would like to conclude by emphasizing the following points.

1. The results of our experiments have indicated that the 13S photophosphorylase (quantasome) particle does not represent the morphological counterpart of the Emerson and Arnold photosynthetic unit, as claimed originally by Park;[3] neither does it participate in electron transport reactions as suggested later.[9]

2. The site of Hill reaction does not appear to be localized on distinct morphological entities in the size range of *ca.* 100 Å. Rather, it appears that photoreduction is a total membrane phenomenon with the photoreducing center either transitory on the membrane or delocalized over the membrane surface.

3. The previous statement does not imply that we view the membrane of chloroplast thylakoids as an amorphous homogeneous lipoprotein layer. On the contrary, our preliminary analysis of the system, as well as work from other laboratories (see ref. 24), reveals a considerable degree of fine structure and organization. The size of these structural elements is considerably smaller than the size of the 13S photophosphorylase (quantasome). It is likely that subunits of this ultrastructure could be described and isolated in the near future. In addition, it is interesting to speculate on a possible association of the electron transport chain with these subunits, and on their organization in some unique pattern within the chloroplast lamellae. It is also possible that the 13S photophosphorylase (quantasome) particle can interact with one or more kinds of these subunits and that this interaction

results in one or another type of array of the photophosphorylase on the chloropast lamellae.

Acknowledgements

We are greatly indebted to Prof. A. T. Jagendorf for his interest and moral support in this study. The technical assistance of Mr. Joel Lamon and Mr. Frank Gafford is greatly appreciated.

REFERENCES

1. Park, R. B., *J. Chem. Ed.*, **39**, 424 (1962).
2. Park, R. B., and Pon, N. G., *J. Mol. Biol.*, **3**, 1 (1961).
3. Lichtenthaler, H. K., and Park, R. B., *Nature*, **198**, 1070 (1963).
4. Emmerson, P., and Arnold, W., *J. Gen. Physiol.*, **16**, 191 (1932).
5. Park, R. B., and Biggins, J., *Science*, **144**, 1900 (1964).
6. Park, R. B., and Pon, N. G., *J. Mol. Biol.*, **6**, 105 (1963).
7. Moudrianakis, E. N., *J. Cell Biol.*, **23**, 63A (1964).
8. Howell, S. H., and Moudrianakis, E. N., *J. Mol. Biol.*, **27**, 323 (1967).
9. Park, R. B., and Branton, D., *Brookhaven Symposia in Biology*, No. 19, p. 341 (1966).
10. Sedar, A. W., Rosa, C. G., and Tsou, K. C., *J. Histochem. and Cytochem.*, **10**, 506 (1962).
11. Rosa, C. G., and Tsou, K. C., *J. Cell Biol.*, **16**, 445 (1963).
12. Sedar, A. W., Rosa, C. G., and Tsou, K. C., *Electron Microscopy*, 2, L-7. 5th International Congress for Electron Microscopy, Academic Press, New York, 1962.
13. Atkinson, E., Melvin, S., and Fox, S. W., *Science*, **111**, 385 (1950).
14. Howell, S. H., and Moudrianakis, E. N., *Proc. Natl. Acad. Sci., U.S.A.*, **58** (1967).
15. Hind, G., and Jagendorf, A. T., *Proc. Natl. Acad. Sci., U.S.A.*, **49**, 715 (1963).
16. McCarty, R. E., and Racker, E., *Brookhaven Symposia in Biology*, No. 19, p. 202 (1966).
17. Avron, M., *J. Biol. Chem.*, **237**, 2011 (1962).
18. Vambutas, V. K., and Racker, E., *J. Biol. Chem.*, **240**, 2660 (1965).
19. Karu, A. E., and Moudrianakis, E. N., in preparation.
20. Trown, P. W., *Biochemistry*, **4**, 908 (1965).
21. Hasselkorn, R., Fernandez-Moran, H., Kieras, F. J., and van Bruggen, E. F. J., *Science*, **150**, 1598 (1965).
22. Park, R. B., *J. Cell Biol.*, **27**, 151 (1965).
23. Sedar, A. W., and Burde, R. M., *J. Cell Biol.*, **27**, 53 (1965).
24. Branton, D., *Proc. Natl. Acad. Sci., U.S.A.*, **55**, 1048 (1966).

On the Nature of the Particles Attached to the Thylakoid Membrane of Spinach Chloroplasts

S. MURAKAMI

Department of Biology, University of Tokyo, Komaba, Meguro, Tokyo

Various particles have been described as constituent units of the thylakoid membrane as investigated by different techniques of electron microscopy.[1–6] The particulate structure of the thylakoid membrane was also deduced from the results of low-angle X-ray scattering analysis.[7] However, interpretations of the electron microscopic images so far obtained are widely divergent, thus leading to various different structural models of the thylakoid membrane.[8–12] Besides these particles in the membrane, several particles have been discovered to be attached to the thylakoid membrane as observed by negative staining[13, 14] or by freeze-etching techniques.[15] The structure and biochemical properties of all these particles, however, have not yet been fully elucidated. This paper deals with the structure and function of the particles observed by the negative staining technique.

Chloroplasts were isolated from spinach leaves (*Spinacia oleracea*) in 0.02 *M* Tris-HCl buffer, pH 8.0, containing 0.4 *M* sucrose and 0.01 *M* NaCl. The chloroplasts were ruptured by osmotic shock in 0.01 *M* NaCl, and the thylakoids were separated from water-soluble substances in the stroma by centrifugation. All these procedures were carried out at temperatures lower than 5°C.

For electron microscopic observation the thylakoids were placed, at room temperature, on a grid covered with a collodion film coated with a thin layer of carbon, and negatively stained with one per cent phosphotungstic acid neutralized by the addition of KOH. Fig. 1 shows a part of the thylakoid membrane. Numerous spherical particles, 80–100 Å in diameter, will be found uniformly covering the whole surface of the membrane. In slightly under-focused images these particles show a granular appearance, suggesting a structure consisting of several subunits. No pseudo-crystalline array of particles was observed under the conditions of preparation employed. The average number of particles was about 30 per 0.1 μ square of thylakoid surface(i.e., 3,000 per 1 μ^2). For convenience in carrying out the experiment, counting and computation were made on the basis of 0.1 μ square. At the edge of the thylakoid, particles will be seen closely

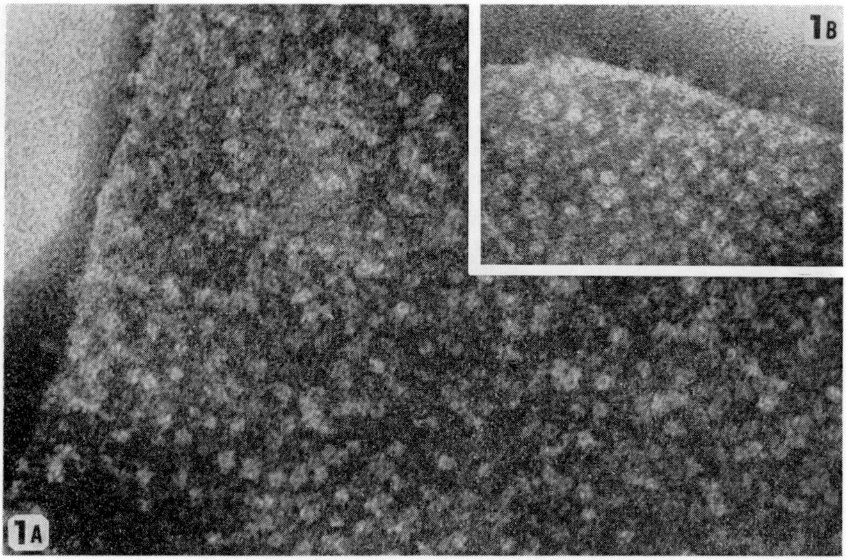

Fig. 1. Part of the thylakoid membrane of spinach chloroplast, showing the particles attached to the outer surface of the membrane (A). At the edge of the thylakoid, particles are seen closely attached to the outer surface of the membrane (B). ×200,000.

attached to the outer surface of the membrane; no stalk-like figure was observed (see Fig. 1B).

The association of the particles with the thylakoid was affected by various treatments. Washing the thylakoids with 0.05 *M* Na-K phosphate buffer, pH 7.4, or distilled water reduced the number of particles on the membrane. When the thylakoids were extracted with aqueous acetone, the particles almost completely disappeared, no particles of similar appearance being observed either on the membrane surface or in the supernatant solution. On the other hand, after treatment with non-polar sovents like ether, hexane and heptane, a large portion of the particles remained attached to the membrane. In experiments using EDTA solutions of varied concentrations as washing media, a remarkable concentration-dependent effect of the chelating agent was discovered. In these experiments, thylakoids (100 μg chlorophyll per ml) were treated for 30 minutes at 4°C with the washing medium containing 0.1–4.0 m*M* EDTA and 0.01 *M* NaCl. At concentrations between 0.3 m*M* and 1.0 m*M*, EDTA caused a marked removal of the particles from the thylakoid surface (Fig. 2), while at concentrations of 0.1 m*M* and 2.0 m*M* or higher, there was but little effect (Fig. 3). No

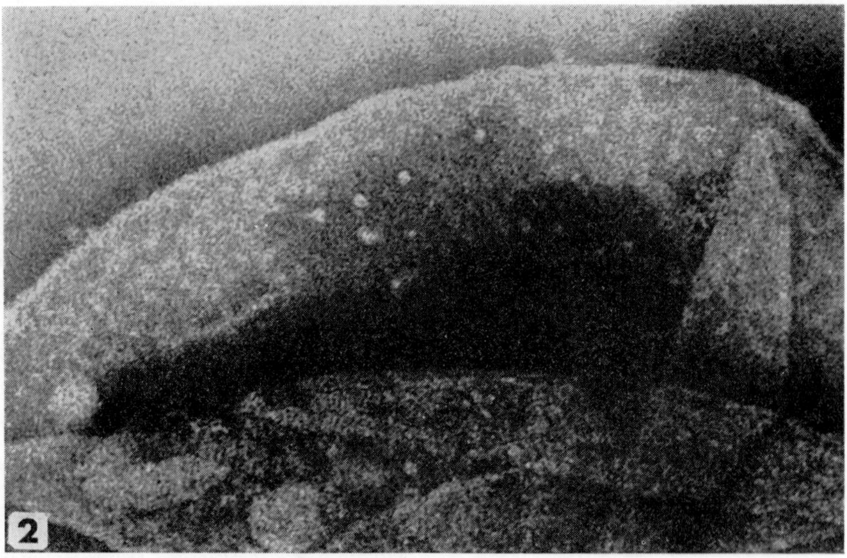

Fig. 2. Thylakoid membrane treated for 30 minutes with 1.0 m*M* EDTA solution containing 0.01 *M* NaCl. Most of the particles on the membrane had been removed. ×200,000.

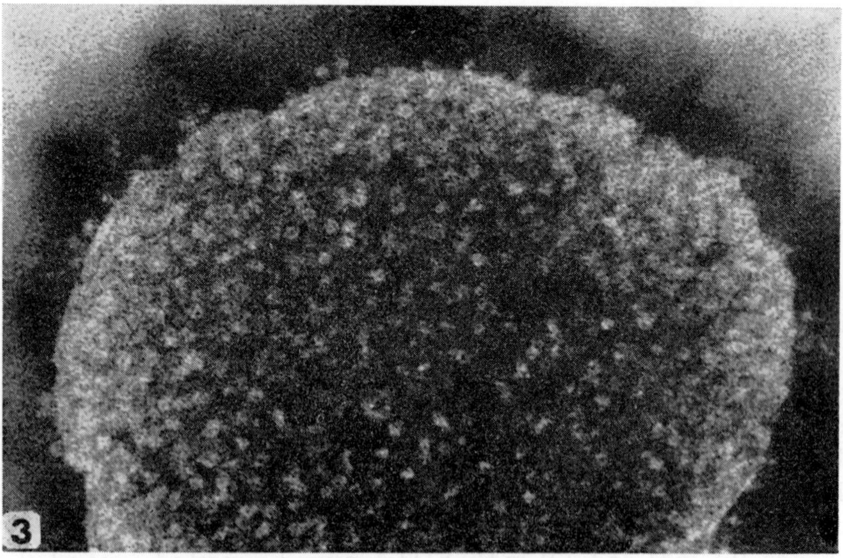

Fig. 3. Thylakoid membrane treated for 30 minutes with 4.0 m*M* EDTA solution containing 0.01 *M* NaCl. The particles were not affected by EDTA extraction. ×200,000.

significant change in the general shape and magnitude of the individual thylakoids was noticed before and after treatment with EDTA. The same effects of varied concentrations of EDTA were found with respect to the population of the particles at the edge of the thylakoid. The average number of particles per unit area ($0.1\ \mu \times 0.1\ \mu$) of the washed and non-washed thylakoids was plotted against the concentration of EDTA in the extracting medium (Fig. 4). The effectiveness of EDTA at the limited concentration range of 0.3–1.0 m*M* can be clearly seen.

The above-described behavior of the particles was reminiscent of a similar finding reported by McCarty and Racker[16] pertaining to the uncoupling effect of EDTA in photophosphorylation of spinach chloroplasts. These authors reported that EDTA was effective in this respect just at the same concentration range (0.5 to 1.0 m*M*) as in the present study. An experiment was therefore designed to compare the effects of various EDTA concentrations on the photophosphorylation, on the one hand, and on the number of particles on the membrane, on the other hand. After treatment at various EDTA concentrations (see above), the thylakoids were collected by centrifugation, and the number of remaining particles and activity of photophosphorylation were determined. For the assay of photophosphorylation, the thylakoids were incubated for 3 minutes at 20°C with 50,000 lux incandescent light. The reaction mixture contained 40 μM Tris-HCl (pH 7.8), 90 μM NaCl, 5 μM $MgCl_2$, 0.1 μM PMS, 5 μM (^{32}P)Pi, 1 μM ADP and chloroplasts equivalent to 30 μg chlorophyll per 3 ml. ATP formation was measured by the method of Avron.[17] As shown in Fig. 4, the effects of EDTA on the particles and on photophosphorylation were similar; at concentrations between 0.3 m*M* and 1.0 m*M* EDTA causes nearly complete uncoupling of photophosphorylation (cf. Avron[18,19]) and the number of particles on the membrane was also minimal. The similarity of the two curves in Fig. 4 clearly indicates a close correlation between the presence of particles and the activity of photophosphorylation.

An attempt was made to purify the particles under investigation. The chloroplasts were treated with 0.8 m*M* EDTA for 30 minutes. The supernatant solution was passed through Hyflo Super Cel and then applied to a DEAE cellulose column equilibrated with 0.02 *M* Tris-HCl buffer, pH 8.0, containing 0.1 m*M* ATP. The particles were recovered in a colorless fraction obtained on elution with 0.02 *M* Tris-HCl buffer, pH 8.0, containing 2 m*M* EDTA, 30 m*M* ammonium sulfate and 0.1 m*M* ATP, which was further purified by precipitation with 40% ammonium sulfate. These procedures were devised after the prescription of McCarty used for the preparation of the coupling factor of photophosphorylation from spinach

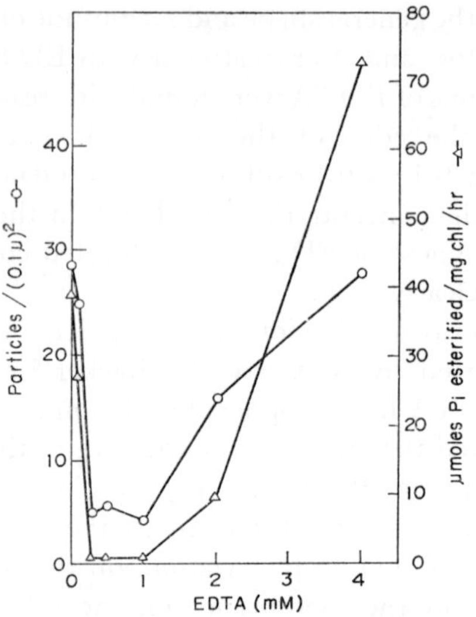

Fig. 4. Effects of various EDTA concentrations on the average number of particles and the activity of photophosphorylation.

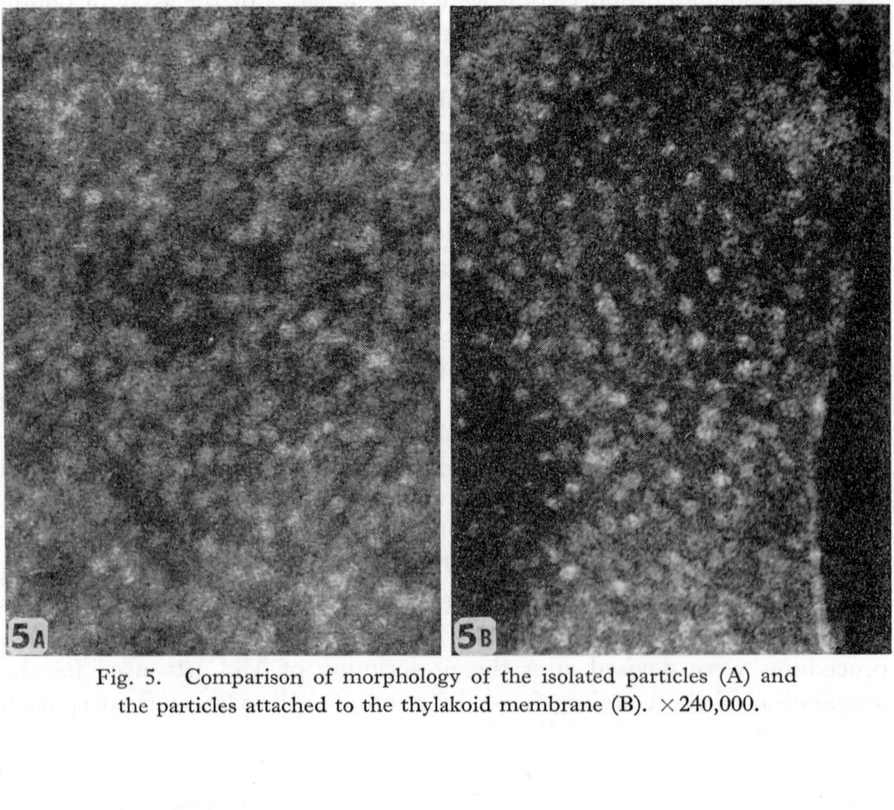

Fig. 5. Comparison of morphology of the isolated particles (A) and the particles attached to the thylakoid membrane (B). ×240,000.

chloroplasts. In Fig. 5 the isolated particles (5A) are compared with the particles on the thylakoid membrane (5B) at the same magnification. The similarity in morphology of the two particles is worth noting; both are 80–100 Å in diameter and have a similar granular appearance. The morphology of the particles either in the isolated state or attached to the thylakoid membrane evidently differs from that of the Fraction I protein particles; the latter are reported to be cubic in shape and 120 Å in edges, and consist of 24 identical subunits.[20]

Dr. H. Sakurai, working with a sample of particles prepared and purified by the same procedure, has tested the coupling effect of the particles in photophosphorylation of spinach chloroplasts and also the activity of Ca^{++} dependent ATPase. He obtained results showing that the particles thus obtained were identical with the "coupling factor" described by Racker and his colleagues.[16,21] It may then be most plausible to assume that the particles observed on the negatively stained thylakoid membrane of spinach chloroplasts represent the site of coupling of photophosphorylation, and that their removal from the membrane was the cause of the uncoupling effect of EDTA at the indicated range of concentrations.

From the average number of particles per unit area, i.e., 30 particles $/(0.1\ \mu)^2$, the area of the thylakoid membrane available for each particle can be calculated to be 180 Å square. According to Wolken,[22] on the other hand, the surface concentration of chlorophyll on the lamellae is of the order of 3×10^{13} molecules/cm^2. From these two values, the ratio of the particle, coupling factor, to chlorophyll molecule is estimated to be approximately 1: 100. This ratio might, however, be somewhat altered when we consider possible swelling of the thylakoids during preparation.

Acknowledgements

The author wishes to thank Dr. H. Sakurai, Department of Biology, Waseda University, for his help in performing the photophosphorylation assay and in preparing the coupling factor particles. This work was supported by a grant in aid for Special Project Research on Biophysics from the Ministry of Education, Japan.

REFERENCES

1. Murakami, S., Katoh, S., and Takamiya, A., in *Intracellular Membrane Structure* (S. Seno and E. V. Cowdry, eds.), pp. 173–189, Japan Soc. for Cellular Chem., Okayama, 1965.
2. Weier, T. E., Engelbrecht, A. H. P., Harrison, A. and Risley, E. B., *J. Ultrastr. Res.*, **13**, 92 (1965).

3. Park, R. B., and Biggins, J., *Science*, **144**, 1009 (1964).
4. Park, R. B., and Branton, D., in *Energy Conversion by the Photosynthetic Apparatus*, pp. 341–352, Brookhaven National Laboratory, Upton, New York, 1967.
5. Menke, W., *Zeitschr. Naturforsch.*, **20b**, 802 (1965).
6. Menke, W., in *Energy Conversion by the Photosynthetic Apparatus*, pp. 328–340, Brookhaven National Laboratory, Upton, New York, 1967.
7. Hosemann, R., and Kreutz, W., *Naturwiss.*, **53**, 298 (1966).
8. Murakami, S., *J. Electron Microscopy*, **13**, 234 (1964).
9. Menke, W., in *Biochemistry of Chloroplasts* (T. W. Goodwin, ed.), Vol. 1, pp. 3–18, Academic Press, London and New York, 1966.
10. Mühlethaler, K., in *Biochemistry of Chloroplasts* (T. W. Goodwin, ed.), Vol. 1, pp. 49–64, Academic Press, London and New York, 1966.
11. Kreutz, W., in *Biochemistry of Chloroplasts* (T. W. Goodwin, ed.), Vol. 1, pp. 83–88, Academic Press, London and New York, 1966.
12. Weier, T. E. and Benson, A. A., in *Biochemistry of Chloroplasts* (T. W. Goodwin, ed.), Vol. 1, pp. 91–113, Academic Press, London and New York, 1966.
13. Oda, T. and Huzisige, H., *Exptl. Cell Res.*, **37**, 481 (1965).
14. Bronchart, R., *C. R. Acad. Sc. Paris*, **260**, 4564 (1965).
15. Mühlethaler, K., Moor, H., and Szarkowski, J. W., *Planta*, **67**, 305 (1965).
16. McCarty, R. E., and Racker, E., in *Energy Conversion by the Photosynthetic Apparatus*, pp. 202–214, Brookhaven National Laboratory, Upton, New York, 1967.
17. Avron, M., *Biochim. Biophys. Acta*, **40**, 257 (1960).
18. Avron, M., *Biochim. Biophys. Acta*, **77**, 699 (1963).
19. Avron, M., in *La Photosynthèse*, pp. 543–555, Colloque International Photosynthèse, C.N.R.S.-C.E.A., Gif-sur-Yvette, France, C.N.R.S., Paris, 1963.
20. Haselkorn, R., Fernández-Morán, H., Kieras, F. J., and van Brugen, E. F. J., *Science*, **150**, 1598 (1965).
21. Vambutas, V. K., and Racker, E., *J. Biol. Chem.*, **240**, 2660 (1965).
22. Wolken, J. J., in *Photosynthetic Mechanisms of Green Plants* (B. Kok and A. T. Jagendorf, eds.), pp. 575–586, Nat. Acad. Sci.-Nat. Res. Council, 1963.

On the Light Scattering Pattern of Spinach Chloroplast Suspension*

YASUO MUKOHATA

Department of Biology, Faculty of Science, Osaka University, Toyonaka, Osaka, Japan

About five years ago, Packer[1] observed that the intensity of the light scattered at 90° from spinach chloroplasts suspended in a hypotonic medium containing photophosphorylating reagents was increased reversibly by actinic illumination. Independently, Itoh, Izawa and Shibata[2] reported that the size distribution of isolated spinach chloroplasts in phosphate buffer shifted on illumination to smaller sizes as measured by a Coulter counter and that the rectilinear attenuance of chloroplast suspensions increased reversibly in accordance with the increase of light scattering. It was concluded from these findings that chloroplasts shrink on illumination under the photophosphorylating condition, which was supported by the results obtained by chlorocrit[2] and viscometry.[3] The apparent viscosity of spinach chloroplast suspensions decreased in the light and was restored in the dark under the photophosphorylating condition. However, there were appreciable differences in the rate between the changes measured by different techniques. The rates of shrinkage measured by the counter or a viscometer were considerably slower (the time required for complete change was about 10 minutes) than those measured by the optical methods (the time was a fraction of a minute).

Following these experiments, Dilley and Vernon[4,5] measured the light scattering change of spinach chloroplasts and the reversible fluxes of cations through the choloroplast membrane, and pointed out the parallelism between them. Hind and Jagendorf[6] also discussed the correlation between the rapid pH response of an unbuffered chloroplast suspension and the light scattering change in relation to the formation of a high energy intermediate. The reversible increase of light scattering in these experiments was observed in conditions where electron transport was occurring but no phosphorylation was taking place. Therefore, if the light scattering increase may be taken as a measure of shrinkage of chloroplasts as predicted by Packer[1,7] and supported

* This work was partly supported by a grant in aid for special project research on biophysics.

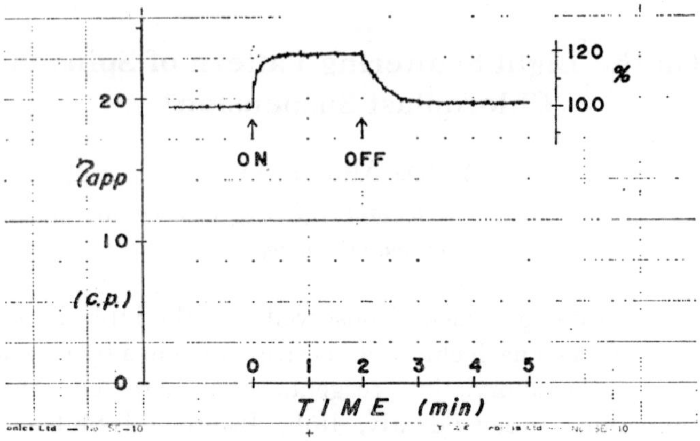

Fig. 1. Change of viscosity of a chloroplast suspension during electron transport. ON and OFF indicate the start and the end of the actinic illumination of white light (750 W projector lamp). The viscosity was measured with a cone-glass plate viscometer[8] at 20±0.2°C under the velocity gradient of 38 sec^{-1} for a suspension of spinach chloroplasts (4.7 mg/ml in chlorophyll concentration) in medium containing 35 mM NaCl, 1 mM $MgCl_2$, 0.1 mM potassium ferricyanide, 10 mM Tris-HCl, pH 7.0. The sample chloroplast suspensions were centrifugal fractions obtained between 400×g (10 min) and 600×g (10 min) in 0.35 M NaCl.

by Dilley and Vernon,[5] the chloroplasts, when placed in circumstances where electron transport is occurring in the light under either photophosphorylating or non-photophosphorylating conditions, could be interpreted as undergoing shrinkage.

Under conditions favorable for electron transport but not for phosphorylation, the viscosity of a concentrated suspension of spinach chloroplasts as examined by a cone-glass plate viscometer[8] increased rather rapidly and reversibly upon actinic illumination, the phenomenon being different from the change observed under phosphorylating conditions.[3] One of the recorded curves is shown in Fig. 1. The viscosity increased on illumination with a rate comparable to that of the scattering increase. The increased viscosity was restored to the initial level when the light was turned off, although in some results the recovery was slower or incomplete. It was also found that, with increasing NaCl concentration, the dark level of the viscosity generally became higher and the viscosity change became smaller. Although it was described[7] that chloropasts swelled slowly in a NaCl medium under the condition favorable for electron transport, this rapid viscosity increase is considered to be due to changes in the electroviscous effects of negatively charged chloroplasts in suspension rather than due to the reversible swelling. Nobel and Mel[9] measured a 15 per cent increase of the electrophoretic mobility of spinach chloroplasts in the light and suggested a change in the surface charge of chloroplasts. Jagendorf, Hind and Neumann[10] reported negative evidence for the size change of chloroplasts during the scattering change.

If the scattering change is due to a structural modification[11,12] caused by a change in the surface charge, a change of pH of the medium may also cause a scattering change. Therefore, changes of scattering on addition of NaOH, KOH or HCl to an unbuffered spinach chloroplast suspension containing an electron carrier or acceptor were observed. The results obtained with potassium ferricyanide were previously reported.[13] The patterns obtained with phenazine methosulfate are summarized in Fig. 2. When the suspension was illuminated, the scattering increased and,

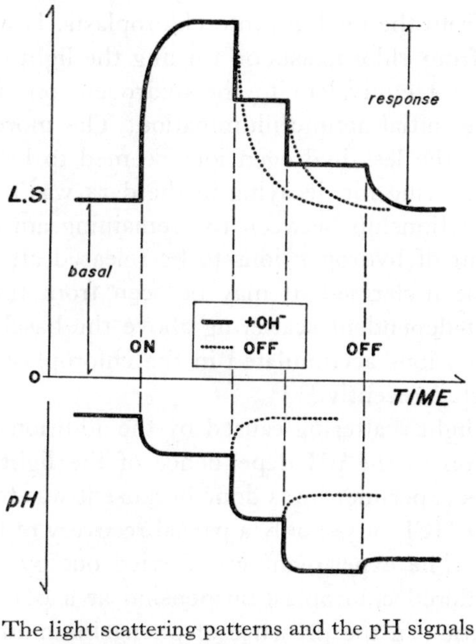

Fig. 2. The light scattering patterns and the pH signals of chloroplasts suspended in an unbuffered medium containing phenazine methosulfate (10 μM) and NaCl at a low concentration.

ON, OFF and OH^- represent the start and the end of the actinic illumination and addition of an alkali solution, respectively. For further details, see text.

simultaneously, the pH value rose as discovered by Neumann and Jagendorf.[14] After both readings of scattering and pH became constant in the light, an alkali (NaOH or KOH) solution was added with a microsyringe. On the addition, the light-induced scattering change decreased instantaneously with a consequent rise of pH. The addition of a large quantity of OH^- ions diminished the scattering intensity to the basal level (the level of scattering in the dark). As shown by the dotted lines, any OH^--reduced levels of scattering in the light decayed to the basal level when the light was turned off. The basal scattering level in the dark was also dependent on the pH of the medium but was not so sensitive to the pH change as the light induced scattering increment. The change of the basal level was approximately 4 per cent per pH unit in the pH range from 6 to 9.5 (ref. 13). Compared with this change of the basal level, the change caused by OH^- ion of the light-dependent scattering was much greater, and therefore the light-dependent and basal scattering were easily distinguishable from each other.

The light-dependent pH increase may be due to the reversible influx of hydrogen ions from the medium into chloroplasts. However, the hydrogen ions released from chloroplasts on turning the light off after the addition of alkali was not equivalent to the hydrogen ions incorporated into chloroplasts on the initial actinic illumination. The more OH^- ions were added in the light, the less hydrogen ions seemed to be released and the less capacity of scattering for decaying in the dark was retained. Although the quantitative relationship between the remaining amplitude of scattering and the amount of hydrogen ions to be released after the addition of alkali has not yet been clarified, it may be seen from the above observations that the light-dependent scattering above the basal level reflects the amount of hydrogen ions accumulated in the chloroplasts during electron transport, as reported recently.[13,15]

The drop of light scattering caused by the addition of OH^- ions was examined in relation to the pH dependence of the light-dependent scattering change. This experiment was done because it was found previously[13] that the addition of HCl caused only a partial recovery of the OH^- reduced level of scattering. The experiment was carried out by adding NaOH or KOH to an unbuffered chloroplast suspension at a pH between 4.5 and 9.0 under illumination in the presence of phenazine methosulfate. In Fig. 3, the pH dependence of the light-dependent scattering above the basal level before the addition of alkali is shown by a curve, and the readings after the addition of alkali are plotted against pH and are shown by open circles. The data shown in the figure are the relative intensities calculated

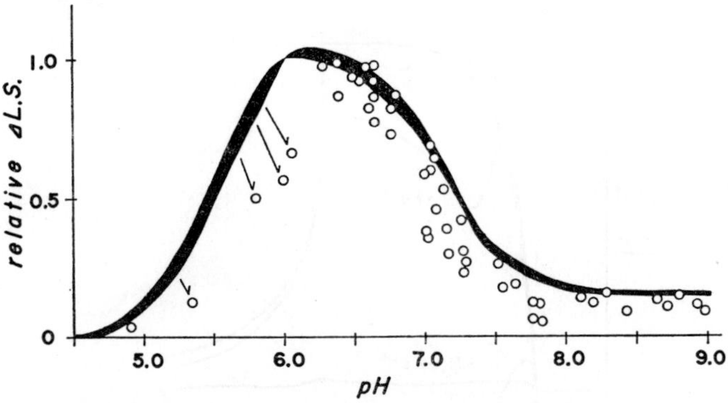

Fig. 3. The pH dependency of the light-dependent scattering (curve) and the effect of the addition of OH^- ions (open circles).

Chloroplasts at a final concentration of 3.5 μg chlorophyll per ml were suspended in 50 ml of a medium containing 0.1 *M* NaCl, 1 m*M* $MgCl_2$ and 10 μM phenazine methosulfate at 30°C. The starting pH of the suspension was adjusted by addition of either NaOH or HCl solution. For further details, see text.

by taking as unity the intensity at pH 6.0 before the addition of alkali, and the relative variations of the intensity observed at a given pH for ten chloroplast preparations before the addition of alkali are shown by the width of the curve. The open circles above pH 6 are nearly on the curve. This appears to suggest that the addition of alkali causes a drop of scattering in the light which can be expected solely from the pH dependence curve obtained before the addition of alkali. As indicated by arrows in the figure, however, the intensities measured below pH 6 also decreased on the addition of alkali, not obeying the pH dependence obtained before the addition of alkali. If the intensity is dependent only on the pH of the medium, we may expect an increase of the intensity on the addition of alkali below pH 6, according to the curve. The unexpected result indicates that OH^- ions added during electron transport interacted with some unknown materials responsible for the light-dependent scattering change. The slight deviations above pH 6 of the open circles from the curve may, therefore, be significant, and the approximate agreement of some of the open circles with the data before the addition of alkali may have been fortuitous.

The time course of the change of light scattering depends on the experimental conditions and is affected by the presence of high valent ions, detergents or denaturants. In Fig. 4 are shown the patterns obtained with spinach chloroplasts suspended in a dilute NaCl (35–175 m*M*) solution

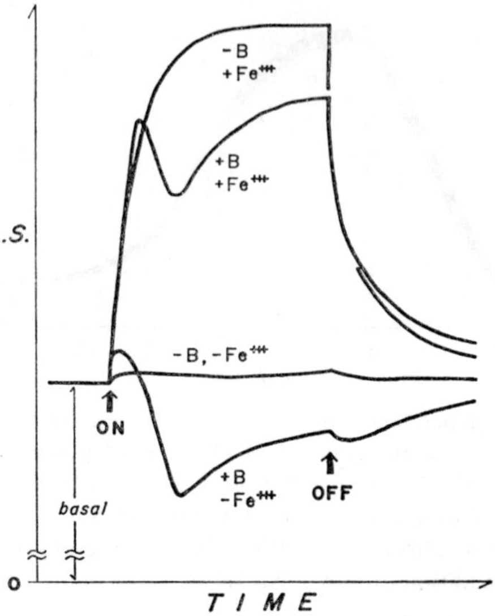

Fig. 4. The light scattering patterns indicating approximate additivity of the electron-transport dependent scattering and the basal scattering.

Chloroplasts (9.6 μg in chlorophyll content) were suspended in 3 ml of a medium containing 35 m*M* NaCl and 1 m*M* $MgCl_2$ with or without a buffer (B: in this case, 10 m*M* Tris-HCl, pH 7.0) and with or without potassium ferricyanide (Fe^{+++}: 0.1 m*M*) at 20°C. The pH of the suspension without the buffer was adjusted to the pH of the corresponding suspension with the buffer by adding to the mixture either NaOH or HCl solution.

with or without a buffer(Tris-HCl, citrate or barbiturate) of about 10 m*M*. In the presence of both potassium ferricyanide(Fe^{+++}, 0.1 m*M*) and the buffer (B), the scattering pattern usually showed a minimum at about 18 seconds after the start of illumination. The scattering pattern obtained without ferricyanide as the control showed a minimum similar to that obtained with ferricyanide, although the scattering level observed without ferricyanide was much lower. In the absence of the buffer and in the presence of ferricyanide, on the other hand, a common pattern, which can be surmised to be a result of combination of two first order reactions,[16] was obtained. Subtraction of the curve obtained with the buffer but without ferricyanide from the curve obtained with both the buffer and ferricyanide provided a pattern similar to the common pattern observable with ferricyanide in the absence of the buffer. In other words, the effect of ferricyanide in the light was not affected by the presence of the buffer. This may imply that the light-induced scattering change superposed on the basal scattering level has a close connection with the electron transport in chloroplasts.

The change of scattering that is dependent on electron transport may be attributed to a conformational modification of the materials inside chloroplast, which will cause a change in birefringent properties. The modification will be triggered and energized by electron transport, and may be closely correlated to hydrogen ion influx,[6] potassium ion efflux[5] and accumulation of hydrogen ions.[13,15] In this connection, it is interesting to note a suggestive work of Pressman[17] who measured, simultaneously, hydrogen ion and potassium ion movements, light scattering, fluorescence and oxygen consumption for a suspension of rat liver mitochondria. His recordings illustrated an oscillatory change of scattering, which decreased in the same phase as an increase of hydrogen ion concentration and in the same phase as a decrease of potassium ion concentration in the medium after addition of valinomycin. The rapid and reversible increase of the light scattering of chloroplasts during electron transport can therefore be considered to be mostly independent of the size change,[16] and rather to be intimately correlated to conformational changes of the materials inside chloroplasts. This scattering change dependent on electron transport will provide a unique means for studying the electron transport system in relation to conformation and orientation of macromolecules in chloroplasts.

Acknowledgement

The author thanks Professor Toshizo Isemura for his discussion and Mr. Hiroshi Maeda and Mr. Masayasu Mitsudo for their technical assistance in some of the experiments.

REFERENCES

1. Packer, L., *Biochim. Biophys. Acta*, **75**, 12 (1963).
2. Itoh, M., Izawa, S., and Shibata, K., *ibid.*, **66**, 319 (1963).
3. Mukohata, Y., and Packer, L., *ibid.*, **79**, 211 (1964).
4. Dilley, R. A., and Vernon, L. P., *Biochemistry*, **3**, 817 (1964).
5. Dilley, R. A., and Vernon, L. P., *Arch. Biochem. Biophys.*, **111**, 365 (1965).
6. Hind, G., and Jagendorf, A. T., *J. Biol. Chem.*, **240**, 3210 (1965).
7. Packer, L., Siegenthaler, P-A., and Nobel, P. S., *J. Cell Biol.*, **26**, 593 (1965).
8. Mukohata, Y., Mitsudo, M., and Isemura, T., *Ann. Rep. Biol. Works, Fac. Sci., Osaka Univ.*, **13**, 93 (1965).
9. Nobel, P. S., and Mel, H. C., *Arch. Biochem. Biophys.*, **113**, 695 (1966).
10. Jagendorf, A. T., Hind, G., and Neumann, J., *Sixth Int. Congr. Biochem.*, New York, X-S-9 (1964).

11. Aden, A. L., and Kerker, M., *J. Appl. Phys.*, **22**, 1242 (1951).
12. Morriss, R. H., and Collins, L. F., *J. Chem. Phys.*, **41**, 3357 (1964).
13. Mukohata, Y., Mitsudo, M., and Isemura, T., *Ann. Rep. Biol. Works, Fac. Sci., Osaka Univ.*, **14**, 107 (1966).
14. Neumann, J., and Jagendorf, A. T., *Arch. Biochem. Biophys.*, **107**, 109 (1964).
15. Deamer, D. W., Crofts, A. R., and Packer, L., *Biochim. Biophys. Acta*, **131**, 81 (1967).
16. Mukohata, Y., *Ann. Rep. Biol. Works, Fac. Sci., Osaka Univ.*, **14**, 121 (1966).
17. Pressman, B. C., *Methods in Enzymology*, **10**, 714 (1967).

Effects of Inhibitors on the Volume Change of Isolated Chloroplasts upon Illumination

K. NISHIDA

Botanical Institute, Faculty of Science, Kanazawa University, Kanazawa, Japan

INTRODUCTION

In recent years, many reports (see reviews 1 and 2) have been published on conformational changes of isolated chloroplasts. Most of the studies were carried out with an apparatus to measure the light scattered by chloroplasts in suspension. Recently, Mukohata *et al.*[3] have reported that the intensity of the scattered light could not be a measure of the volume of chloroplasts and is more or less related to their refractive index or charge. Other methods for observing this phenomenon, measurements of size distribution with a Coulter counter and of packed volume with a hematocrit, have been also applied to the study of conformational changes in chloroplasts. The results obtained by these methods, however, do not agree with those obtained by the optical method, especially in the time course of the change. Inhibitors of photosynthesis have been widely used in the study of conformational changes (4–13). Except for a few experiments,[5,10,13] however, the inhibitory effects were studied by measuring their effects on the light scattering. In the present study, several inhibitors were reexamined and their effects on the volumes of chloroplasts were precisely studied with a Coulter counter. The terms "photoshrinkage" and "photoswelling" are used in this paper. "Photoshrinkage" was first used by Itoh[14] and corresponds to the light-induced low-amplitude shrinkage of Packer *et al.*[15] under photophosphorylating conditions such as observed previously[16] in a 40 m*M* phosphate buffer solution, pH 7.5. "Photoswelling," on the other hand, means the light-induced high-amplitude swelling[15] under non-photophosphorylating conditions such as observed in a 300 m*M* sodium chloride solution buffered with 50 m*M* Tris-HCl, pH 7.5 (ref. 16). Inhibitors were applied in these different conditions in order to observe their effects on photoshrinkage and photoswelling.

Abbreviations: DCMU, 3-(3′,4′-dichlorophenyl)-1,1-dimethylurea; PMS, phenazine methosulfate; DPIP, dichlorophenol indophenol.

MATERIAL AND METHODS

Preparation of sample

Spinach leaves purchased from a commercial source were used as material. About 50 grams of leaves were cut into strips and homogenized at full speed for 15 sec in a blendor with 100 ml of cold 0.4 M sucrose solution buffered with 50 mM Tris-HCl, pH 7.5. Occasionally, 40 mM phosphate buffer, pH 7.5 was used. The green filtrate was first spun at $100 \times g$ for 2 min, and the sediment of whole cells and cell debris was discarded. The green supernatant was centrifuged further for 7 min at $800 \times g$. Sedimented chloroplasts were resuspended in a fresh medium, and this procedure of low and high speed centrifugations at $100 \times g$ and $800 \times g$ was repeated at least twice to make the preparation of whole chloroplasts as uniform as possible. The final preparation of chloroplasts was suspended in a fresh dust-free medium.

Measurement of volume

Measurements of chloroplast volume were made with a Coulter counter model B as described in a previous paper.[16] Readings in a series of experiments were normalized to express the volumes in percentage of the initial volume.

Illumination

The light source used was a 200 W incandescent lamp and the light intensity was about 30,000 lux on the surface of the chloroplast suspension. To eliminate heat radiated from the lamp, the light was passed through a layer of running water, 5 cm in width.

Chlorophyll content

The chlorophyll content was measured by the method of Arnon.[17] The chlorophyll contents of sample suspensions were 0.1–0.05 μg chlorophyll per ml.

RESULTS AND DISCUSSION

As shown in Fig. 1a, the photoshrinkage observed in 40 mM phosphate buffer was completely inhibited by the addition of 5, 10 and 50 μM DCMU before illumination. The inhibitory effect was also observed when 10 μM DCMU was added during shrinkage (Fig. 1b). According to Packer,[4] and Dilley and Vernon,[6] the light scattering change caused by illumination is inhibited by 5 μM DCUM. When DCMU was replaced by 1 mM *o*-phenanthroline, almost the same results were obtained, as reported previously.[5] When 33 μM DPIP and 666 μM ascorbate were added to the

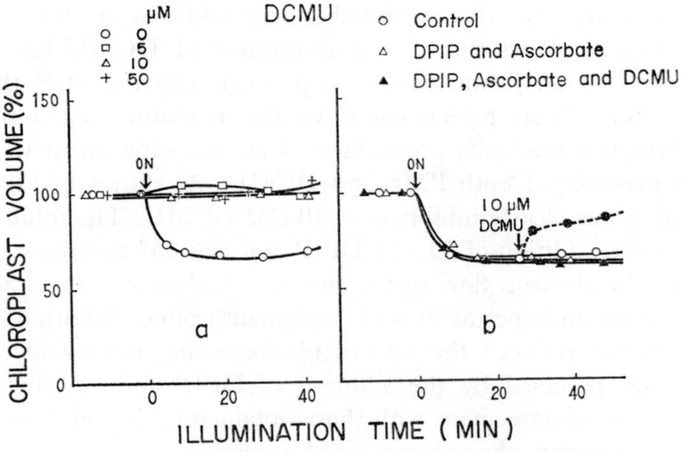

Fig. 1. Effect of DCMU on the photoshrinkage in the presence and absence of 33 μM DPIP and 666 μM sodium ascorbate.

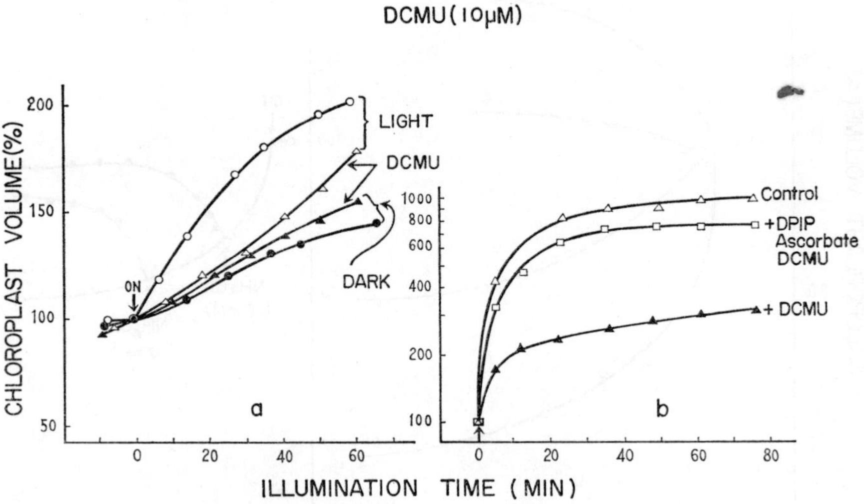

Fig. 2. Effect of DCMU on the photoswelling in the absence of PMS (Fig. 2a) and the effect of DCMU and DCMU+DPIP+ascorbate in the presence of PMS (Fig. 2b). The concentrations of PMS, DPIP and ascorbate were 30, 33 and 666 μM, respectively.

incubation mixture before illumination, the inhibitory effect of DCMU was completely suppressed (Fig. 1b). Probably DPIP and ascorbate supplied electrons to the chloroplasts, and the effect of electron transport thus initiated may have overcome the effect of DCMU.

Photoswelling was also inhibited by the addition of 10 *μM* DCMU (Fig. 2a). Fig. 2a shows that this concentration of DCMU has no effect on the dark swelling. It was shown in previous experiments[16] that when chloroplasts had been preincubated in the medium containing PMS, photoswelling was markedly accelerated. A similar experiment was carried out in the presence of both PMS and DCMU. As shown in Fig. 2b, the photoswelling was clearly inhibited by 10 *μM* DCMU. The inhibition was reversed by the addition of 33 *μM* DPIP and 666 *μM* ascorbate. The importance of the electron flow in the process of photoswelling was further strengthened by an experiment with *o*-phenanthroline. Addition of 1 m*M* *o*-phenanthroline reduced the rate of photoswelling remarkably, but the inhibition was reversed by the addition of DPIP and ascorbate. These results are in good agreement with those obtained by Nobel,[10] who used a hematocrit to measure the volume of chloroplasts.

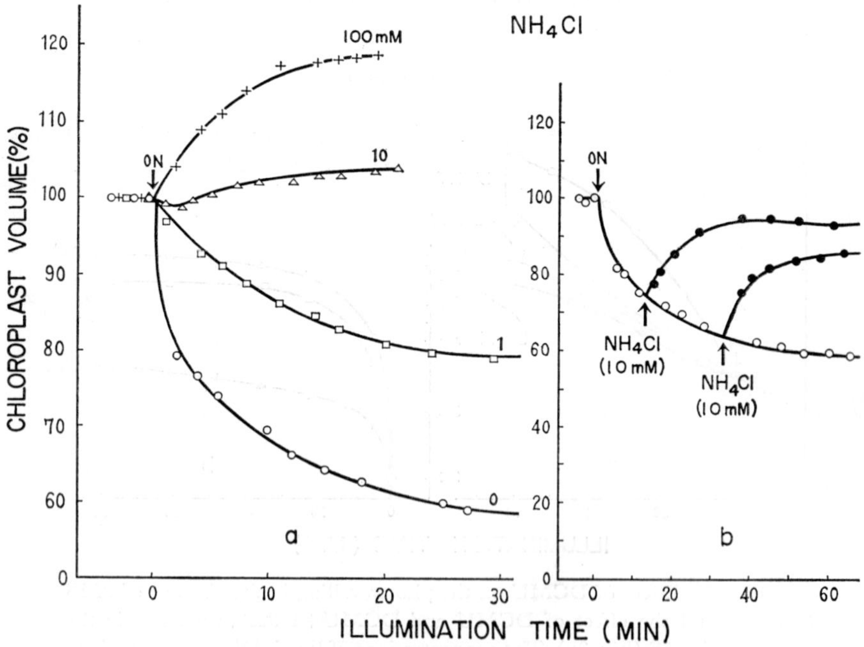

Fig. 3. Effect of NH_4Cl on the photoshrinkage.

The inhibitory effect of NH_4Cl on the photoshrinkage was examined at different concentrations of the reagent. As shown in Fig. 3a, 1 m*M* NH_4Cl noticeably inhibited the shrinkage. At higher concentrations, however, the chloroplasts began to swell when exposed to light. At 100 m*M*, for example, an observable swelling took place when the light was turned on. It was found that this swelling was not reversed when the light was turned off. Such observable and irreversible swelling was not noted in the presence of DCMU or *o*-phenanthroline. When 10 m*M* NH_4Cl was added during the process of shrinkage, the reversal swelling started and proceeded until the volume reached a constant level (Fig. 3b). The inhibitory effect of NH_4Cl on photoshrinkage has been observed by various workers. Itoh *et al.*,[5] using a Coulter counter, showed that photoshrinkage was completely inhibited by preincubation of chloroplasts with 3.3 m*M* NH_4Cl in the dark, and Packer,[4] Hind and Jagendorf,[7] and Dilley and Vernon[6] showed that the intensity of light scattering by chloroplasts was decreased by illumination in the presence of 3–9 m*M* NH_4Cl. They did not, however, test a high concentration of NH_4Cl such as 100 m*M* where the swelling is observable. Recently, Izawa[8] discovered that when 5 m*M* NH_4Cl was added with ferricyanide as an electron acceptor to a chloroplast suspension at pH 7.4, the absorbance decreased in the light, and that this change of absorbance was slowly reversed when the light was turned off. The swelling observed in the present experiment seems to be a similar phenomenon, and may be regarded as a deterioration process, considering its irreversible nature.

The effect of NH_4Cl concentration on the photoswelling is shown by the curves in Fig. 4a, which indicate that the swelling was accelerated by increasing NH_4Cl concentration. The swelling was more noticeably accelerated by NH_4Cl (Fig. 4b) when the chloroplasts were incubated with 33 μM DPIP and 666 μM ascorbate before illumination. This acceleration of swelling by NH_4Cl was probably caused by acceleration of electron transport, by uncoupling the phosphorylation in chloroplasts. Nobel,[10] using the packed volume method, also observed acceleration of photoswelling caused by NH_4Cl.

The inhibitory effect of atebrin depends upon the pH of the medium. At pH 6.0, atebrin scarcely affected the shrinkage, while at pH 8.0 it accelerated the shrinkage at first, but later reversed the change (Fig. 5). This reversal effect was clearly observed at 40 μM of atebrin. The swelling was not affected by light, so that this phenomenon seems to differ from photoswelling. It too, however, may be a deterioration process, judging from its irreversible nature. By measuring the absorbance changes of

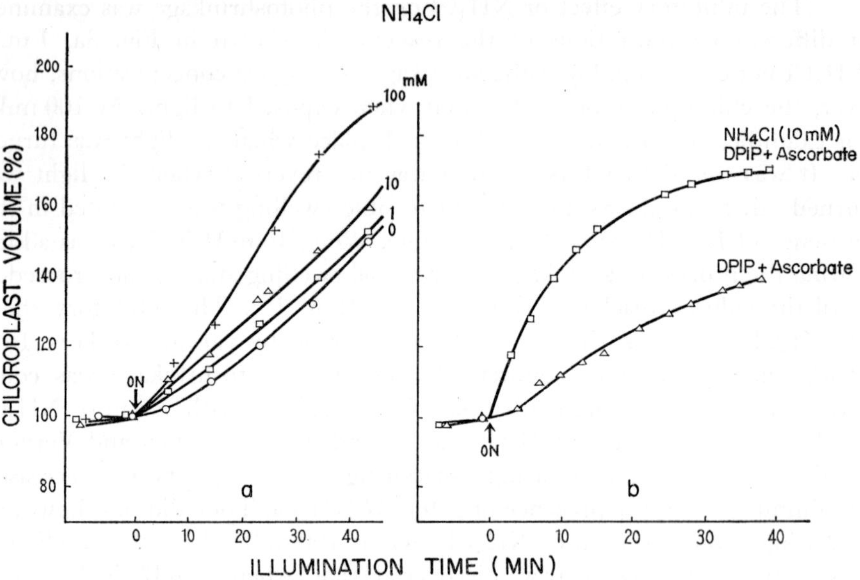

Fig. 4. Effect of NH_4Cl on the photoswelling in the presence and absence of 33 μM DPIP and 666 μM sodium ascorbate.

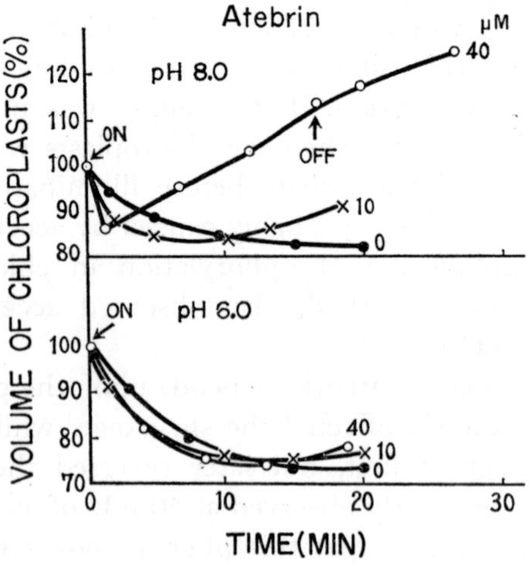

Fig. 5. Comparison of the effects of atebrin at pH 6.0 and 8.0 on the process of photoshrinkage.

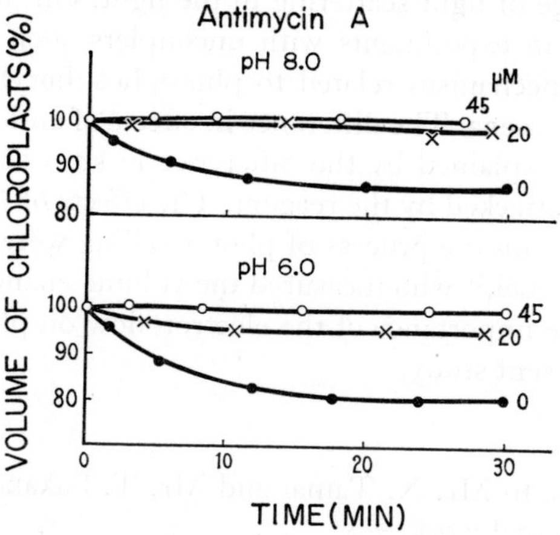

Fig. 6. Effect of antimycin A on the process of photoshrinkage.

chloroplasts, Hind and Jagendorf[7] observed that at pH 8 atebrin increases the absorbance of a chloroplast suspension noticeably, but at pH 6 it decreases the absorbance. Izawa,[8] and Izawa and Good[9] also reported an accelerating effect after measuring the change of absorbance. Dilley and Vernon,[6] on the other hand, could not confirm this effect of atebrin at a concentration of 8×10^{-5} *M*.

A clear inhibitory effect of antimycin A on the photoshrinkage was observed, as shown in Fig. 6. It was shown by Bamberger *et al.*[18] that antimycin A has an inhibitory effect on the reduction of NADP, cyclic and non-cyclic photophosphorylation, and carbon dioxide assimilation, and that its inhibitory effects are independent of the pH of the suspension. The data presented in Fig. 6 do not agree with the result obtained by Dilley and Vernon,[7] who reported that 4.5×10^{-5} *M* antimycin A did not exhibit any inhibitory effect on the change of absorbance of a chloroplast suspension in Tris-acetate buffer, pH 6.0, containing trimethylquinone (TMQ) as an electron acceptor. The disagreement may be caused by a difference in the method or in the suspending medium used.

Most of the data presented above are in good agreement with those obtained by previous workers who studied conformational changes with the light-scattering method, except for the antimycin A experiments discussed above. The experimental results obtained with inhibitors of electron transport such as DCMU and *o*-phenanthroline suggested that the electron flow is a condition necessary for the occurrence of photoshrinkage or the accompanying change of light scattering in the light. On the other hand, it may be deduced from experiments with uncouplers such as NH_4Cl and atebrin, that some mechanism related to photophosphorylation is also involved in these phenomena. The difference in effects of antimycin A on these phenomena may be explained by the difference in sites of the photophosphorylation process attacked by the reagent. The effects of NH_4Cl, DCMU, and *o*-phenanthroline on the process of photoswelling were coincident with those observed by Nobel,[10] who measured the volume change by the packed volume method. The importance of the electron flow on photoswelling was confirmed in the present study.

Acknowledgement

Thanks are due to Mr. N. Tamai and Mr. T. Sakata for their assistance in the experimental work.

SUMMARY

Investigations were made on the effects of various inhibitors, DCMU, *o*-phenanthroline, NH_4Cl, atebrin and antimycin A, on the processes of photoshrinkage and photoswelling of isolated spinach chloroplasts, which were measured using a Coulter counter, and the results were compared with those obtained previously by the light scattering method. It was found that the photoshrinkage—as measured by the present method—was inhibited by the addition of DCMU, *o*-phenanthroline, NH_4Cl, and atebrin in the same manner as when measured by the method of light scattering. The two methods gave different results, however, when antimycin A was added to the chloroplast suspension. Based on these results the mechanism of photoshrinkage was discussed. Photoswelling was also inhibited by the addition of DCMU and *o*-phenanthroline, while it was accelerated by NH_4Cl. The importance of the electron flow in the process of photoswelling was thus confirmed.

REFERENCES

1. Packer, L., and Siegenthaler, P. A., *Int. Rev. Cytology*, **20**, 97 (1966).
2. Nishida, K., *Bot. Mag.* (Tokyo), **79**, 830 (1966). (in Japanese).
3. Mukohata, Y., Mitsudo, M., and Isemura, T., *Ann. Rep. Biol. Works Fac. Sci. Osaka Univ.*, **14**, 107 (1966).
4. Packer, L., *Biochem. Biophys. Res. Comm.*, **9**, 355 (1962).
5. Itoh, M., Izawa, S., and Shibata, K., *Biochim. Biophys. Acta*, **66**, 319 (1963).
6. Dilley, R. A., and Vernon, L. P., *Biochemistry*, **3**, 817 (1964).

7. Hind, G., and Jagendorf, A. T., *J. Biol. Chem.*, **240**, 3202 (1964).
8. Izawa, S., *Biochim. Biophys. Acta*, **102**, 373 (1965).
9. Izawa, S., and Good, N. E., *Plant Physiol.*, **41**, 533 (1966).
10. Nobel, P. S., *Doctoral Thesis,* University of California, Berkeley.
11. Siegenthaler, P. A., and Packer, L., *Plant Physiol.*, **40**, 785 (1965).
12. Siegenthaler, P. A., *Physiol. Plantarum*, **19**, 437 (1966).
13. Nobel, P. S., *Biochim. Biophys. Acta*, **131**, 127 (1967).
14. Itoh, M., *Plant and Cell Physiol.*, **6**, 221 (1965).
15. Packer, L., Siegenthaler, P. A., and Nobel, P. S., *J. Cell Biol.*, **26**, 593 (1965).
16. Nishida, K., Tamai, N., and Ryoyama, K., *Plant and Cell Physiol.*, **7**, 415 (1966).
17. Arnon, D. I., *Plant Physiol.*, **24**, 1 (1949).
18. Bamberger, E. S., Black, C. C., Fewson, C. A., and Gibbs, M., *Plant Physiol.*, **38**, 483 (1963).

Chairman's Remarks

Some progress in the sub-microscopic morphology of the chloroplast and in the separation of particles containing each of the photosynthetic systems was reported and followed by vigorous discussion. Two limitations in the current technology of such investigations became apparent. One of these is the unfortunate fact that electron microscopy does not show the location of chlorophyll in the structures. Another serious difficulty is in the disintegration of chloroplasts into small enough particles to separate the two photochemical systems. Nearly all of the more or less satisfactory disintegration procedures use detergents. There was general agreement that although detergents may be essential, there is always liable to be danger of confusion from their use. Possible difficulties are the chance of disturbing the relation of pigments to their natural carriers, leading to both spectroscopic and activity changes, the chance of making larger-than-actual particles by addition of detergent micelles, and the chance of differentially inhibiting the chemical reactions.

In spite of these limitations, the hopes of separating System 1 and System 2 particles and of finding out how they are arranged in relation to each other in the active chloroplast has attracted many workers to this important subject.

The work on separation of particles having System 1 or System 2 activity with corresponding differences in pigments and other components of the photosynthetic apparatus fortunately still shows considerable diversity of techniques. The different methods of physical disruption—supersonic treatment, extrusion through a needle valve, shaking with glass beads, grinding with fine abrasives, or rapid extrusion of frozen preparations through a small orifice—all have their enthusiastic partisans. We seem, however, to be in great need of more comprehensive comparative studies of the different methods by themselves and in combination with each other. Detergents may be used alone or with physical methods to break the components apart, and the hope is to do so without spoiling the activity or shifting the absorption spectra of the pigments.

The discussions showed that most workers hoped for successful disruption without detergents, but that physically broken particles without detergents, when separated by centrifugation or electrophoresis, do not differ much from each other in their content of the two photosystems. Therefore, the use of detergents usually seems to be a necessary evil. Consensus of opinion during the discussions favored digitonin even though good results are also often reported with Triton X-100, SDS, and others. As with the physical methods, each usable detergent has its group of supporters, and systematic investigation of the comparative merits and dangers of various detergents used at various concentrations and pH values are needed.

Similarly, a great deal of comparative work on the separation procedures, following the disruptive steps, would seem to be profitable. Studies of the separation and disruption procedures must go on together, a fact that makes it difficult to plan a logical program of investigation. Considering the present state of the art, more would appear to be gained in the long run by greater emphasis on comparative investigation of possible techniques than on attempts to get immediate and detailed information about the nature of the particles prepared by any of the procedures so far developed.

The separation procedures—gel electrophoresis, differential centrifugation, and density gradient centrifugation—each have their supporters, but little comparative information as to their relative effectiveness was presented. It would be very good to know if the fractions obtained from any of these procedures could be further purified by later use of other separation methods. Separation methods based on selective precipitation are little used because of the difficulty of resuspending the precipitates, but may nevertheless be worth further study because of their ease and speed.

Ogawa, Kanai, and Shibata reported on the kinds and quantities of carotenoids found in two electrophoretic fractions of chloroplast particles in a red alga, a blue-green alga, and in a diatom. The results of their elegant and comprehensive study from the Tokugawa Institute confirms their previous conclusions based on similar fractionation of spinach chloroplasts. The more reduced carotenoids, the carotenes, prevailed in System 1 particles, while the carotenoids, with more oxygen in the side chain, predominated in System 2 particles. This work gives strong support to the idea that carotenoid transformations are an immediate part of the mechanism of oxygen evolution by photosynthesis.

The chloroplasts were solubilized by SDS at pH 10.3 and separations made by gel electrophoresis. Recordings of chlorophyll and protein content at different positions in the electrophoresis tube gave good evidence for still a third chlorophyll-protein fraction present as a shoulder, but strongly overlapped by the major fractions. We may hope for further reports by this group on these less well known chlorophyll-protein fractions. Perhaps they can be more highly resolved by electrophoresis in longer tubes. Their results suggest that treatment of the chloroplasts by light, with and without various electron carriers, before the disintegration and separation might be used to localize the carotenoid changes within the two photochemical systems of photosynthesis. A review of such carotinoid changes found by other workers is included in their paper.

Another comprehensive paper on chloroplast fractionation was given by Huzisige in a different session, as well as the work of Vernon, Garcia, Mollenhauer, and Ke. They used sonication followed by Triton X-100 and made separations by sucrose density gradient centrifugation. Their important conclusion is that the detergent splits off some small particles from membranes. The heavier particles are pieces of the membranes. This most informative report compared the fractions from four purple bacteria and from spinach chloroplasts. In addition to numerous electron micrographs, they showed the absorption changes on illumination of "B890" and ubiquinone in the "light" and the "heavy" fractions.

In *Chromatium* and *R. palustris*, the small particles were in the heavy centrifugal fractions, while in *R. rubrum*, the small particles were in the lighter fractions. In *Chromatium*, the cytochromes were mainly in the small particles. The small particles contained the photo-active bacteriochlorophyll and ubiquinone, while the larger membraneous particles were inactive. *R. palustris* gave small particles similar to those for *Chromatium* except that they aggregated in strands.

With *R. rubrum* and *R. sp.* NHTC-113 (Eimhjellen), the photooxidation of "B890" or its equivalent took place in both heavy and light particles and there was no separation of the bacteriochlorophyll forms. With *R.* NHTC, the ubiquinone reduction was present in the light but not in the heavy particles. With this bacterium, a stacked disc formation, not present in the original chromatophore, was found to be caused by the detergent. Electron microscopists beware!

Homan, Schmid, and Gaffron talked about the photosynthetic abilities of chloroplasts from a yellow leaf, the yellow part of a variegated mutant, and the normal leaves of tobacco. The yellow mutant, in sufficiently bright light, can give photosynthetic rates on a leaf area basis even exceeding that of the normal green leaf, although they have only 1/5 to 1/10 the normal amount of chlorophyll. Their chloroplasts have long and widely separated lamellae that either touch or are folded at the ends. Their rate of O_2 evolution from ferricyanide was twice that of the normal chloroplasts on a chlorophyll basis. Both NADP photoreduction and PMS photophosphorylation also exceeded that of the normal chloroplasts. These yellow mutants had chlorophyll units, measured as moles of CO_2 per mole of chlorophyll per flash, of about 800 in a majority of the 60 mutants. Values of 2,500 and of 400 were also found. The green chloroplasts gave more of the larger values, up to 5,500, in a series of 74 measurements. This statistical study of the size of the photosynthetic unit from many determinations on the same material is in itself a most important contribution to the understanding of the significance of the photosynthetic unit. Obviously, the apparent unit size depends on some as yet uncontrolled variable, thus giving the impression that the functional unit size may not actually represent the real number of chlorophyll molecules attached to a reaction center. They propose that a variable number of reaction centers may be attached to a cluster of chlorophyll molecules of a standard size.

By contrast with the normal and the yellow mutants, their chloroplasts from yellow leaves occurring in patches on green plants gave almost no photosynthesis or Hill-reaction activity. These chloroplasts, however, had high System 1 activity. Their submicro-structure showed widely separated lamellae that did not touch, overlap, or fold. These authors, therefore, suggest that the contact between layers of lamellae may be essential for proper function.

Mukohata gave a very penetrating and unifying discussion of old and new work on the various changes, other than catalysis of chemical reactions, taking place in chloroplasts exposed to light or given other treatments. These reversible changes were light scattering, shrinkage as seen by the Coulter counter or by centrifuged volume, viscosity of the suspension, reversible fluxes of cations, and electrophoretic mobility. New measurements were presented of light induced increases in apparent viscosity of the suspension, and of light scattering and pH changes caused by addition of OH^- to illuminated suspensions. The maximum of the light scattering response was at about pH 6.2.

The light scattering change consists of two parts. The internal scattering depends on the internal pH and may be rapidly altered by OH^- addition. The other part of the change, related to the size of the chloroplasts, changes more slowly.

Some interesting work on light-induced volume changes of chloroplasts as measured with a Coulter counter was presented by Nishida. Shrinkage, swelling, or both together but with different rates, can result from light exposure. The shrinkage measured in 40 μM phosphate pH 7.5 with 30 K lux amounted to about 30% of the initial volume and had a half time of about three minutes. The swelling to about twice the original volume, measured in 0.3 M NaCl+0.05 M Tris at pH 7.5 at the same intensity, had a half time exceeding 20 minutes. Both processes were inhibited by DCMU unless DPIP and ascorbate were added. NH_4Cl increased the swelling rate, thus appearing to inhibit shrinking. Atebrin at 40 μM and pH 8, but not pH 6, also increased swelling but less spectacularly than NH_4Cl. Antimycin A stopped the shrinkage, but without increasing the slower swelling. PMS greatly increased the rate and amplitude of swelling if added before the light. It seems as though the shrinkage is an immediate effect of light while the swelling comes from the buildup of photoproducts over a long period.

The structure of thylakoids was discussed by Park and Shumway, who pointed out that their disagreement with Muhlethaler *et al.* is due to differing interpretations of electron microscope photographs. Park presented evidence for his contention that the particles he calls quantasomes are inside the membranes of the thylakoids rather than outside as believed by the Swiss group. Moudrianakis, however, takes these particles to be carboxydismutase rather than chlorophyll-bearing structures. Thus we are left with the conclusion that the nature of the structures visible by electron microscopy is still worth further study.

To go below the size of visible particles to the molecular arrangement of chlorophyll, carotenoid, protein, cytochrome, and lipid molecules requires speculation on the basis of analytical composition. A model of possible arrangements at the molecular level proposed by Benson has far greater artistic merit than the older models and does not appear to violate any known facts. No doubt this model will shortly appear in elementary textbooks as have so many of its predecessors.

C. S. FRENCH

PART II

PHYSICAL AND CHEMICAL ASPECTS

Separation of Two Photochemical Systems of Photosynthesis by Fractionation of Chloroplasts

DANIEL I. ARNON,* HARRY Y. TSUJIMOTO, BERAH D. MCSWAIN AND RICHARD K. CHAIN

Department of Cell Physiology, University of California, Berkeley, California

INTRODUCTION

Professor Tamiya opened this Conference with reminiscences about the first Photosynthetic Conference at Gatlinburg in 1955. We find it interesting that there is a certain similarity in the reports from our laboratory at the Gatlinburg Conference and here. In both instances our contributions might be characterized as being unorthodox in nature. At Gatlinburg my colleagues, M. B. Allen, F. R. Whatley, and I reported on the then fresh discoveries of photophosphorylation and carbohydrate synthesis by isolated chloroplasts.[1] These two discoveries ran counter to the views widely held in 1955 when ATP in photosynthesis was still thought to come from a collaboration between chloroplasts and mitochondria,[2,3] and isolated chloroplasts were thought to be capable only of oxygen evolution, not of CO_2 assimilation.[4]

At this conference our paper is unorthodox in that it presents evidence in support of the hypothesis that the two light reactions in chloroplasts are cyclic and noncyclic photophosphorylation operating in parallel. We recently adopted this hypothesis[5,6] in place of the one which we embraced in 1961[7] and which is now very popular; namely, that two light reactions operating in series are required for noncyclic photophosphorylation and that cyclic photophosphorylation is a variant of one of them. The reasons for abandoning the earlier hypothesis and adopting the present one are discussed in detail elsewhere.[5,8] Here we will only describe the main outlines of our present hypothesis, enough to facilitate the presentation and interpretation of our new data.

* Aided by grants from the National Institute of General Medical Sciences, Office of Naval Research and the National Science Foundation.

RESULTS AND DISCUSSION

The Two Light Reactions in Chloroplasts

According to our present hypothesis, cyclic and noncyclic photophosphorylation in chloroplasts are two photochemical reactions, each of which involves a single photoact.[8] Noncyclic photophosphorylation would involve only what is generally called photosystem II and cyclic photophosphorylation (and its variants discussed below) would involve only photosystem I.

In noncyclic photophosphorylation ATP formation is stoichiometrically coupled with the photoreduction of ferredoxin and evolution of oxygen (ATP: 2Fd:$\frac{1}{2}O_2$=1: 1: 1) and occurs at a "noncyclic" site different from that in cyclic photophosphorylation. By using substrate amounts of ferredoxin, its photoreduction, accompanied by oxygen evolution and ATP formation, can be measured directly.[9] Usually, however, it is operationally more convenient to follow the photoreduction of ferredoxin indirectly. Catalytic amounts of ferredoxin are photoreduced and reoxidized, independently of light, by ferredoxin-NADP reductase and substrate amounts of NADP.[10,6]

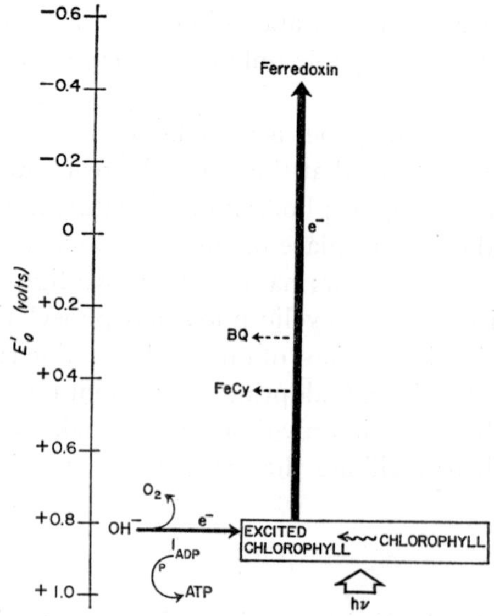

Fig. 1. Scheme for noncyclic photophosphorylation in chloroplasts.

Ferredoxin, which has an oxidation-reduction potential about equal to that of hydrogen gas, is the strongest reductant that has been isolated from the photochemical apparatus of photosynthesis.[11,12] As shown in Fig. 1, our hypothesis envisages that nonphysiological electron acceptors such as ferricyanide replace ferredoxin in noncyclic photophosphorylation, but with a great loss of photochemically generated reducing potential. Nevertheless, it follows from our hypothesis that the same photoact is involved in the photoreduction of ferricyanide by water as in the photoreduction of ferredoxin by water.[5,8] We shall see later how this interpretation was experimentally verified.

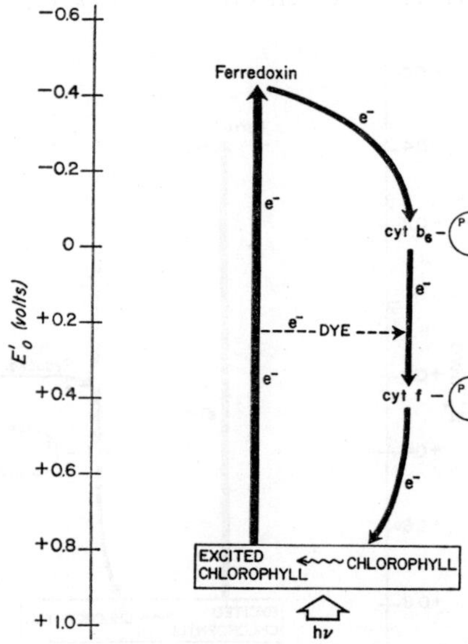

Fig. 2. Scheme for cyclic photophosphorylation in chloroplasts.

In cyclic photophosphorylation ferredoxin is also the physiological catalyst[13,6] but here ATP formation occurs at a cyclic site(s) and is coupled with a cyclic electron flow that is initiated by the photoreduction of ferredoxin by electrons supplied by the bound components of chloroplasts (Fig. 2). Thus, in cyclic photophosphorylation there is no unidirectional transfer of electrons from water to ferredoxin (and thence via NADP to the carbon cycle) but a cycling of "internal" electrons derived from the constituents of the photosynthetic apparatus. Consequently, cyclic photophosphorylation produces no oxygen and contributes no reducing equivalents to carbon assimilation. The sole product of cyclic photophosphorylation is ATP.

Recent evidence that the rates of ferredoxin-catalyzed cyclic photophosphorylation are comparable with maximum rates of photosynthesis *in vivo* strengthens the view that ferredoxin is the physiological catalyst in that process.[6] It is well known, however, that in experiments with chloroplasts *in vitro*, cyclic photophosphorylation can be catalyzed by such nonphysiological agents as phenazine methosulfate (PMS) which, in our view, are substitutes for ferredoxin. Unlike ferredoxin-catalyzed cyclic photophosphorylation, the PMS-catalyzed process is insensitive to antimycin A,[14,6] suggesting a bypass by PMS around a portion of the physiological cyclic pathway, probably that involving cytochromes of the *b* type.

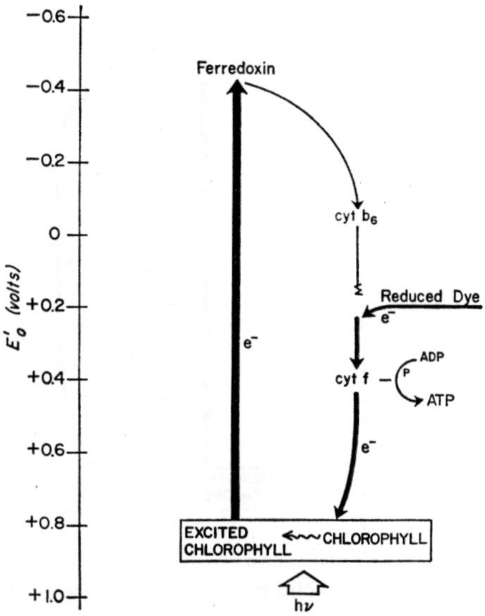

Fig. 3. Scheme for photoreduction of ferredoxin by reduced dyes and the accompanying phosphorylation.

An important difference between our present views and those of 1961[7] is the interpretation of the photoreduction of NADP by reduced dyes and the concomitant ATP formation. In 1961 we interpreted this reaction as a part of noncyclic photophosphorylation.[7] According to our present hypothesis, this reaction is a variant of cyclic photophosphorylation.[5] As shown schematically in Fig. 3, electrons from the reduced dye flow to ferredoxin via a portion of the cyclic chain and ATP is formed at a cyclic site. However, the electrons from ferredoxin do not cycle but are transferred, via ferredoxin-NADP reductase, to a NADP "trap."

A consequence of this concept of cyclic photophosphorylation is that PMS-catalyzed cyclic photophosphorylation and the phosphorylation that accompanies the photoreduction of NADP by reduced dyes are both measures of at least a partially functioning cyclic photophosphorylation system in chloroplasts. Any chloroplast preparation that is able to give a phosphorylation linked to photoreduction of NADP by reduced dyes must also give a cyclic photophosphorylation catalyzed by PMS. We shall see later how this interpretation was experimentally verified.

Evidence that cyclic and noncyclic photophosphorylation are two distinct photochemical reactions that differ in mechanisms and chlorophyll pigment systems came from contrasting response to inhibitors and monochromatic light.[15,6] Inhibitors like *o*-phenanthroline and DCMU, which strongly inhibit noncyclic photophosphorylation, stimulate cyclic photophosphorylation. Low concentrations of antimycin A, 2,4-dinitrophenol, and desaspidin, which strongly inhibit cyclic photophosphorylation, have no effect on the noncyclic process.[6]

A sensitive physical method to differentiate between cyclic and noncyclic photophosphorylation is the use of monochromatic light. Red and far-red monochromatic light give a "red rise" for cyclic photophosphorylation and a "red drop" for noncyclic photophosphorylation. The rate of noncyclic photophosphorylation declines as the wavelength is increased from 664 to 679 mμ; the reaction stops almost completely at 708 and 714 mμ. By contrast, the rate of cyclic photophosphorylation increases with an increase in wave-length, most strikingly about 700 mμ.[6]

We have concluded from this evidence that noncyclic photophosphorylation is the short wavelength, and cyclic photophosphorylation the long wavelength, reaction of plant photosynthesis. It follows that the enhancement effect in photosynthesis *in vivo* results not from the collaboration of two partial reactions of noncyclic photophosphorylation as called for by the older hypothesis but from the collaboration of cyclic and noncyclic photophosphorylation to produce sufficient ATP for CO_2 assimilation. The carbon cycle requires ATP and $NADPH_2$ in a ratio of 3: 2. Noncyclic photophosphorylation produces ATP and $NADPH_2$ in a ratio of 1: 1. The additional ATP requirement must be provided by cyclic photophosphorylation. Reports[16,17] that the ATP/2e ratio in noncyclic photophosphorylation is greater than one and might, therefore, provide the extra ATP needed for CO_2 assimilation, have not been substantiated in this laboratory.[18]

Fractionation of Chloroplasts by Digitonin

The experiments of Wessels[19,20] and those of Anderson and Boardman[21] showed that treatment of chloroplasts with digitonin may bring about at least a partial separation of the two photosystems. In the experiments of Wessels,[19,20] chloroplast fragments obtained by a digitonin treatment had a capacity for cyclic photophosphorylation and for photoreduction of NADP by a reduced dye (DPIP) but lost completely their capacity to photoreduce NADP with water as the electron donor.

The photochemical activity of the digitonin-treated particles obtained by Wessels could be explained by our current hypothesis as a physical separation of the cyclic photophosphorylation from the noncyclic photophosphorylation system. However, the results of Anderson and Boardman[21] were more complicated, from our point of view. Contrary to our hypothesis, which envisages that the same photoact is involved in the photoreduction of NADP (strictly speaking, ferredoxin) by water as in the photoreduction of any Hill reagent, their results showed that the digitonin treatment of chloroplasts reduced the rate of photoreduction of NADP much more severely than that of ferricyanide or trichlorophenol indophenol. (Since these two Hill reagents produced similar effects, we will henceforth deal only with ferricyanide.)

They observed first that incubation of chloroplasts with digitonin produced with time a much greater decline in the rate of photoinduced electron flow from water to NADP than from water to ferricyanide.[21] Second, their digitonin treatment of chloroplasts yielded particles of different size, separable by differential centrifugation, that differed in composition and in photochemical activity.[21] Of particular interest was the comparison of the composition and chemical activity of the original chloroplasts with two types of particles: heavy particles sedimented by centrifugation at $10{,}000 \times g$ (D-10 particles) and light particles, sedimented by centrifugation at $144{,}000 \times g$ (D-144 particles).

The D-10 fraction showed an enrichment in chlorophyll *b* relative to chlorophyll *a* whereas the D-144 fraction showed an enrichment of chlorophyll *a* relative to chlorophyll *b*. The D-144 fraction lost completely the capacity to use water as an electron donor for the photoreduction of either ferricynide or NADP (in the presence of ferredoxin and ferredoxin-NADP reductase). The only photochemical activity retained by the D-144 fraction was the photoreduction of NADP by reduced dye (DPIP). By contrast, the D-10 fraction retained a diminished but still appreciable capacity to photooxidize water. However, its rate of photoreduction of NADP by water was lower than the rate of photoreduction of ferricyanide by water.[21]

Anderson and Boardman[21] explained their results on the premise that the photoreduction of ferricyanide by water requires only photosystem II[22,23] whereas the photoreduction of NADP by water requires the collaboration of both photosystems I and II. They interpreted the high activity of the D-144 preparation in the photoreduction of NADP by reduced dye as evidence that the NADP reducing system proper always resides in photosystem I but that without photosystem II the photoreduction of NADP by water is impossible. They considered that the D-144 preparation with the high chl *a*/chl *b* ratio is representative of photosystem I whereas the D-10 preparation, showing an enrichment in chlorophyll *b*, contained an enrichment of particles representative of photosystem II.

Separation of Cyclic from Noncyclic Photophosphorylation by Fractionation of Chloroplasts Treated with Digitonin

In view of these results we undertook recently to investigate the effect of digitonin on cyclic and noncyclic photophosphorylation by chloroplasts.

TABLE I. Effect of time of incubation of spinach chloroplasts in 0.5% digitonin on photoreduction of ferricyanide and NADP with water as the electron donor.

Incubation time (hrs)	Q_{2e} FeCy	Q_{2e} NADP
Control	87	127
0.5	47	40
1	32	38
2	29	23
3	21	16

Q_{2e} FeCy=1/2 μmoles ferricyanide reduced/mg chlorophyll/hr.

Q_{2e} NADP=μmoles NADP reduced/mg chlophyll/hr.

The reaction mixture (final volume 1.5 ml) contained spinach chloroplasts (C_{1s}, equivalent to 100 μg of chlorophyll) and the following in μmoles : Tricine buffer, pH 8.2, 50 ; $MgCl_2$, 2.5 ; ADP, 5 ; $K_2HP^{32}O_4$, 5. In addition, the FeCy system contained potassium ferricyanide, 10 μmoles ; and the NADP system, ferredoxin, 0.015 μmole ; NADP, 5 μmoles ; and a saturating amount of NADP reductase. The reaction mixtures were equilibrated with argon for 5 minutes in rubber-capped test tubes and 0.7 ml was transferred anaerobically with syringes to rubber-capped, 2 mm thick cuvettes previously flushed with argon. The reactions were carried out for 5 minutes at 20°C with uniform illumination by incandescent light, filtered through a Corning 2–60 filter to cut off wavelengths below 620 mμ and through a Corning 1–69 filter and 3 cm of water to cut off the infra-red radiation. Illumination intensity was 21 milliwatts/cm^2 at the cuvette surface. The FeCy or NADP reduced in the light was measured by change in absorbance at 420 or 340 mμ, respectively.

We followed, with a few modifications, the procedure of Anderson and Boardman.[21] Spinach chloroplasts, prepared by the methods used in our laboratory,[24] were suspended at 0° in digitonin (final concentration 0.5 per cent) for 30 min. Two fractions were isolated by differential centrifugation: one sedimented between 1,000 and 10,0000 $\times g$ (D-10 fraction) and another sedimented between 50,000 and 144,000 $\times g$ (D-144 fraction). Chlorophyll was determined as described by Arnon[25] and ATP by the method of Hagihara and Lardy.[26] Other experimental particulars are included with the experimental data.

Our results show, on several significant points, agreement as well as disagreement with those of Anderson and Boardman.[21] Table 1 shows that with increasing time of incubation of chloroplasts in 0.5 per cent of digitonin there was a decrease in the rate of photoreduction of ferricyanide and NADP with water as the electron donor. However, Table I gives no evidence that the rate of decline of photoreduction of NADP was significantly greater than the rate of decline of photoreduction of ferricyanide.

TABLE II. Effect of digitonin treatment on noncyclic photophosphorylation.

System	Chloroplasts		D-10 fraction		D-144 fraction	
	Q_{2e}	Q_{ATP}	Q_{2e}	Q_{ATP}	Q_{2e}	Q_{ATP}
Ferricyanide	121	106	58	37	0	0
NADP	133	157	62	51	2	0

Q_{2e} = μmoles NADP or 1/2 μmoles ferricyanide reduced/mg chlorophyll/hr.
Q_{ATP} = μmoles ATP formed/mg chlorophyll/hr.
The experimental conditions were as given in Table I. In this experiment the chlorophyll *a* : chlorophyll *b* ratios in the untreated, D-10 and D-144 preparations were 2.9, 2.3 and 6.1, respectively.

Next, we undertook to compare the photochemical activity of untreated chloroplasts and the activity of the D-10 and D-144 fractions on noncyclic photophosphorylation, i.e. under conditions when the electron flow from water is coupled with ATP formation. The concomitant phosphorylation substantially enhances the rate of electron flow from water to ferricyanide and NADP[27-29] and hence the presence of phosphorylation was deemed especially desirable in investigating the effect of digitonin on noncyclic electron transport. Table II shows the complete loss of noncyclic electron transport and phosphorylation in the D-144 fraction. In the D-10 fraction, the rate of electron transport and phosphorylation was cut about in half but again there was no significant difference in this decline of photo-

TABLE III. Effect of digitonin treatment on cyclic photophosphorylation.

System	Chloroplasts Q_{ATP}	D-10 fraction Q_{ATP}	D-144 fraction Q_{ATP}
Ferredoxin	154	20	0
Menadione	118	22	0
Phenazine methosulfate	314	76	46

Q_{ATP} = μmoles ATP formed/mg chlorophyll/hr.

The reaction mixture (final volume 1.5 ml) included spinach chloroplasts (C_{1s}, equivalent to 100 μg of chlorophyll) and the following in μmoles: Tricine buffer, pH 8.2, 50; $MgCl_2$, 2.5; ADP, 5; $K_2HP^{32}O_4$, 5. In addition, the ferredoxin system contained ferredoxin, 0.15 μmole, and 3-(3, 4-dichlorophenyl)-1, 1-dimethyl urea (DCMU) to give a final concentration of 2×10^{-7} *M*. The other two systems contained, respectively, menadione, 0.15 μmole, and phenazine methosulfate, 0.05 μmole. Other experimental conditions as given in Table I. In this experiment the chlorophyll *a*: chlorophyll *b* ratios in the untreated D-10 and D-144 preparations were 2.6, 2.3 and 5.7, respectively.

chemical activity whether the terminal electron acceptor was NADP or ferricyanide.

Table III shows the effect of digitonin treatment of chloroplasts on cyclic photophosphorylation with three different catalysts, ferredoxin, menadione and PMS. Cyclic photophosphorylation in the D-10 fraction was sharply reduced with all three catalysts to a greater degree than noncyclic photophosphorylation (Table II). The rate of cyclic photophosphorylation was reduced about 87 per cent with ferredoxin, 82 per cent with menadione and 76 per cent with PMS. The D-144 fraction had no photophosphorylating activity at all when ferredoxin or menadione was used as a catalyst but had an appreciable capacity for ATP synthesis when PMS was the catalyst. These results are consistent with earlier views[30] that PMS catalyzes cyclic photophosphorylation by giving rise to a shortened cyclic electron pathway which bypasses one or more electron carriers. It appears that the digitonin treatment results in a loss of a component(s) of chloroplasts that is indispensable when ferredoxin or menadione catalyze cyclic photophosphorylation but is not essential when the catalyst is PMS.

To recapitulate, the light chloroplast particles (D-144) obtained by the digitonin treatment were completely devoid of the noncyclic photophosphorylation system but retained a significant portion of the cyclic system, measurable with PMS as a catalyst. In the heavy particles (D-10) both cyclic and noncyclic photophosphorylation were reduced by the digitonin treatment but the decline was relatively greater for the cyclic than for the noncyclic system—consistent with the idea that the D-10 fraction was enriched with the noncyclic system relative to the cyclic one. The activity of the noncyclic photophosphorylation system in the D-10 fraction was, relative to the activity in the untreated chloroplasts, approximately the same with either NADP or ferricyanide as the terminal electron acceptor. There was no evidence that the digitonin treatment damaged the photochemical electron transfer from water to ferredoxin (and thence to NADP) more than it damaged the photochemical electron transfer from water to ferricyanide.

Effects of Reduced Dye and Plastocyanin

Table IV summarizes experiments on the effect of the digitonin treatment on the photoreduction of NADP by reduced dye (DPIP) and the coupled phosphorylation. To avoid the complication that NADP might be simultaneously reduced by electrons from water, all these experiments were carried out in the presence of DCMU.

TABLE IV. Effect of plastocyanin (PC) and 2,6-dichlorophenol indophenol (DPIP) on photoreduction of NADP and coupled phosphorylation by chloroplasts treated with digitonin.

Treatment	Chloroplasts		D-10 fraction		D-144 fraction	
	Q_{NADP}	Q_{ATP}	Q_{NADP}	Q_{ATP}	Q_{NADP}	Q_{ATP}
Asc+DPIP	65	28	47	20	48	16
Asc+DPIP+PC	64	30	96	26	224	14
Asc+PC	18	7	66	4	209	2
Ascorbate	22	7	10	5	1	1

Q_{NADP}=μmoles NADP reduced/mg chlorophyll/hr.

Q_{ATP}=μmoles ATP formed/mg chlorophyll/hr.

The reaction mixture was as given in Table I for the NADP system except that 3-(3, 4-dichlorophenol)-1, 1-dimethyl urea (DCMU) at a final concentration of 1×10^{-6} M was added in all treatments. Other additions, where indicated, were ascorbate, 10 μmoles ; plastocyanin (PC), 10 mμmoles ; and 2, 6-dichlorophenol indophenol (DPIP), 0.1 μmole.

Table IV shows that, consistent with previous discussion, the D-144 fraction, which had appreciable PMS-catalyzed cyclic photophosphorylation activity, has also retained over half of the phosphorylating activity that is coupled with the photoreduction of NADP by reduced DPIP (ascorbate+DPIP). The activity of the corresponding system in the D-10 fraction was almost the same as in the D-144 fraction.

Table IV also shows a striking effect of the digitonin treatment on the response of chloroplasts to added plastocyanin. With ascorbate-DPIP as the electron donor system, the addition of plastocyanin had no effect on untreated chloroplasts but doubled the rate of NADP photoreduction by the D-10 preparation and increased more than four-fold the rate of NADP photoreduction by the D-144 fraction.

Another striking effect of plastocyanin shown in Table IV is that it acted directly as a carrier for electrons from ascorbate, independently of DPIP, in the digitonin-treated chloroplast fractions, particularly in D-144. In the untreated chloroplasts, ascorbate was an effective electron donor only in the presence of DPIP (whether plastocyanin was added or not) but in the D-144 and, to a lesser extent, in the D-10 fraction, the addition of catalytic amounts of plastocyanin gave a higher rate of NADP reduction by ascorbate than the addition of catalytic amounts of DPIP. Davenport[31] observed similar effects of ascorbate, plastocyanin and DPIP on chloroplasts disrupted by sonication. It appears that disruption of chloroplasts, either by digitonin or sonication, makes electrons from reduced plastocyanin readily accessible to the electron transport chain.

Another interesting effect of plastocyanin was that while it strikingly increased the rate of NADP photoreduction, it had no effect on the concomitant phosphorylation (Table IV). These results suggest that electrons from reduced plastocyanin enter the electron transport chain at a point past

TABLE V. Effect of plastocyanin (PC) on electron transport and photophosphorylation by digitonin-treated chloroplasts.

System	Chloroplasts		D-10 fraction		D-144 fraction	
	Q_{2e}	Q_{ATP}	Q_{2e}	Q_{ATP}	Q_{2e}	Q_{ATP}
1. NADP, noncyclic	133	157	62	51	2	0
2. 1+PC	140	177	85	52	13	0
3. Ferricyanide, noncyclic	121	106	58	37	0	0
4. 3+PC	140	118	61	42	11	0
5. Ferredoxin, cyclic	—	154	—	20	—	0
6. 5+PC	—	161	—	20	—	0
7. Menadione, cyclic	—	118	—	22	—	0
8. 7+PC	—	127	—	19	—	0
9. Phenazine methosulfate, cyclic	—	314	—	76	—	46
10. 9+PC	—	307	—	68	—	40

Q_{2e} = μmoles NADP or 1/2 μmoles FeCy reduced/mg chlorophyll/hr.
Q_{ATP} = μmoles ATP formed/mg chlorophyll/hr.

The experimental conditions for the noncyclic systems were as given in Table I. The experimental conditions for the cyclic systems were as given in Table III. Plastocyanin (PC), 10 mμmoles, was added where indicated.

the cyclic phosphorylation site. By contrast, electrons from reduced DPIP must enter the electron transport at or near the cyclic phosphorylation site which is possibly identified with cytochrome *f*.[7]

Table V shows that added plastocyanin had no effect on cyclic and non-cyclic photophosphorylation, not only on untreated chloroplasts but also on the D-10 and D-144 fractions.

Chemical Characterization of Digitonin-Treated Chloroplasts

A chemical characterization of our digitonin-treated chloroplast fractions was made with respect to chlorophyll $a\!:\!b$ ratios, P-700 (ref. 32), plastocyanin,[33] cytochrome *f*, b_6 and 559 (ref. 34, 35). Cytochrome 559 is a tentative designation, pending further characterization, of what appears to be a cytochrome *b* component that may prove to be the same as the cytochrome b_3 with an α-peak at 559 mμ described by Hill and Scarisbrick[36] and Lundegardh.[35]

As shown in Table VI, the average chlorophyll *a*: chlorophyll *b* ratios in untreated chloroplasts, D-10, and D-144 fractions were, respectively, 2.83, 2.38 and 5.98. The comparable average chl *a*/chl *b* ratios in the experiments of Anderson and Boardman[21] were 2.83, 2.27 and 5.34. It appears, therefore, that based on chlorophyll ratios, our D-10 and D-144 preparations were comparable to those of Anderson and Boardman.[21]

Fig. 4 shows that, in agreement with results of Anderson *et al.*,[37] the concentration of P-700 is lower in the D-10 fraction but is markedly increased in the D-144 fraction relative to untreated chloroplasts. The high

TABLE VI. Effect of digitonin on chlorophyll *a*/chlorophyll *b* ratios of spinach chloroplasts.

Experiment	Chlorophyll *a*/chlorophyll *b* ratios		
	Chloroplasts	D-10 fraction	D-144 fraction
A	2.50	2.00	5.05
B	3.05	2.54	5.73
C	3.18	2.38	6.08
D	2.74	2.34	6.41
E	2.68	2.30	5.55
F	2.94	2.30	6.09
G	3.04	2.55	6.80
H	2.61	2.52	6.70
I	2.90	2.56	5.72
J	2.67	2.30	5.68
Average	2.83	2.38	5.98

concentration of P-700 in the D-144 fraction accompanies its enrichment in chlorophyll *a* and the decrease of P-700 in the D-10 fraction accompanies the decrease in the chlorophyll *a* concentration (relative to chlorophyll *b*) in that fraction.

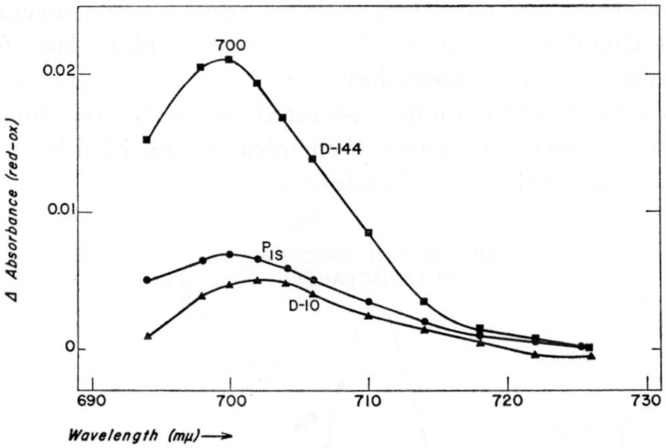

Fig. 4. P-700 in digitonin-treated chloroplasts. The difference spectra (reduced minus oxidized) were obtained by adding sodium ascorbate to the sample cuvette and potassium ferricyanide to the reference cuvette. The cuvettes (10 mm light path) contained, in a final volume of 3 ml, chloroplasts (equivalent to 300 μg of chlorophyll); Tris-HCl buffer, pH 8.2, 100 μmoles; and $MgCl_2$, 5 μmoles. Temperature, 20°C; gas phase, air.

A striking difference in plastocyanin content between the untreated and digitonin-treated chloroplasts is shown in Fig. 5. The plastocyanin content of the D-10 fraction is sharply reduced and drops almost to zero in the D-144 fraction. It will be recalled that the increase in the rate of NADP photoreduction by ascorbate (without DPIP) that resulted from the addition of plastocyanin was larger with the D-144 fraction (Table IV).

Fig. 6 shows the effect of digitonin treatment on the cytochrome components of chloroplasts. In general agreement with the results of Boardman and Anderson,[38] the digitonin treatment reduced the cytochrome *f* and b_6 content in the D-10 and the D-144 fractions. The most pronounced change occurred with respect to cytochrome 559. This cytochrome component was lost almost completely from the D-144 fraction but remained tightly bound in the D-10 fraction.

A summary of the relative abundance of the several chloroplast constituents and their molar ratios relative to total chlorophyll is given in

Table VII. A comparison of these data for cytochromes with those of Boardman and Anderson[38] show, on the whole, good agreement, considering differences in techniques and varietal and seasonal differences in leaf material. This similarity of constituents coupled with similar chlorophyll ratios (Table VI) between our D-10 and D-144 fractions and those of the Australian workers[21, 38] are consistent with the agreement on several points as to the photochemical activity of digitonin-treated chloroplasts found in the two laboratories. It also follows, however, that the divergencies as to the photochemical activity of the digitonin-treated chloroplast fractions cannot be explained by assumed differences of chemical composition between corresponding fractions in the two laboratories.

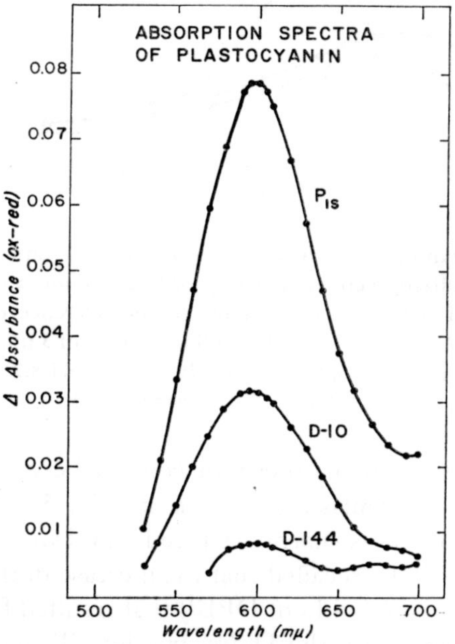

Fig. 5. Difference spectra (oxidized minus reduced) of plastocyanin extracted from digitonin-treated chloroplasts. The difference spectra were obtained by adding potassium ferricyanide to the sample cuvette and sodium ascorbate to the reference cuvette. In each case, plastocyanin was extracted from the chloroplast preparation to give 1 ml of plastocyanin extract equivalent to 5 mg of chlorophyll. Temperature, 20°C; gas phase, air; light path, 10 mm.

Effect of Sonication

Apart from digitonin, sonication of chloroplasts has been widely used as a method for separating the photochemical activities of chloroplasts. Of

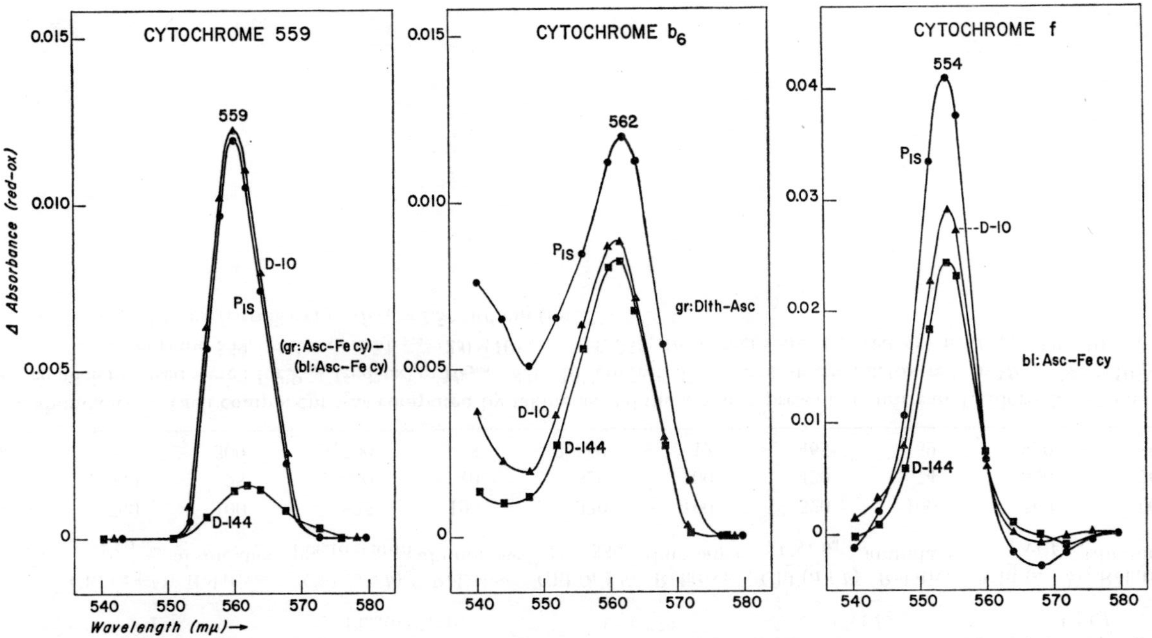

Fig. 6. Difference spectra (reduced minus oxidized) of three cytochrome components in chloroplasts treated with digitonin. The cytochrome *f* difference spectra were obtained on bleached (extracted with 80% acetone) chloroplast preparations by adding sodium ascorbate to the sample cuvette and potassium ferricyanide to the reference cuvette. The cytochrome b_6 difference spectra were obtained on green (unbleached) chloroplast preparations by adding sodium dithionite to the sample cuvette and ascorbate to the reference cuvette. The cytochrome 559 difference spectra were obtained by subtracting the cytochrome *f* difference spectrum from bleached chloroplast preparations (see above) from a difference spectrum obtained from a green chloroplast preparation by adding ascorbate to the sample cuvette and ferricyanide to the reference cuvette. The bleached chloroplast preparation was equivalent, on the basis of its original chlorophyll content, to the green chloroplast preparation. For the determination of cytochrome b_6 and cytochrome 559, chloroplast preparations containing 0.2 mg total chlorophyll/ml were used. Chloroplast preparations containing 0.9 mg total chlorophyll/ml were used for cytochrome *f* determination. Temperature, 20°C; gas phase, air; light path, 10 mm.

TABLE VII. Relative abundance and molar ratios of total chlorophyll (Chl a+Chl b) to P_{700}, plastocyanin and cytochromes f, b_6 and 559 in digitonin-treated chloroplasts.

	P_{700}		Plastocyanin		Cyt 559		Cyt b_6		Cyt f	
	Chl (a+b) / P_{700}	Relative abundance	Chl (a+b) / Plastocyanin	Relative abundance	Chl (a+b) / Cyt 559	Relative abundance	Chl (a+b) / Cyt b_6	Relative abundance	Chl (a+b) / Cyt f	Relative abundance
Chloroplasts	760	100	625	100	320	100	320	100	560	100
D-10 fraction	1,000	75	1, 600	40	320	100	430	74	800	70
D-144 fraction	255	300	13,500	5	2,650	12	465	69	950	59

The relative abundance of each component was computed by taking as 100 the amount present in untreated chloroplasts. The extinction coefficients used were: for P_{700}, (red-ox) $mM\epsilon_{700}=8.0\times10^1$/cm (ref. 32); for plastocyanin, (ox-red) $mM\epsilon_{597}=9.8\times10^3$/cm (ref. 33); for cytochrome 559, (red-ox) $mM\epsilon_{559}=2.0\times10^4$/cm (ref. 35); for cytochrome b_6, (red-ox) $mM\epsilon_{562}=2.0\times10^4$/cm (ref. 35); and for cytochrome f, (red-ox) $mM\epsilon_{554}=2.5\times10^4$/cm (ref. 35).

special interest to us was the paper of Katoh and Takamiya[39] in which a sonication of chloroplasts produced with time a much greater inactivation of photoreduction of NADP than of photoreduction of ferricyanide by water. As shown in Fig. 7, we found that, under our conditions, the decline with increasing time of sonication in the rate of photoreduction of NADP was about the same as the decline in the rate of ferricyanide. However, this investigation is still in a preliminary stage and we cannot draw final conclusions until we have investigated the effect of sonication more extensively, particularly in the light of the recent results of Katoh and San Pietro.[40]

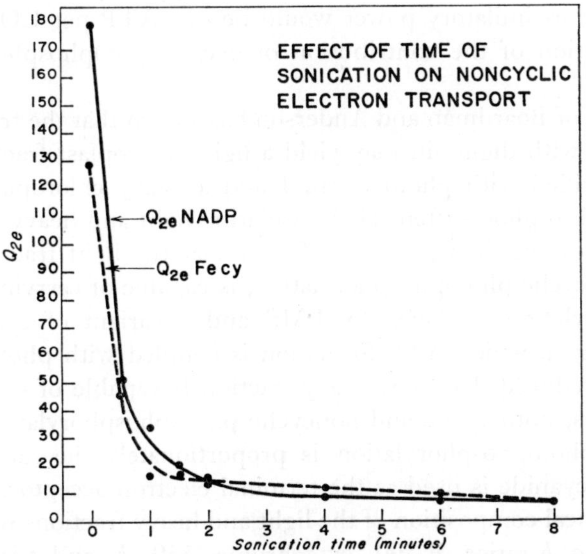

Fig. 7. Effect of time of sonication of chloroplasts on photoreduction of NADP and ferricyanide with water as the electron donor. Q_{2e} = μmoles NADP or 1/2 μmoles FeCy reduced/mg chlorophyll/hr. Experimental conditions were as given in Table I.

CONCLUDING REMARKS

There is now wide agreement among investigators of plant photosynthesis that the assimilatory power for CO_2 assimilation, which consists of two moles of $NADPH_2$ and three of ATP per mole of CO_2, is supplied by two photochemical reactions. There is some disagreement, however, about the identity of the two photochemical reactions. A currently popular hypothesis identified the two photochemical reactions as two component reactions of noncyclic photophosphorylation (photosystems I and II) operating in series. The proponents of this hypothesis either question the operation of cyclic photophosphorylation *in vivo* (despite the evidence to the contrary reviewed in ref. 8) or identify cyclic photophosphorylation as a variant of photosystem I.

By contrast, the hypothesis that now guides the work of this laboratory holds that the two photochemical reactions of chloroplasts are cyclic and noncyclic photophosphorylation, operating in parallel. Noncyclic photophosphorylation would involve only photosystem II and its contribution to the assimilatory power (per mole of CO_2) would be 2 $NADPH_2$ and 2 ATP. Cyclic photophosphorylation would involve only photosystem I and its contribution to the assimilatory power would be one ATP per CO_2. (A more refined elaboration of the contribution of cyclic photophosphorylation is given in ref. 8.)

The work of Boardman and Anderson has shown that the fractionation of chloroplasts with digitonin may yield a light chloroplast fraction that is primarily identified with photosystem I and a heavy chloroplast fraction that is enriched in photosystem II. By isolating light and heavy chloroplast fractions by these methods, we have found that the light fraction, totally inactive in noncyclic photophosphorylation, is capable of carrying on cyclic photophosphorylation catalyzed by PMS and a variant of cyclic photophosphorylation in which ATP formation is coupled with photoreduction of NADP by reduced dye. The heavy fraction is capable of carrying out, at reduced rates, both cyclic and noncyclic photophosphorylation. Its rate of noncyclic photophosphorylation is proportionately the same whether NADP or ferricyanide is used as the terminal electron acceptor.

The chemical composition of the light and heavy fractions with respect to chlorophyll $a:b$ ratios, P-700, cytochromes 559, b_6 and f is similar to that reported for these fractions by Boardman and Anderson. Plastocyanin has appreciably decreased in the heavy fraction and has almost totally disappeared in the light fraction. The addition of plastocyanin to the light fraction gives a striking increase in the rate of electron flow from ascorbate to NADP (via ferredoxin) without, however, any corresponding increase in ATP formation.

The photochemical activity of the digitonin-treated chloroplasts is consistent with the hypothesis that cyclic and noncyclic photophosphorylation are the two photochemical reactions of chloroplasts.

REFERENCES

1. Gaffron, H. (ed.), *Research in Photosynthesis,* Interscience Publishers, New York, 1957.

2. Vishniac, W., and Ochoa, S., *J. Biol. Chem.*, **195**, 75 (1952).
3. Ohmura, T., *Arch. Biochem. Biophys.*, **57**, 187 (1955).
4. Lumry, R., Spikes, J. D., and Eyring, H., *Ann. Rev. Plant Physiol.*, **5**, 271 (1954).
5. Arnon, D. I., Tsujimoto, H. Y., and McSwain, B. D., *Nature*, **207**, 1367 (1965).
6. Arnon, D. I., Tsujimoto, H. Y., and McSwain, B. D., *Nature*, **214**, 562 (1967).
7. Losada, M., Whatley, F. R., and Arnon, D. I., *Nature*, **190**, 606 (1961).
8. Arnon, D. I., *Physiol. Rev.*, **47**, 317 (1967).
9. Arnon, D. I., Tsujimoto, H. Y., and McSwain, B. D., *Proc. Natl. Acad. Sci. U.S.A.*, **51**, 1274 (1964).
10. Shin, M., and Arnon, D. I., *J. Biol. Chem.*, **240**, 1405 (1965).
11. Tagawa, K., and Arnon, D. I., *Nature*, **195**, 537 (1962).
12. Arnon, D. I., *Science*, **149**, 1460 (1965).
13. Tagawa, K., Tsujimoto, H. Y., and Arnon, D. I., *Proc. Natl. Acad. Sci. U.S.A.*, **49**, 567 (1963).
14. Losada, M., and Arnon, D. I., in *Metabolic Inhibitors* (R. M. Hochster and J. H. Quastel, eds.), Vol. II, p. 559, Academic Press, New York, 1963.
15. Tagawa, K., Tsujimoto, H. Y., and Arnon, D. I., *Proc. Nat. Acad. Sci. U.S.A.*, **50**, 544 (1963).
16. Winget, G. D., Izawa, S., and Good, N. E., *Biochem. Biophys. Res. Comm.*, **21**, 438 (1965).
17. Lynn, W. S., and Brown, R. H., *J. Biol. Chem.*, **242**, 412 (1967).
18. Ramirez, J. M., Del Campo, F. F., and Arnon, D. I., *Federation Proc.*, **26**, 866 (1967).
19. Wessels, J. S. C., *Biochim. Biophys. Acta*, **65**, 561 (1962).
20. Wessels, J. S. C., *Proc. Roy. Soc.* (*London*), **B 157**, 345 (1963).
21. Anderson, J. M., and Boardman, N. K., *Biochim. Biophys. Acta*, **112**, 403 (1966).
22. Horio, T., and San Pietro, A., *Proc. Natl. Acad. Sci. U.S.A.*, **51**, 1226 (1964).
23. Biggins, J., and Sauer, K., *Biochim. Biophys. Acta*, **88**, 655 (1964).
24. Whatley, F. R., and Arnon, D. I., in *Methods in Enzymology* (S. P. Colowick and N. O. Kaplan, eds.), Vol. VI, p. 308, Academic Press, New York, 1963.
25. Arnon, D. I., *Plant Physiol.*, **24**, 1 (1949).
26. Hagihara, B., and Lardy, H. A., *J. Biol. Chem.*, **235**, 889 (1960).
27. Arnon, D. I., Whatley, F. R., and Allen, M. B., *Science*, **127**, 1026 (1958).
28. Arnon, D. I., Whatley, F. R., and Allen, M. B., *Biochim. Biophys. Acta*, **32**, 47 (1959).
29. Davenport, H. E., *Nature*, **184**, 524 (1959).
30. Arnon, D. I., in *Light and Life* (W. D. McElroy and B. Glass, eds.), p. 361, Johns Hopkins Press, Baltimore, 1961.
31. Davenport, H. E., in *Non-Heme Iron Proteins: Role in Energy Conversion* (A. San Pietro, ed.), p. 129, Antioch Press, Yellow Springs, Ohio, 1965.
32. Kok, B., in *Photosynthetic Mechanisms of Green Plants* (B. Kok and A. T. Jagendorf, eds.), p. 35, Nat. Acad. Sci.-Nat. Res. Council, 1963.
33. Katoh, S., Shiratori, I., and Takamiya, A., *J. Biochem.*, **51**, 32 (1962).
34. Davenport, H. E., and Hill, R., *Proc. Roy. Soc.*, **B 139**, 327 (1952).
35. Lundegardh, H., *Physiol. Plantarum*, **15**, 390 (1962).
36. Hill, R., and Scarisbrick, *New Phytologist*, **50**, 98 (1951).
37. Anderson, J. M., Fork, D. C., and Amesz, J., *Biochem. Biophys. Res. Comm.*, **23**, 874 (1966).
38. Boardman, N. K., and Anderson, J. M., *Biochim. Biophys. Acta*, **143**, 187 (1967).

39. Katoh, S., and Takamiya, A., *Biochim. Biophys. Acta*, **99**, 156 (1965).
40. Katoh, S., and San Pietro, A., in *The Biochemistry of Copper* (J. Peisach, P. Aisen and W. E. Blumberg, eds.), p. 407, Academic Press, New York, 1966.

Possibility of Differentiation of Three Photoreaction Systems in *Chromatium* D

SHIGEHIRO MORITA

Department of Biophysics and Biochemistry, Faculty of Science, University of Tokyo, Tokyo

INTRODUCTION

One of the fundamental differences between bacterial photosynthesis and photosynthesis of green plants pertains to differentiation of photochemical reaction systems. Two successive light reactions (systems I and II) are operating in green plants, while in photosynthetic bacteria only one light reaction has been generally assumed to be working.

On the other hand, evidence is accumulating that there are cyclic and non-cyclic systems of electron transport in *Chromatium*.[1–3] Furthermore, in *Chromatium* it has been shown that the photoinduced oxidations of three different hemoproteins, cytochromes c_{553}, cc' and c_{555}, are participating in three distinct paths of electrons.[1,4,5] These three reactions were found to differ from each other in several respects: reactivity at low temperatures,[2] kinetics of dark reduction of the oxidized forms,[1,4] dependency on light intensity,[4] sensitivity towards various inhibitors,[1] sensitivity to oxygen,[1,4] as well as the effects of endogenous or added substrates.[1] In this report these three reactions will be named reactions I, II and III. Reaction I is saturated at a low light intensity, reactions I and II are sensitive to oxygen while reaction III is insensitive. Without addition of substrate, only reaction III is observed.

Furthermore, there have been several reports describing experimental facts which seem to suggest the possibility of differentiation of photochemical reaction systems in *Chromatium*. On the basis of the finding of two kinds of photoinduced absorption changes of bacteriochlorophyll-like pigments, i.e. bleaching of P890 and red or blue shift of the absorption band of bacteriochlorophyll, Clayton[6] and Vredenberg *et al.*[7,8] have suggested the possibility of two kinds of photoreactions in photosynthetic bacteria. Amesz and Vredenberg,[9] investigating the quantum yields of fluorescence, the oxidation of cytochrome and the shift of absorption bands of carotenoids, suggested compartmentation of light harvesting pigments in photosynthetic bacteria. In this connection, it has to be noted that it was only the oxidation of one component of the hemoproteins in *Chromatium* (i.e. cytochrome c_{553}) that they actually measured. Vernon[10,11] has reported, for *Chromatium*, separation of two kinds of pigmented particles by means of degradation of the chromatophores with Triton X-100, followed by separation by density gradient centrifugation. The two kinds of fractions thus obtained showed different ratios of the bacteriochlorophyll forms, and it was only in the "heavy particles" that the photoinduced absorption changes at 890 nm and 270 nm occurred. His results are of special interest in connection with the possible compartmentation of bacteriochlorophyll.

In the present study, the photoinduced oxidations of the three hemoproteins were investigated to discover a correlation between these reactions and the forms of light-harvesting bacteriochlorophyll. The action spectra for the oxidation of the three hemoproteins were measured separately under individually specified experimental conditions, using intact cells of *Chromatium*.

EXPERIMENTAL

The absorption changes were measured with an Amino-Chance Dual-Wavelength Spectrophotometer, in a modified sample compartment. The diagram of the sample compartment is shown in Fig. 1. The sample cuvette was illuminated from the same side as the measuring light beams by actinic light furnished from a Bausch and Lomb monochromator.

The reactions were measured separately in the following way:

1) Under anaerobic conditions and in the presence of substrate in the

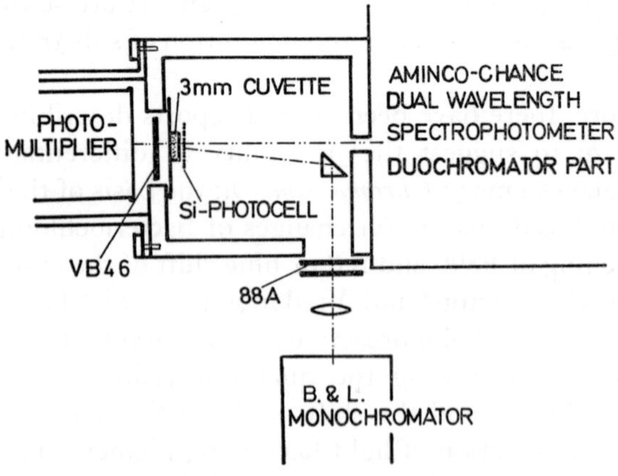

Fig. 1. Diagram of the sample compartment.

reaction medium (the culture medium was used as substrate), illumination at lower intensities of actinic light causes reaction I, without inducing reactions II and III in the bacterial cell.[1,4]

2) On illumination under aerobic conditions, and without addition of the substrate, the starved cells only show reaction III.[1,4]

3) In the presence of a high concentration of 1,10-phenanthroline ($10^{-3}M$), reactions I and III were inhibited, and on illumination at higher intensities of actinic light only reaction II was observed.[1] The absorbancy changes on turning on and turning off the actinic light were measured at 423.5 nm (reference wavelength: 440 nm). The measurements were made at three different light intensities for each wavelength of actinic light. The light intensities were varied by insertion of appropriate neutral density filters and measured with a Si-photocell placed in the place of the sample cuvette at the beginning and at the end of each measurement. A curve was drawn for the changes in level of steady state of absorbancy as a function of intensities of actinic light of a given wavelength. The light intensities causing a given amount of absorbancy change were estimated mostly by interpolating, but in a few cases also by extrapolating, the plotted curves. The reciprocals of such light intensities were plotted *vs.* wavelengths to obtain the action spectrum.

RESULTS

The resulting action spectra are shown in Figs. 2 and 3. The action spectra approximately followed the absorption spectrum of the intact cells, showing three peaks in the near-infrared region. On closer examination, however, meaningful differences were discovered with respect to the shapes of the spectral curves. In Fig. 2 the action spectra for reactions I and III were superimposed. A difference will be seen in relative efficiencies at 810 nm and 890 nm; 890 nm light was more effective for reaction III, while 810 nm light was more effective for reaction I.

In Fig. 3 the action spectra for reactions I and II were superimposed. Small but detectable differences will be seen between the two spectral curves, especially at 850 nm. In the oxidation of cytochrome *cc'*, participating in reaction II, the discrepancy between the action and the absorption spectra is not very marked. However, it may be stated that 850 nm light is more effective for reaction II, as compared with the light of the other absorption bands of bacteriochlorophyll.

The precise shapes of the action spectra, as well as absorption spectra, differed from bacterial sample to sample. In spite of such differences, the

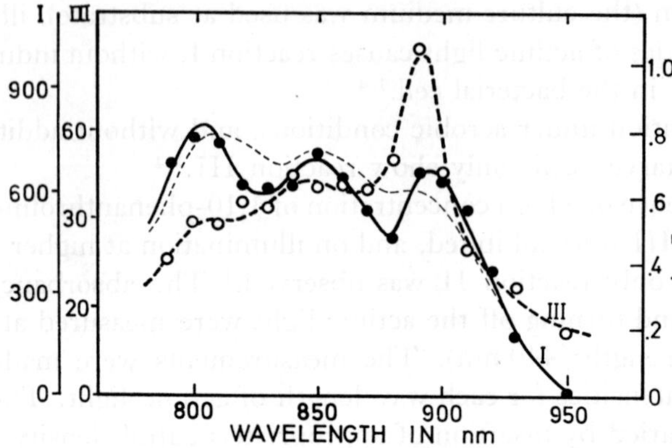

Fig. 2. Action spectra for photo-induced oxidation of cytochromes c_{553} and c_{555}. Thin dashed curve, absorption spectrum. Open circles, aerobic without substrate (reaction III). Solid circles, anaerobic with substrate (reaction I). The action spectra are normalized to make the averages of the heights of three peaks in each spectral curve coincide. Activities are expressed by the reciprocals of the flux of quanta (10^8 einstein^{-1}. sec cm^2), inducing 0.8% change (reaction I) or 0.15% change (reaction III) in $\Delta T_{423.5\ \mathrm{nm}-440\mathrm{nm}}$. Half-widths of actinic light were 5 nm (reaction I) and 20 nm (reaction III). Amounts of bacterial cells used, 1.86 mg Folin protein/ml.

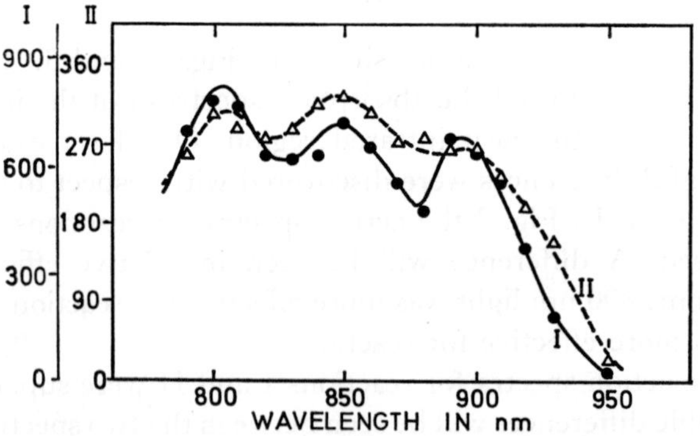

Fig. 3. Action spectra for photo-induced oxidation of cytochromes c_{553} and cc'. Triangles, anaerobic with substrate and 1,10-phenanthroline ($10^{-3}M$) added 15 minutes prior to measurement (reaction II). Solid circles, anaerobic with substrate (reaction I). The action spectra are normalized to make the averages of the heights of three peaks in each spectral curve coincide. Activities are expressed by the reciprocals of the flux of quanta (10^8 einstein^{-1} sec cm^2), inducing 0.8% change (reaction I) or 1.0% change (reaction II) in $\Delta T_{423\cdot5\mathrm{nm}-440\mathrm{nm}}$ Half-widths of actinic light were 5 nm (reaction I) and 10 nm (reaction II). Amounts of bacterial cells used, 1.86 mg Folin protein/ml.

TABLE I.

Exp. no.	Relative efficiency at 810 nm*		Ratio reaction III/reaction I
	Reaction I (anaerobic, with substrate)	Reaction III (aerobic, without substrate)	
1	1.15	0.85	0.75
2	0.89	0.75	0.85
3	1.02	0.75	0.7
4	1.18	0.51	0.4

* Activity at 810 nm/activity at 890 nm.

relationship pertaining to the relative heights of the three peaks under various conditions remained unaltered, at least in so far as the action spectra of reactions I and III were concerned. The relative heights between 810 nm and 890 nm peaks in the action spectra for reactions I and III were compared, using the results of three independent experiments, in which the samples from different cultures were used. In these cases (experiments 1, 2 and 3 in Table I), the efficiency for each wavelength of actinic light was calculated by dividing the observed absorbancy change with the light intensity. The ratios of the relative efficiencies at 810 nm for reactions III and I were 0.75, 0.85 and 0.7. In the same table, the corresponding values for the experiment given in Fig. 2 (experiment 4) are also included (Table I). It will be seen that in each case the efficiency at 810 nm is higher for reaction I than for reaction III.

DISCUSSION

The action spectra for the oxidation of the hemoproteins obtained under various conditions appreciably differ from each other as well as from the absorption spectrum of the intact cells of *Chromatium*. The results of another series of experiments, in which the effects of light intensity on the steady state level of oxidations were compared at various wavelengths, also led to the same conclusion.* This finding shows that the observed differences in shape of action spectra obtained under various conditions were not a result of artifact due to the difference in experimental conditions under which the reaction was measured. Previous studies by various investigators and also the present author have shown that separate electron transporting paths are operating in the photoinduced oxidation of the hemoproteins in *Chro-*

* To be published elsewhere.

matium.[1-5] The result of the present study suggests that each one of these three electron transport reactions is correlated to one of the three forms of bacteriochlorophyll showing absorption maxima at 810, 850 and 890 nm. The interrelation between these reaction systems is still obscure. It cannot be decided with certainty whether they are working in parallel or functioning in a successive series like the "two light reactions" in green plants. At present the author would like to assume that these three reactions are operating rather independently. Some support for this tentative scheme will be found in the well known fact that the turnover rate of cytochrome c_{555} is ten times higher than that of cytochrome c_{553}. Such a big difference in rates cannot occur in a serial arrangement of the two photoreactions.

There has been no indication of any clear "red drop" in the action spectra in *Chromatium*, nor was any single-peaked action spectrum obtained for the oxidation of hemoproteins. These facts may indicate that a strong interaction is working between the photochemical systems in question.

Another possibility is that there may be a differentiation of the electron transport systems in three compartments, in each one of which all the three bacteriochlorophyll forms are involved, but in different ratios of amounts. The combination of these interactions and of a separate package of bacteriochlorophyll is also possible. Some support for the separate package of light harvesting systems will be found from the studies of the fluorescence of bacteriochlorophyll.[12] In any condition, with a weak exciting light or strong exciting light, with or without substrate, only one peak of the emission spectrum of fluorescence has been observed in the intact cells of *Chromatium*. In the light harvesting system, one form (890 nm) out of the three forms of bacteriochlorophyll emits fluorescence *in vivo*. The fact suggests that, in each pigment system, the 890 nm absorbing form of bacteriochlorophyll must be involved.

In summary, the author is inclined to believe in the occurrence of three photosynthetic reactions, which are probably operating in parallel rather than in series.

SUMMARY

The action spectra for the oxidations of cytochromes c_{553}, cc' and c_{555} in *Chromatium* were measured.

The experimental results indicated that 810 nm light was favorable for the oxidation of cytochrome c_{553}; 850 nm light was more favorable for the oxidation of cytochrome cc'; and 890 nm light was more favorable for the oxidation of cytochrome c_{555}.

Summarizing these results, possible differentiation of three photochemical reaction systems in *Chromatium* was proposed.

REFERENCES

1. Morita, S., Edwards, M., and Gibson, J., *Biophys. Biochim. Acta*, **109**, 45 (1965).
2. Vredenberg, W. J., and Duysens, L. N. M., *Biophys. Biochim. Acta*, **79**, 456 (1964).
3. Chance, B., Nishimura, M., Roy, S. B., and Schleyer, H., *Biochem. Z.*, **338**, 654 (1963).
4. Olson, J. M., and Chance, B., *Arch. Biochem. Biophys.*, **88**, 26 (1960).
5. Olson, J. M., and Chance, B., *Arch. Biochem. Biophys.*, **88**, 40 (1960).
6. Clayton, R. K., *Proc. Natl. Acad. Sci.*, **50**, 583 (1963).
7. Vredenberg, W. J., and Amesz, J., *Biophys. Biochim. Acta*, **126**, 244 (1966).
8. Vredenberg, W. J., Amesz, J., and Duysens, L. N. M., *Biochem. Biophys. Res. Comm.*, **18**, 435 (1965).
9. Amesz, J., and Vredenberg, W. J., *Biophys. Biochim. Acta*, **126**, 254 (1966).
10. Garcia, A., Vernon, L. P., and Mollenhauer, H., *Biochemistry* (Am. Chem. Soc.), **5**, 2399 (1966).
11. Vernon, L. P., *this volume.*
12. Tamaki, T., *personal communication.*

Effect of Hill Reaction Inhibitors on Photosystem I

S. Izawa

Department of Biology, Queen's University, Kingston, Ontario, Canada

INTRODUCTION

Scores of compounds represented by DCMU and CMU are known to inhibit the Hill reaction at very low concentrations. The site of inhibition is believed to be located very close to photosystem II.[1] Photosystem I-dependent reactions such as the photoxidation of $DCPIH_2$ with NADP serving as electron acceptor[2] and cyclic photophosphorylation[3,4] are unaffected by these inhibitors at concentrations which are high enough to suppress the Hill reaction completely. However, system I reactions are not completely immune to these system II inhibitors. The best-documented example is the fact that PMS- or pyocyanine-mediated cyclic photophosphorylation is inhibited by high concentrations ($\geqq 0.1$ mM) of DCMU, CMU and similar compounds.[4,5] It has been postulated that the electron transport on the reducing side of pyocyanine (i.e. system I) involves another DCMU-sensitive site.[5] The apparent existence of a DCMU-sensitive component in system I has tended to give an impression that the two photosystems may have more in common than would otherwise have been thought.[6] These interpretations are based on the seemingly reasonable but unproven assumption that the inhibition of cyclic photophosphorylation by Hill reaction inhibitors means the inhibition of cyclic electron transport. The crucial experiment, i.e., direct measurement of the cyclic electron flow, is unfortunately nearly impossible. However, it seems obvious that if photosystem I involves a DCMU-sensitive site, the inhibition should also show up in the system I-dependent noncyclic electron flow. These considerations led us to the reexamination of the secondary high concentration effect of DCMU by using the reaction system described below.

Abbreviations: CMU, 3-(4-chlorophenyl)-1,1-dimethylurea; CTCA, N-(4-chlorophenyl) trichloracetamide; 3-CTCA, 3-chlorophenyl analog of CTCA; DCCA, N-(3,4-dichlorophenyl) chloracetamide; DAD, 2,3,5,6-tetramethylphenylenediamine; DCMU, 3-(3,4-dichlorophenyl)-1,1-dimethylurea; DCPI, 2,6-dichlorophenolindophenol; $DCPIH_2$, reduced form of DCPI; MV, methylviologen; MVH, reduced form of MV; PMS, phenazine methosulfate (N-methylphenazonium methosulfate); TMPD, N,N,N′,N′-tetramethylphenylenediamine.

EXPERIMENTAL

Measurement of photosystem I-dependent electron flow

Connolly and Good[7] have recently developed a simple method of measuring the rate of system I-dependent electron transport based on the previous findings of Krogmann,[8] Vernon and Zaugg[2] and Trebst *et al.*[9,10] In the presence of DCMU in an amount which is sufficient to block system II, electrons from artificial donors (e.g. $DCPIH_2$ with excess asscorbate) are transported to a low potential, autoxidizable acceptor such as methylviologen through system I, and the reduced acceptor is reoxidized by air to produce H_2O_2. The hydrogen peroxide formed is reduced quantitatively by the same electron donor, probably via a chloroplast-bound peroxidase (N. E. Tolbert, personal communication), thus enabling the measurement of electron flow as oxygen uptake:

$$DCPIH_2 + 2MV \xrightarrow{\text{Photosystem I}} DCPI + 2MVH$$

$$2MVH + O_2 \xrightarrow{\text{spontaneous}} 2MV + H_2O_2$$

$$H_2O_2 + DCPIH_2 \xrightarrow{\text{peroxidase?}} DCPI + 2H_2O$$

$$DCPI + \text{ascorbate} \xrightarrow{\text{spontaneous}} DCPIH_2 + \text{dehydroascorbate}$$

$$\overline{\text{Ascorbate} + O_2 \longrightarrow \text{Dehydroascorbate} + H_2O}$$

Interestingly, this $DCPIH_2$ photoxidation represents a type of electron transport which is only weakly, if at all, coupled to phosphorylation but is nevertheless very sensitive to uncouplers, the otherwise rather slow rate greatly increasing in their presence.* The reaction with TMPD (non-phosphorylating) or DAD (partially coupled) as electron donor is exceedingly fast and uncoupler-insensitive. Thus the $DCPIH_2$ reaction is suitable for detecting an uncoupler effect, and the TMPD (or DAD) system is well qualified for testing electron transport inhibition. We have examined the effect of high concentrations of various Hill reaction inhibitors† on the above reaction system and obtained the unexpected results described below.

* With ferredoxin-NADP as electron acceptor the reaction is slower and the uncoupler effect is much less pronounced, suggesting that the teminal reduction step is more or less rate-limiting.[7] Therefore, the NADP system is less satisfactory for the present purpose.

† The inhibitors and electron donors used in this study were a generous gift from Dr. N. E. Good.

RESULTS AND DISCUSSION

It should be noted first that in the following experiments photosystem II was always blocked completely by 3 μM DCMU. High concentrations of inhibitors including DCMU were applied to the reaction mixtures in addition to this basic low concentration of DCMU.

The results are summarized in Table I. Any of the six inhibitors tested, even at 0.5 mM, has little effect on TMPD photoxidation, a very fast non-phosphorylating process. Surprisingly, however, some of these compounds remarkably enhance the slow, weakly coupled $DCPIH_2$ reaction (see also Fig. 1), a typical uncoupler effect.[7] Moreover, their potency in stimulating $DCPIH_2$ photoxidation very clearly parallels their potency to inhibit the PMS-mediated cyclic photophosphorylation. It therefore seems very probable that the inhibition of the cyclic photophosphorylation by these Hill reaction inhibitors is not a blocking of the system I-driven electron flow but

TABLE I. Effect of high concentration of Hill reaction inhibitors on system I reactions.

Compound	pI_{50} for Hill reaction[a]	TMPD photo-oxidation[b] Rate	TMPD photo-oxidation[b] Effect (%)	$DCPIH_2$ photo-oxidation[b] Rate	$DCPIH_2$ photo-oxidation[b] Effect (%)	Cyclic photo-phosphorylation[c] Rate	Cyclic photo-phosphorylation[c] Effect (%)	Partition ratio[d]
(None)	...	1850	...	370	...	440	...	...
Atrazine	6.6	1950	(+ 4)	430	(+ 16)	490	(+11)	—
CMU	6.3	1600	(−14)	440	(+ 19)	340	(−23)	5
DCMU	7.5	1850	(0)	500	(+ 35)	280	(−36)	26
CTCA	4.6	1600	(−14)	960	(+160)	4	(−99)	53
3-CTCA	5.0	1450	(−22)	1120	(+200)	20	(−95)	—
DCCA	6.5	2000	(+ 8)	1310	(+250)	34	(−92)	180

a. Log of concentration causing 50% inhibition of the Hill reaction (from ref. 4).

b. For experimental conditions, see the legend for Fig. 1. The rates given are μequiv/hour/mg chlorophyll.

c. The reaction mixture contained: 3 μM DCMU, 50 mM tricine buffer (pH 8.1), 2 mM $MgCl_2$, 1 mM ADP, 10 mM $^{32}P\text{-}Na_2HPO_4$, 20 μM PMS, and 0.15 mM of the inhibitors listed. Chlorophyll concentration was 13 μg/ml. The reaction mixtures (2 ml) in a cuvette were illuminated for 2 min and analyzed for radioactive ATP formed. Other conditions were the same as described for Fig. 1. Rates, μmoles ATP/hour/mg chlorophyll.

d. The ratio of inhibitor concentrations in the chloroplast phase and in the medium phase at an overall inhibitor concentration of 7×10^{-5} M. Values were obtained by determining (as absorbance decreases in UV region) the amount of inhibitor disappeared from the medium on addition of a known packed volume of chloroplasts. At lower concentrations where the Hill reaction is specifically inhibited the ratios are much higher.[11]

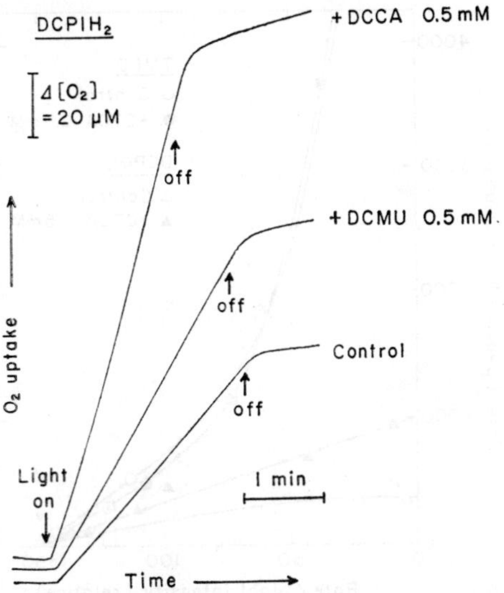

Fig. 1. Enhancement of $DCPIH_2$ photoxidation by high concentrations of Hill reaction inhibitors. The reaction mixtures contained: 3 μM DCMU, 50 mM TES buffer (pH 7.0), 3 mM sodium ascorbate, 2 mM $MgCl_2$, 0.1 mM MV, 0.1 mM DCPI (0.5 mM TMPD or DAD, when used) and the high concentrations of inhibitors indicated. Chlorophyll concentration was 10 μg/ml. The reaction mixtures (4 ml) in a small vial were stirred with a magnetic stirrer and oxygen uptake was measured with a Clark oxygen electrode. The actinic light used was broad-band orange light (540–700 mμ). Temperature was 18°C. The curves shown are recorder-tracings.

an uncoupling of electron transport from the phosphorylation mechanism.

The uncoupling potency of these inhibitors does not seem to be at all related to their potency to inhibit system II (as indicated by their ability to inhibit the Hill reaction). For instance, the most potent Hill reaction inhibitor DCMU is only a mild uncoupler, while the poorest inhibitor CTCA is one of the strongest uncouplers. However, a correlation seems to exist between uncoupling potency and binding to the chloroplast material. CTCA and DCCA, the strongest uncouplers, are the substances absorbed by chloroplasts most strongly, as indicated by the partition ratios. CMU is only weakly absorbed by chloroplasts and is a poor uncoupler.

The light intensity experiments shown in Figs. 2 and 3 indicate the following facts: (a) Enhancement of the system I electron transport by uncoupling concentrations of the Hill reaction inhibitors is due to a release of some rate-limiting dark process. The quantum yield remains unchanged. (b) The three reaction systems ($DCPIH_2$, DAD and TMPD) have the same

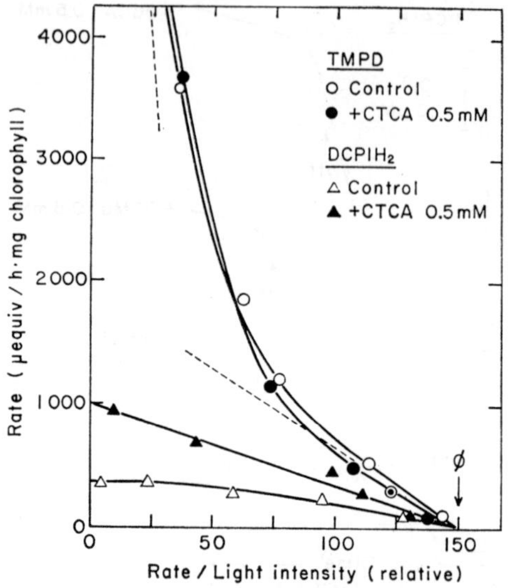

Fig. 2. Effect of high concentrations of Hill reaction inhibitors on photoxidations of $DCPIH_2$ and TMPD at varied light intensities. Experimental conditions were as described for Fig. 1 except for variation in light intensity. The stimulation (uncoupling) of $DCPIH_2$ photoxidation by the inhibitors is an enhancement of a rate-limiting dark process (intercepts on the ordinate) and involved no change in relative quantum efficiency (ϕ).

quantum efficiencies, except that the TMPD reaction appears to be complex, comprising in part a low-efficiency process (Fig. 3).

Fig. 4 depicts the kinetics of cytochrome *f* response in the presence and absence of DCCA ($DCPIH_2$ as electron donor). Upon illumination the cytochrome is rapidly oxidized and re-reduced in the dark. (With ascorbate alone as electron donor, the dark reduction proceeds extremely slowly.) Evidently $DCPIH_2$ provides electrons for the transport chain between system II and the cytochrome.[2,12] In the presence of DCCA the reduction of cytochrome proceeds markedly faster, indicating that the dark electron flow from $DCPIH_2$ to the cytochrome is accelerated by the compound. This is consistent with the conclusion drawn from the light intensity experiments mentioned above. The typical uncoupler ammonia similarly stimulates the dark reduction of cytochrome *f* (*cf.* ref. 13). Fig. 5 diagrams our interpretation of these observations.

At concentrations which are completely uncoupling, the amount of inhibitors bound to chloroplasts is roughly 1 inhibitor molecule for each 10 chlorophyll molecules. The significance of this value in relation to the

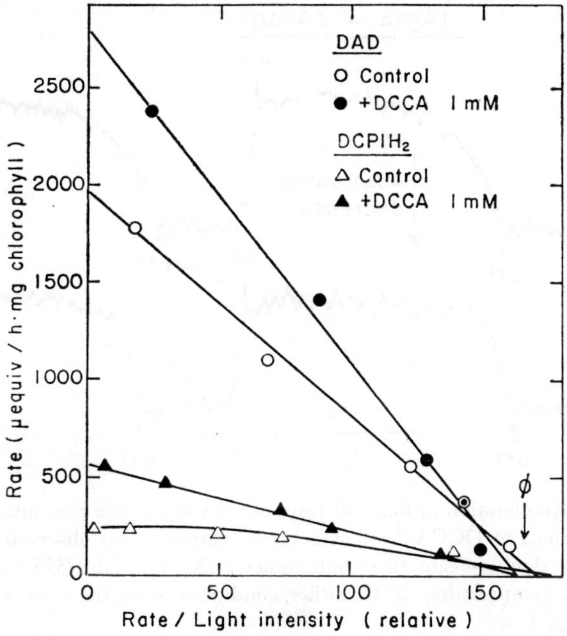

Fig. 3. Effect of high concentrations of Hill reaction inhibitors on photoxidation of DAD and $DCPIH_2$ at varied light intensities. Experimental conditions were as described for Fig. 1 except for variations in light intensity. Chloroplasts (+DCCA) were treated with the compound drastically (1 hour preincubation at room temperature). There is still no indication of electron transport inhibition. The quantum efficiency (ϕ) is only negligibly affected in these pretreated chloroplasts.

uncoupling mechanism is not clear since we do not know what proportion of the chloroplast-bound inhibitors is actually at the uncoupling site.

Mukasa *et al.*[14] reported an interesting observation that high concentrations of CMU affect changes of cytochromes in respiring yeast cells as antimycin A does. The effect of DCCA on the kinetics of changes of cytochrome *f* in chloroplasts (Fig. 4), however, is opposite to the antimycin effect.[15]

The significant finding in this study is that photosystem I seems to lack a DCMU-inhibitable site completely. Neither the quantum efficiency nor the electron transport potential is affected by the highest concentrations (0.5–1 m*M*) of the system II inhibitors. The action of DCMU on system II is generally believed to be a blocking of the reoxidation of the as yet unidentified quencher "Q".[1] If a similar mechanism of quantum conversion is

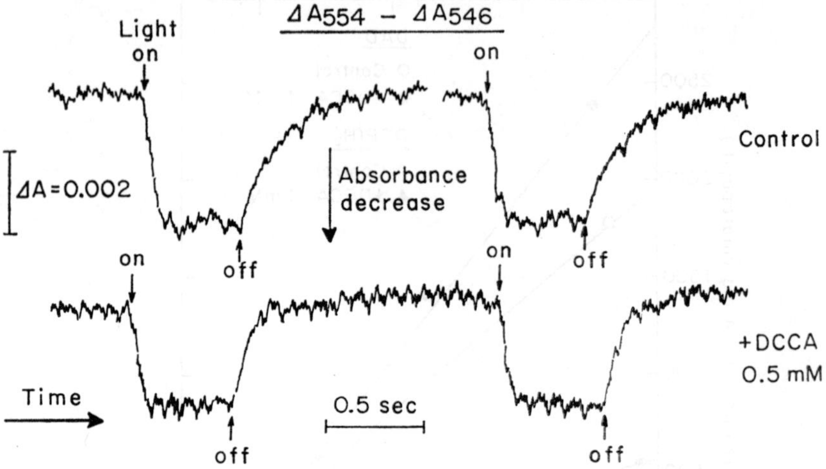

Fig. 4. Acceleration of light-off response of cytochrome *f* by uncoupling concentrations of DCCA. The cytochrome response was observed with a Chance-type double-beam spectrophotometer. Actinic light, 575 mμ (60 Kerg/sec cm²), temperature, 20°C. Other conditions were the same as described for Fig. 1.

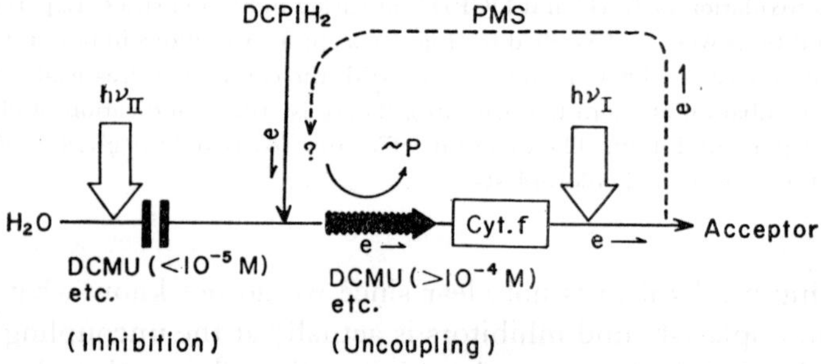

Fig. 5. Scheme proposed to illustrate the dual effect of Hill reaction inhibitors on the electron transport system in chloroplasts. The non-phosphorylating by-pass electron transport (see ref. 7) is omitted from the diagram.

involved in photosystem I, the chemical nature of the components participating therein may be quite different from those in system II.

Acknowledgements

The author wishes to thank Dr. G. Hind, Brookhaven National Laboratory, for his help in double-beam spectroscopy. Thanks are also due to Dr. N. E. Good and Mr. T. N. Connolly, Michigan State University, for valuable discussions. This work was supported by a grant (A-3743) from the National Research Council, Ottawa, awarded to the author at Queen's University, and in part by a grant (GB4568) from the National Science Foundation, Washington, D. C., awarded to Dr. N. E. Good and the author at Michigan State University.

SUMMARY

Photosystem I-dependent cyclic photophosphorylation is severely inhibited by high concentrations ($\geqq 0.1$ mM) of inhibitors of the Hill reaction (photosystem II) such as 3-(3,4-dichlorophenyl)-1,1-dimethylurea (DCMU). Evidence is presented that this inhibition is not due to a blocking of electron flow but is rather an uncoupling effect. The quantum conversion mechanism involved in photosystem I and the electron transport driven by the system are completely resistant to the highest concentrations (0.5–1 mM) of these system II inhibitors.

REFERENCES

1. Duysens, L. N. M., and Amesz, J., *Biochim. Biophys. Acta*, **64**, 243 (1962).
2. Vernon, L. P., and Zaugg, W. S., *J. Biol. Chem.*, **235**, 2728 (1960).
3. Jagendorf, A. T., and Margulies, M. M., *Arch. Biochem. Biophys.*, **90**, 184 (1960).
4. Good, N. E., *Plant Physiol.*, **36**, 788 (1961).
5. Asahi, T., and Jagendorf, A. T., *Arch. Biochem. Biophys.*, **100**, 531 (1963).
6. Kok, B., and Cheniae, G. M., in *Current Topics in Bioenergetics* (D. R. Sanadi, ed.), Vol. 1, p. 1, Academic Press, New York, 1966.
7. Izawa, S., Winget, D. G., Connolly, T. N., and Good, N. E., *Brookhaven Symposia in Biology*, No. 19, 169 (1966).
8. Krogmann, D. W., *Biochim. Biophys. Acta*, **30**, 655 (1958).
9. Trebst, A., and Eck, H., *Z. Naturforsch.*, **16 b**, 455 (1961).
10. Trebst, A., and Pistorius, E., *Z. Naturforsch.*, **20b**, 143 (1965).
11. Izawa, S., and Good, N. E., *Biochim. Biophys. Acta*, **102**, 20 (1965).
12. Witt, H. T., Müller, A., and Rumberg, B., *Nature*, **192**, 967 (1961).
13. Avron, M., and Chance, B., *Brookhaven Symposia in Biology*, No. 19, 149 (1966).
14. Mukasa, H., Itoh, M., and Nosoh, Y., *Plant Cell Physiol.*, **7**, 683 (1966).
15. Hind, G., and Olson, J. M., *Brookhaven Symposia in Biology*, No. 19, 188 (1966).

Photoreaction of Chloroplasts: NADP Photoreduction by *Euglena* Chloroplasts*

SAKAE KATOH AND A. SAN PIETRO

Department of Biophysics and Biochemistry, Faculty of Science, University of Tokyo, Tokyo
and
Charles F. Kettering Research Laboratory, Yellow Springs, Ohio

INTRODUCTION

The current representation of photosynthesis in green plants and algae[1,2] involves two photosystems: a long wavelength system (photosystem I) and a shorter wavelength system (photosystem II). It is visualized that the action of photosystems I and II is the conversion, presumably at their reaction centers, of electromagnetic energy into chemical energy in the form of separated oxidizing and reducing entities. The weak reductant produced by photosystem II serves as the electron donor for photosystem I; the strong oxidant generated by photosystem II interacts with water to yield molecular oxygen. The role of water in this hypothesis is, therefore, as an electron donor for photosystem II. The resultant of the absorption of light by photosystem I is the formation of a strong reductant and the concomitant oxidation of the weak reductant produced by photosystem II. Electron flow between the two photosystems occurs via an electron transfer pathway which interconnects them. The strong reductant produced by photosystem I serves to reduce ferredoxin or NADP. In concert with these reactions is the formation of ATP; however, the exact mechanism of ATP formation is unknown. It is presumed that electron flow between the two photosystems serves ultimately as the energy source for ATP formation.

In the present paper, the results obtained from experiments in which chloroplasts of *Euglena gracilis*, strain Z, were mainly employed are described. Firstly, evidence will be presented that a cytochrome of the *c*-type, cytochrome-552 (cytochrome *f*), functions in the electron transfer system connecting the two photosystems in the manner predicted by Hill and Bendall.[1] Secondly, it will be shown that the mechanism of inactivation

* Contribution No. 285 of the Charles F. Kettering Research Laboratory. This research was supported in part by a Research Grant (GM-10129) to one of us (A.S.P.) from the National Institutes of Health, United States Public Health Service.

caused by heat-treatment of chloroplasts differs from that of inhibition by poisons. Ascorbate or cysteine can serve in place of water as the electron donor for photosystem II with heated chloroplasts but not with the poisoned chloroplasts. Finally, a comparison will be made of the various physical and chemical treatments of chloroplasts which inactivate preferentially the oxygen evolution system.

RESULTS AND DISCUSSION

Requirement of Cytochrome-552 for NADP Photoreduction

Smillie[3] and Perini *et al.*[4] showed that *Euglena* cytochrome-552 was very easily solubilized during preparation of chloroplasts in ordinary aqueous medium. This observation was confirmed in the present investigation. A difference spectrum (reduced with dithionite minus oxidized with ferricyanide) of an acetone powder of *Euglena* chloroplasts showed the presence only of a *b*-type cytochrome but no appreciable amount of cytochrome-552. These chloroplasts supported the Hill reaction with ferricyanide and DPIP as oxidant but were incapable of photoreducing NADP or other substances of low redox potential.[5] Presumably, the loss of cytochrome-552 from the electron transfer chain was the reason why *Euglena* chloroplasts were unable to photoreduce NADP. If such were the case, the addition of cytochrome-552 to the chloroplasts should restore the Hill activity with NADP as oxidant. This was found to be the case; a substantial rate of NADP photoreduction with water as the electron donor was obtained when cytochrome-552 was added to the chloroplasts. In addition, cytochrome-552 was effective in restoring the Hill reaction with cytochrome *c* and methyl viologen as oxidants, as well as the photoreduction of NADP with reduced DPIP and TMPD as electron donors. These observations are consistent with the scheme for the photosynthetic electron transport system proposed by Hill and Bendall, which consists of two photoreactions connected by an electron transport chain involving cytochrome(s). In the absence of cytochrome-552, only substances of high redox potential such as ferricyanide or DPIP were reduced via the weak reductant produced by photosystem II. On addition of cytochrome-552, a cooperation of the two photosystems was possible such that the chloroplasts could photoreduce NADP with the strong reductant produced by photosystem I.

Evidence in support of this view was obtained from a study of the spectral behavior of the added cytochrome-552 during NADP photoreduction.[6] When either reduced or oxidized cytochrome-552 was added to a reaction system containing ferredoxin and NADP, it was rapidly oxidized or reduced, respectively, on illumination until a steady state level corresponding to about 90 per cent reduction was reached. Thereafter, the redox level of the cytochrome-552 remained constant during the continued photoreduction of NADP by the regular Hill reaction system. It is evident that cytochrome-552 underwent repeated oxidation-reductions in the light, thereby transporting electrons from photosystem II to photosystem I during NADP photoreduction.

In the absence of any added electron acceptor, *Euglena* chloroplasts catalyzed a rapid and complete photoreduction of oxidized cytochrome-552 and this photoreduction was inhibited by poisons such as DCMU. Furthermore, even in the presence of the poison, the chloroplasts catalyzed the photooxidation of reduced cytochrome-552 with molecular oxygen as oxidant. The rate of this photooxidation was markedly stimulated by the addition of ferredoxin plus NADP or methyl viologen. In contrast, the photooxidation of reduced cytochrome-552 was negligible in the absence of any added electron acceptor including molecular oxygen. A good stoichiometry between the photooxidation of reduced cytochrome-552 and the concomitant photoreduction of NADP was obtained under anaerobic conditions.

It was found, however, that both the photooxidation and photoreduction of the added cytochrome-552 required a higher concentration of cytochrome for maximal activity than that required to support the maximal rate of NADP photoreduction. In addition, the rate of the NADP Hill reaction was always higher than either that of the photooxidation or photoreduction of the added cytochrome-552. It was postulated, therefore, that only a small portion of the cytochrome-552 provided was reincorporated into the original reaction site in the chloroplasts, whereas the bulk of the cytochrome-552 remained in solution. The cytochrome-552 bound at the appropriate reaction site would function as an electron carrier between the two photosystems. On the other hand, the photooxidation-reduction of the cytochrome-552 remaining in solution might be carried out secondarily and involve a diffusion process. In agreement with this assumption was the observation that the kinetics of the photooxidation-reduction of cytochrome-552 during NADP photoreduction, measured by sensitive spectrophotometry, were biphasic: an initial small but very rapid absorbance change followed by a large but slower absorbance change. These two phases of absorbance change might be a reflection of the two physical states of the added cytochrome-552.

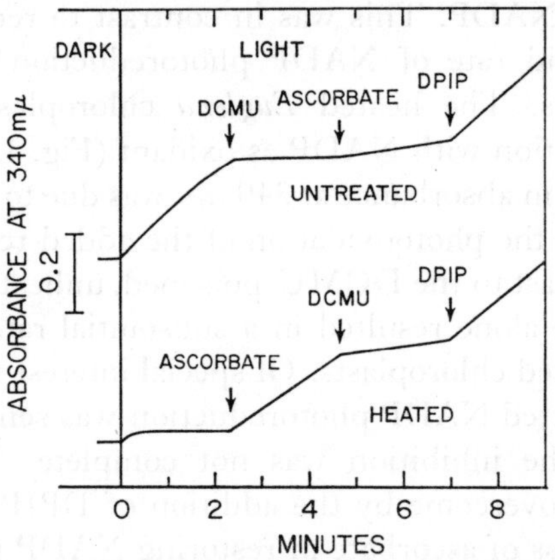

Fig. 1. Time course of NADP photoreduction by untreated chloroplasts and chloroplasts incubated at 45° for 5 minutes.

Absorbance changes at 340 mμ were determined in a Beckman spectrophotometer (Model DB). A cuvette (0.5 × 1 × 4 cm) was illuminated with a Unitron lamp through a Corning glass filter (No. 2403) from one side of the photo-tube compartment. Scattered light was removed with a glass filter inserted between the cuvette and phototube. Each reaction mixture contained in a final volume of 0.7 ml: 50 m*M* phosphate, pH 6.0; 10 m*M* NaCl; 0.86 m*M* NADP; 20 μ*M* reduced *Euglena* cytochrome-552; a saturating amount of spinach ferredoxin and *Euglena* chloroplasts corresponding to 11 μg of chlorophyll. Where indicated with arrows, the following additions were made: 10 μ*M* DCMU; 14 m*M* neutralized ascorbate; or 20 μ*M* DPIP.

Heat Inactivation

The oxygen evolution system of the Hill reaction is one of the most labile reactions in photosynthesis. Hinkson and Vernon[7] reported that treatment of chloroplasts at 51° for 10 minutes resulted in complete inhibition of the Hill activity, whereas the photoreduction of indigo carmine with reduced DPIP as electron donor and the photooxidation of ascorbate in the presence of DPIP were suppressed only partly. The photoreduction of ferredoxin by mildly heated chloroplasts at the expense of reduced DPIP was also observed by Whatley *et al.*[8]

Euglena chloroplasts lost the Hill activity with ferricyanide, DPIP or oxidized cytochrome-552 as oxidants on heating at 45° for 5 minutes. The time course of NADP photoreduction with untreated *Euglena* chloroplasts and *Euglena* chloroplasts heated at 45° for 5 minutes is shown in Fig. 1. The upper curve shows that the NADP Hill reaction with untreated chloroplasts is highly sensitive to DCMU. Ascorbate was a poor electron donor in the DCMU-poisoned system and supported a steady but only very slow rate of reduction of NADP. This was in contrast to reduced DPIP which could support a rapid rate of NADP photoreduction with the DCMU-poisoned chloroplasts. The heated *Euglena* chloroplasts were unable to catalyze the Hill reaction with NADP as oxidant (Fig. 1, lower curve). The initial, small increase in absorbance at 340 mμ was due to the photoreduction of NADP coupled to the photooxidation of the added reduced cytochrome-552. In marked contrast to the DCMU-poisoned, unheated chloroplasts, the addition of ascorbate alone resulted in a substantial rate of NADP photoreduction by the heated chloroplasts. Of special interest was the finding that this ascorbate-supported NADP photoreduction was sensitive to the poison DCMU, although the inhibition was not complete. This inhibition by DCMU was largely overcome by the addition of DPIP.

The effectiveness of ascorbate in restoring NADP photoreduction was most significant with chloroplasts which were inactivated by incubating at 40–45° for 5 minutes; the extent of restoration amounted usually to 70 per cent of the original activity. Ascorbate was less effective when the chloroplasts were heated at higher temperature and was without effect when the chloroplasts were heated at 60°. NADP photoreduction supported by reduced DPIP was more resistant to heat treatment but again was inactivated completely by heating the chloroplasts at 60° for 5 minutes.

It was found that cysteine could replace ascorbate in supporting NADP photoreduction by heated chloroplasts. The cysteine-supported reaction was also sensitive to the poison, thereby indicating that cysteine acted in a manner similar to ascorbate in supporting NADP photoreduction by heated chloroplasts.

The ascorbate-supported NADP photoreduction by the heated chloroplasts resembled in many respects the regular Hill reaction.[9] NADP photoreduction by the ascorbate system was dependent upon cytochrome-552, and the maximal activity was obtained with a concentration of cytochrome-552 which gave maximal activity of the Hill reaction. The pH-dependency of the ascorbate system was also similar to that of the Hill reaction; the optimum was at a pH slightly higher than 6. The most interesting feature of the ascorbate-supported NADP photoreduction was that it was inhibited by all the poisons of the Hill reaction so far tested.[9] These include DCMU, simazine, atrazine, *o*-phenanthroline, hydroxylamine and CCCP. The concentration of the poisons required for 50 per cent inhibition was the same with both the ascorbate system and the normal Hill reaction system. It seems, therefore, that the inhibitory site of the poisons in the ascorbate-supported reaction is the same as that in the regular Hill reaction.

Ascorbate Photooxidation

There are two possible ways to account for the effect of ascorbate on NADP photoreduction by the heated chloroplasts. First, ascorbate simply reverses the heat-inactivation of the Hill reaction so that NADP photoreduction in the presence of ascorbate is catalyzed by the regular Hill reaction system. This seems unlikely because pre-treatment of the heated chloroplasts with ascorbate failed to restore the activity.[9] The second possibility is that, although the oxygen evolution system is irreversibly inactivated in the heated chloroplasts, ascorbate can serve in place of water as the electron donor for NADP photoreduction. In the latter case, ascorbate should be photooxidized stoichiometrically during the photoreduction of NADP. Of interest in this connection is the ability of chloroplasts prepared from higher plants to photooxidize ascorbate with molecular oxygen as oxidant. Ikeda,[10] Chiba and Okayama[11] and Trebst *et al.*[12] reported that the photooxidation of ascorbate in the absence of DPIP was sensitive to poisons of the Hill reaction.

In order to study the quantitative aspects of ascorbate photooxidation, similar experiments were performed with *Euglena* chloroplasts, wherein NADP was replaced by methyl viologen as the electron acceptor. The

TABLE I. Ascorbate Photooxidation by *Euglena* Chloroplasts.

Treatment	Omission/Addition	μmoles O_2 absorbed / mg chlorophyll hr
Untreated	−ascorbate	44
	−cytochrome-552	11
	None	170
	+DCMU	94
	+DPIP	242
	+DPIP, DCMU	245
Heated	−ascorbate	3
	−cytochrome-552	18
	None	144
	+DCMU	110
	+DPIP	217
	+DPIP, DCMU	222

Oxygen uptake was determined with a Clark electrode in a semi-closed vessel which was illuminated by a 1,000 watt tungsten lamp. Each reaction mixture contained in a final volume of 4.9 ml: 50 m*M* phosphate buffer of pH 6, 10 m*M* NaCl, 14 m*M* neutralized ascorbate, 4 mg of catalase, 0.1 *M* ethanol, 1 m*M* methyl viologen, 20 μ*M* *Euglena* cytochrome-552, 1 m*M* ethylenediamine tetraacetate and *Euglena* chloroplasts corresponding to 115 μg chlorophyll.

Where indicated, 10 μ*M* DCMU and 20 μ*M* DPIP were added.

amount of methyl viologen reduced in the light was determined by measuring oxygen uptake, since the reduced dye was rapidly reoxidized by molecular oxygen. The hydrogen peroxide formed was trapped with the ethanol and catalase couple. The results obtained are presented in Table I. In the presence of cytochrome-552, untreated *Euglena* chloroplasts showed a significant rate of oxygen uptake with methyl viologen as electron acceptor. The rate of oxygen uptake was markedly stimulated when ascorbate was added to the reaction system. This ascorbate-enhanced reaction was largely inhibited by DCMU. The effect of ascorbate alone might be different from that of the ascorbate-DPIP couple, which could support a higher rate of oxygen consumption, in that the latter reaction was much less susceptible to the poison.

A considerable activity of light-induced oxygen uptake survived after heat-treatment of *Euglena* chloroplasts at 45° for 5 minutes, which inactivated the Hill reaction completely, provided that cytochrome-552, methyl viologen and ascorbate were added to the chloroplasts. The effect of DCMU on this reaction, however, was unexpectedly small and variable with different preparations and ranged from 40 to 0 per cent inhibition. In some experiments, the inhibitory action of the poison on the ascorbate-supported NADP photoreduction and the ascorbate-stimulated photoconsumption of oxygen was compared with the same chloroplast preparation. The results obtained were difficult to evaluate; whereas NADP photoreduction was

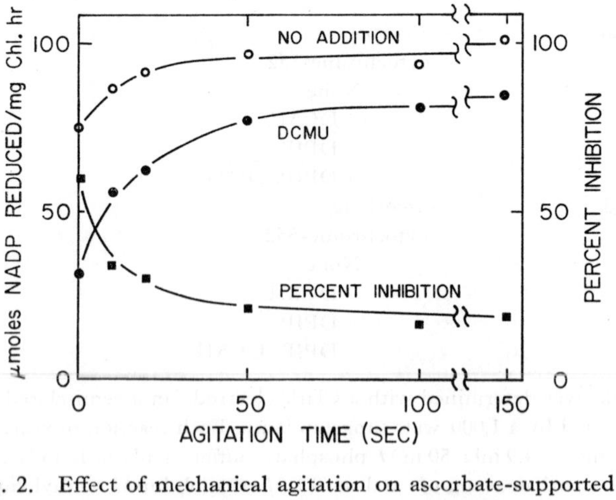

Fig. 2. Effect of mechanical agitation on ascorbate-supported NADP photoreduction by heated *Euglena* chloroplasts in the absence and presence of DCMU.

Experimental conditions were as described in Fig. 1.

strongly suppressed by the poison, the photoconsumption of oxygen was much less sensitive to it.

It was found, however, that the sensitivity of the ascorbate-supported NADP photoreduction to the poison decreased markedly by mechanical agitation of the heated chloroplasts. As shown in Fig. 2, vigorous agitation of the heated chloroplasts with a Volto mixer caused an increase in the activity, slight in the absence of the poison and more marked in its presence. For example, we observed a decrease in the extent of inhibition from 60 percent before agitation to 20 per cent after 50 seconds of the treatment. Presumably, the chloroplasts became so structurally labile on heating that mechanical agitation, with the mixer or a magnetic stirrer in the polarographic experiments, caused a modification in the chloroplasts so as to make ascorbate more accessible to photosystem I as electron donor. An additional observation in this regard was that upon repeated evacuation and flushing with nitrogen, the heated chloroplasts progressively lost their sensitivity to the poison. The unheated chloroplasts were much more re-

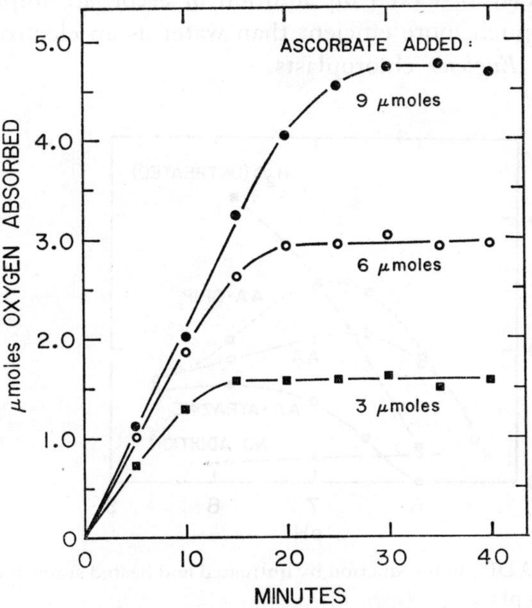

Fig. 3. Light-induced oxygen uptake by *Euglena* chloroplasts in the presence of different amounts of ascorbate.

Oxygen uptake was determined with a Warburg manometer. Flasks were illuminated with a bank of fluorescent lamps from the bottom. Each reaction mixture of 2.7 ml contained the components indicated in Table I, with the exception that catalase and ethanol were omitted.

sistant to the mechanical agitation in this respect. Because of these experimental limitations, further study of ascorbate photooxidation by heated chloroplasts was not pursued.

It was found that the light-induced oxygen uptake discussed above was accompanied by a stoichiometric photooxidation of ascorbate. The results obtained from manometric experiments in which oxygen uptake was measured in the presence of different amounts of ascorbate are shown in Fig. 3. Oxygen uptake ceased when about one mole of oxygen to two moles of ascorbate added was consumed in the absence of catalase and ethanol. Titration with DPIP revealed that there was no detectable amount of ascorbate remaining in the reaction mixture after the reaction ceased.

These observations indicate that ascorbate is actually serving as an electron donor for photosystem II, not only with the heated chloroplasts but also with the untreated chloroplasts. The effectiveness of ascorbate as an electron donor for photosystems I and II would be governed mainly by the structural characteristics of the chloroplasts. A marked increase in the rate of oxygen uptake observed on addition of ascorbate implies that ascorbate might be much more efficient than water as an electron donor for photosystem II in *Euglena* chloroplasts.

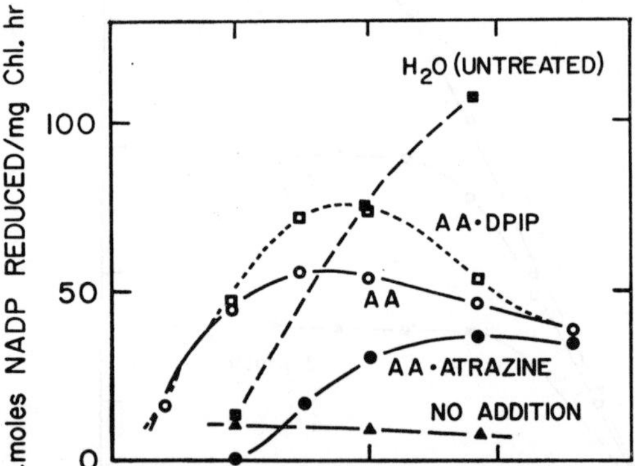

Fig. 4. NADP photoreduction by untreated and heated spinach chloroplasts at various pH's.

NADP photoreduction was determined as described in Fig. 1. Each reaction mixture contained 50 m*M* buffer of various pH, 10 m*M* NaCl, 0.86 m*M* NADP, a saturating amount of ferredoxin and ferredoxin NADP oxido-reductase and spinach chloroplasts corresponding to 8 μg of chlorophyll. Where indicated, 10 μ*M* atrazine, 14 m*M* neutralized ascorbate or 20 μ*M* DPIP was added.

Spinach Chloroplasts

A related question was whether or not this ascorbate-supported, poison-sensitive photoreduction of NADP could be observed with chloroplasts other than *Euglena* chloroplasts. Spinach chloroplasts were used to answer this question. It was found firstly that with untreated spinach chloroplasts, ascorbate had a stimulatory effect on the NADP Hill reaction at alkaline pH's where the activity dropped sharply and became negligible at a pH higher than 9. The effect of ascorbate on the activity at acidic and neutral pH's was small and variable, often being even slightly inhibitory.

A complete inactivation of the Hill activity of spinach chloroplasts was attained by heating them at 50° for 5 minutes. As shown in Fig. 4, ascorbate, like the ascorbate-DPIP couple, restored the activity of the heated spinach chloroplasts to photoreduce NADP. In contrast to *Euglena* chloroplasts, however, cysteine was not as effective as ascorbate with spinach chloroplasts. It was observed that this ascorbate effect was variable from preparation to preparation, both with respect to the magnitude of restoration and to the pH-dependency of the restored activity. In addition, the kinetics of the ascorbate-supported reduction by the heated spinach chloroplasts showed a significant lag before a steady state of reduction was attained. The duration of the lag period was found to be pH-dependent, lasting longer at higher pH's and disappearing at pH's lower than 7. It was found that the lag was effectively eliminated by the addition of a catalytic amount of ferredoxin-NADP oxidoreductase without any change in the rate of the steady state reduction. Presumably, the endogenous enzyme was inactivated reversibly in the heated chloroplasts. The present experiments were, therefore, carried out in the presence of exogenously added enzyme.

The ascorbate-supported reaction was invariably sensitive to poisons such as DCMU, although the extent of inhibition again varied with different preparations. Two features of the inhibition were consistently observed; first, in contrast to the *Euglena* system, the effect of the poison was pH-dependent, being marked at acidic pH's but not at higher pH. Secondly, the concentration of the poison required to inhibit the ascorbate-supported reaction was in the same range as that required to inhibit the regular Hill reaction, except that the inhibition of the ascorbate system was always incomplete. It appears, therefore, that ascorbate, but no longer water, can serve as an electron donor for photosystem II with the heated spinach chloroplasts. The poison-insensitivity of the reaction at high pH, as well as the stimulating effect of ascorbate at higher pH with the untreated chloroplasts, could be accounted for in terms of the effect of alkaline pH on chloroplast structure which is discussed in a later section.

Aging of Chloroplasts

There are a number of physical and chemical treatments of chloroplasts which inactivate preferentially the oxygen evolution system of the Hill reaction. In the light of the above findings, a study was made to compare the inactivation mechanisms involved in these treatments by determining the effect of ascorbate and the poisons on the NADP photoreduction activity of the inactivated chloroplasts.

Vernon and Zaugg[13] showed that aging of chloroplasts resulted in a loss of the Hill activity; however, with NADP as the electron acceptor the inhibition was overcome by the addition of the ascorbate-DPIP couple. They assumed, therefore, that aging of the chloroplasts preferentially inactivated the oxygen evolution system.

Euglena chloroplasts were stored in the dark at 4° and the changes in the NADP photoreduction activity were followed for 4 days. The results presented in Fig. 5 show that after two days' storage the Hill activity was markedly decreased, and it was only about 10 per cent of the original activity after 4 days of aging. The stimulatory effect of ascorbate alone and the ascorbate-DPIP couple was apparent after one day of aging and about one-half of the activity was restored with 4 day-old chloroplasts. This ascorbate-supported NADP photoreduction by the aged chloroplasts was highly sensitive to DCMU, whereas the ascorbate-DPIP system was not. Thus,

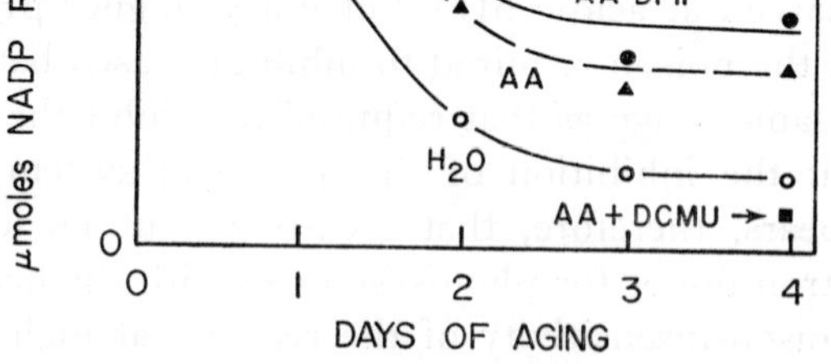

Fig. 5. NADP photoreduction by aged *Euglena* chloroplasts. Experimental conditions were as described in Fig. 1.

the inactivation of the oxygen evolution system in aged chloroplasts resembles that in heated chloroplasts.

Other Treatments

Similar experiments showed that the mechanism of inactivation by acid treatment was the same as that of heating or aging. Chloroplasts, wherein the Hill activity was inactivated by incubation at pH 4, catalyzed an ascorbate-supported NADP photoreduction which was sensitive to the poison. However, the rate of NADP photoreduction with either ascorbate or reduced DPIP was quite low, presumably because photosystem I was also severely damaged by acid incubation. On the other hand, the activity of the ascorbate-supported NADP photoreduction with the pH 9-treated chloroplasts was relatively high and resistant to the poison. It seems that ascorbate can serve as a direct and efficient electron donor for photosystem I, presumably because of a structural change around photosystem I caused by incubation of the chloroplasts at high pH. Because of the functionality of this short-circuit reaction which is not sensitive to the poison, it was difficult to determine whether or not ascorbate still served as an electron donor for photosystem II in the alkaline-treated chloroplasts. The insensitivity of NADP photoreduction to the poison, as well as the stimulatory effect of ascorbate at high pH with the heated and untreated spinach chloroplasts could be accounted for by an effect of alkaline pH on chloroplast structure.

A similar situation was observed with detergent-treated chloroplasts. Whereas the Hill activity of *Euglena* chloroplasts decreased with increasing concentration of Triton X-100 or deoxycholate, the activity of photosystem I, assayed by NADP photoreduction or photooxidation of reduced cytochrome-552, was stimulated before it was severely inhibited by a higher concentration of the detergent. Ascorbate could support a significant rate of NADP photoreduction but this reduction was not sensitive to the poison. This was also the case with the chloroplasts treated with linolenic acid. On the other hand, chloroplasts inactivated by irradiation with ultraviolet light showed a high activity of NADP photoreduction in the presence of reduced DPIP but no appreciable activity in the presence of ascorbate alone. It is concluded that ultraviolet irradiation inactivates the Hill reaction in a manner different from heat treatment.

SUMMARY

This study of the characterization of the electron transport system of

Euglena may be summarized as follows: (*a*) *Euglena* chloroplasts require cytochrome-552 for the photoreduction of NADP or other substances of low redox potential, but not for the photoreduction of ferricyanide or DPIP; (*b*) oxidized cytochrome-552 was photoreduced by photosystem II, and reduced cytochrome-552 was photooxidized by photosystem I with a concomitant photoreduction of NADP; (*c*) the effectiveness of ascorbate or cysteine in supporting NADP photoreduction with the poison- and heat-inactivated chloroplasts was compared. The inhibitory effect of the poison on the ascorbate-supported NADP photoreduction by the heated chloroplasts was studied; (*d*) *Euglena* chloroplasts catalyze the poison-sensitive photooxidation of ascorbate with methyl viologen as electron acceptor; (*e*) the ascorbate-supported and the poison-sensitive photoreduction of NADP was also observed with heated spinach chloroplasts; (*f*) a comparison was made of several physical and chemical treatments of chloroplasts which inactivate the Hill reaction with regard to restoration of NADP photoreduction by ascorbate and the effect of the poisons on the restored activity.

The results are discussed briefly in terms of the Hill and Bendall scheme of photosynthesis.

REFERENCES

1. Hill, R., and Bendall, F., *Nature,* **186**, 136 (1960).
2. Duysens, L. N. M., *Science*, **121**, 210 (1955).
3. Smillie, R. M., *Can. J. Botany,* **41**, 123 (1963).
4. Perini, F., Kamen, M. D., and Schiff, J. A., *Biochim. Biophys. Acta,* **88**, 74 (1964).
5. Katoh, S., and San Pietro, A., *Arch. Biochem. Biophys.*, **118**, 488 (1967).
6. Katoh, S., and San Pietro, A., *Arch. Biochem. Biophys.*, **121**, 211 (1967).
7. Hinkson, J. W., and Vernon, L. P., *Plant Physiol.*, **34**, 268 (1959).
8. Whatley, F. R., Tagawa, K., and Arnon, D. I., *Proc. Natl. Acad. Sci.*, **49**, 266 (1963).
9. Katoh, S., and San Pietro, A., *Arch. Biochem. Biophys.*, **122,** 144 (1967)
10. Ikeda, S., *Mem. Res. Inst. Food Sci. Kyoto Univ.*, **18**, 57 (1959).
11. Chiba, Y., and Okayama, S., *Plant and Cell Physiol.*, **3**, 379 (1962).
12. Trebst, A., Eck, H., and Wagner, S., in *Photosynthetic Mechanisms of Green Plants* (B. Kok and A. T. Jagendorf, eds.), p. 174, Nat. Acad. Sci.-Nat. Res. Council, 1963.
13. Vernon, L. P., and Zaugg, W. S., *J. Biol. Chem.*, **235**, 2728 (1960).

Occurrence of Back Flow of Electrons against the Action of Photochemical System I in Sonicated Lamellar Fragments

YOSHIHIKO FUJITA* AND FUMIO MURANO†

Institute of Applied Microbiology, University of Tokyo, Tokyo

INTRODUCTION

The photosynthetic electron flow system *in vivo* has been known to involve cyclic electron flow(s) induced by the action of photochemical system I (longer wavelength photochemical system).[1,2] In isolated systems, therefore, the oxidation-reduction reaction induced by photochemical system I may be affected by such cyclic electron flows. Kok *et al.*[3] reported that a back reaction of the P 700-photooxidation occurs in the O_2-linked photooxidation of cytochromes or plastocyanin by detergent-treated chloroplasts. In the photooxidation of mammalian cytochrome *c* by sonicated lamellar fragments of *Anabaena cylindrica*, we also found an occurrence of a back flow of electrons to the added cytochrome *c*.[4]

Our further investigation on the back reaction has been carried out mainly with the *Anabaena* system and partly with systems of other algae and spinach. Results reported herein indicate that the back flow of electrons to the added electron donor commonly occurs in the reaction induced by the action of photochemical system I. Further, it was suggested that the back reaction is mediated by CRS, a redox substance, which is contained in the sonicated and washed lamellar fragments.

MATERIALS AND METHODS

Methods for algal cultures and preparations of sonicated lamellar fragments were the same as those in our previous work.[5–7]

Abbreviations: CM-, carboxymethyl; CRS, cytochrome *c* reducing substance; DEAE-, diethylaminoethyl; DPIP, 2,6-dichlorophenol indophenol; MV, methyl viologen; NADP, nicotinamide adenine dinucleotide phosphate; TMPD, tetramethyl phenylene-diamine.

* Present address: Ocean Research Institute, University of Tokyo, Nakano-ku, Tokyo, Japan.

† Present address: Nomura Research Institute for Technology and Economics, Kamakura, Kanagawa, Japan.

Anabaena cytochrome 553 was prepared from the soluble protein fraction of cell-sonicates as described in the previous paper.[7] The cytochrome preparation was further purified by chromatography on a CM-cellulose column. Spectroscopic examination indicated that the preparation did not contain plastocyanin. Ferredoxin and ferredoxin-NADP reductase were prepared from spinach leaves according to the methods of Tagawa *et al.* and of Shin *et al.*, respectively.[8] Spinach plastocyanin was obtained according to the method of Katoh *et al.*[9] CRS was prepared from *Anabaena* sonicated lamellar fragments as reported previously.[10]

The O_2-linked photooxidation reaction[7] and NADP-photoreduction reaction[6] were measured as reported previously. The reaction mixture for both reactions contained a large amount of ascorbate ($2\times10^{-2}M$). The reactions strictly required the addition of the electron carrier between ascorbate and the photochemical system, indicating that ascorbate cannot serve as the direct electron donor for the photochemical system. Light-induced anaerobic redox reactions of cytochromes and DPIP were followed spectrophotometrically by a Hitachi Perkin-Elmer spectrophotometer 139. The sample was illuminated with 20,000 lux red light (longer than 650 mμ), and the detector was protected from the actinic light by combined use of a color filter and an interference filter.

RESULTS AND DISCUSSION

Effect of added MV on the O_2-linked photooxidation of various redox substances

In our previous work on the spinach system,[6] we found that MV stimulates strongly the plastocyanin-photooxidation but only slightly the DPIP-photooxidation. A similar insignificant stimulation by MV was generally observed with the DPIP-photooxidation by sonicated lamellae of other organisms (Tables I and II). The plastocyanin-photooxidation, on the other hand, was far slower than the DPIP-photooxidation, when MV was not included. On addition of MV, the reaction rate was markedly enhanced up to the same level as (or slightly higher than) that of the DPIP-photooxidation (Table I). The same slow reaction without MV and the strong stimulation by MV were observed with the photooxidation of mammalian cytochrome *c* (Table I), *Anabaena* cytochrome 553 and TMPD (Table II). This implies that MV does not simply stimulate the electron flow from the primary reductant of the photochemical system to O_2, but also suppresses a back flow which lowers the over-all rate of the photooxidation. This is supported by the evidence that, even when the rate of over-all reaction was limited at the step of the electron transfer from ascorbate to the photochemical system, MV-stimulation was still marked (Table II).

As the MV-stimulation of the photooxidation of cytochrome *c* or plastocyanin occurs in various organisms, the occurrence of the back reac-

TABLE I. Effect of added MV on O_2-linked photooxidation of DPIP, plastocyanin and horse-heart cytochrome *c* by sonicated lamellar fragments of spinach and various algae.[a]

Organisms	DPIP (10^{-3} *M*)			Plastocyanin (2×10^{-5} *M*)			Horse-heart cytochrome *c* (2×10^{-5} *M*)		
	−MV (μmoles/mg chl/hr)	+MV (μmoles/mg chl/hr)	Stimulation (%)	−MV (μmoles/mg chl/hr)	+MV (μmoles/mg chl/hr)	Stimulation (%)	−MV (μmoles/mg chl/hr)	+MV (μmoles/mg chl/hr)	Stimulation (%)
Spinach I	195	242	24	116	234	102	—	—	—
II	266	270	2	94	294	218	—	—	—
Porphyra yezoensis	240	269	12	108	358	232	—	—	—
Porphyra sp. Yamamoto II	309	405	47	106	347	228	—	—	—
Anabaena cylindrica	161	199	24	—	—	—	96	193	102
Anabaenopsis circularis	144	206	43	—	—	—	51	179	254
Tolypothrix tenuis	181	165	−9	—	—	—	134	255	90
Plectonema sp.	273	405	48	—	—	—	167	484	189

a. Light-induced O_2-uptake was manometrically measured under illumination with 20,000 lux incandescent light at 26°. The reaction mixture (2.4 ml) contained 48 μmoles sodium ascorbate, 2.4 μmoles KCN, lamellar fragments (30 to 60 μg chlorophyll), 240 μmoles Tris buffer (pH 7.6) and electron-carriers as indicated. In the center well of each flask, 0.1 ml of KOH (20%) was placed.

TABLE II. O_2-linked photooxidation of DPIP, TMPD and *Anabaena* cytochrome 553 by sonicated lamellar fragments of *Anabaena*.[a]

Added electron feeder	O_2-uptake (μmoles/mg chl/hr) −MV	O_2-uptake (μmoles/mg chl/hr) +MV (5×10^{-5} *M*)
None	12	14
DPIP (10^{-3} *M*)	174	236
TMPD (10^{-4} *M*)	21	58
TMPD (2×10^{-3} *M*)	30	236
Anabaena cytochrome 553 (2×10^{-7} *M*)	20	62
Anabaena cytochrome 553 (3×10^{-6} *M*)	44	272

a. For experimental procedures, see Table I.

tion is probably a common phenomenon in the O_2-linked photooxidation reaction by sonicated lamellae. In the DPIP-photooxidation, however, the back reaction was not significant. This suggests that electrons are transferred back to the added electron donor.

Anaerobic redox reactions of mammalian cytochrome c, Anabaena cytochrome 553 and DPIP by sonicated lamellae of Anabaena

As reported previously,[4] a sonicated and well washed lamellar preparation, which retains the activity of photochemical system I, contains a relatively large amount of a redox substance tentatively named CRS. The redox substance is photoreduced when the lamellae are illuminated and reoxidized by horse-heart cytochrome *c*.[4] Thus, the sonicated lamellae can photooxidize the reduced cytochrome *c* even under anaerobic conditions and reduce the cytochrome up to the original reduced level in the subsequent dark period, as shown in Fig. 1A (also *cf.* ref. 4). If the isolated CRS is added to the reaction mixture, the light oxidation and the dark reduction are extended markedly (curve b in Fig. 1A). This indicates that CRS can mediate the back reaction in the photooxidation of mammalian cytochrome *c*.

Figs. 1B and 1C show the patterns of light-induced anaerobic redox reactions of *Anabaena* cytochrome 553 and DPIP by *Anabaena* lamellae. *Anabaena* cytochrome 553 was photooxidized in the light and reduced in the subsequent dark period in a similar way to mammalian cytochrome *c* (Fig. 1B). In the presence of added CRS, the light oxidation and the dark reduction were enhanced (curve b in Fig. 1B). This indicates that the cytochrome is reduced by CRS which had been photoreduced. However, neither the light oxidation nor the dark reduction was observed with reduced DPIP even in the presence of added CRS (Fig. 1C). On the first illumina-

---

Fig. 1. Light-induced anaerobic redox reactions of horse-heart cytochrome *c* (A), *Anabaena* cytochrome 553 (B) and DPIP (C) by sonicated *Anabaena* lamellae. The reaction mixture (4 ml) contained lamellar preparation (80 μg chlorophyll), 4 μmoles KCN, 200 μmoles Tris buffer (pH 7.6) and 0.1 μmole horse-heart cytochrome *c* in chart A, 21 mμmoles *Anabaena* cytochrome 553 in chart B or 4 μmoles DPIP in chart C. The reaction mixture was placed in a Thunberg-type cuvette, and the air was replaced with pure N_2. Horse-heart cytochrome *c* and *Anabaena* cytochrome 553 were reduced by sodium hydrosulfite, and excess hydrosulfite was removed by a Sephadex G 25 column. DPIP was reduced by titration with sodium hydrosulfite solution. Downward arrows indicate light on, and upward arrows light off. Reactions were run at room temperature.

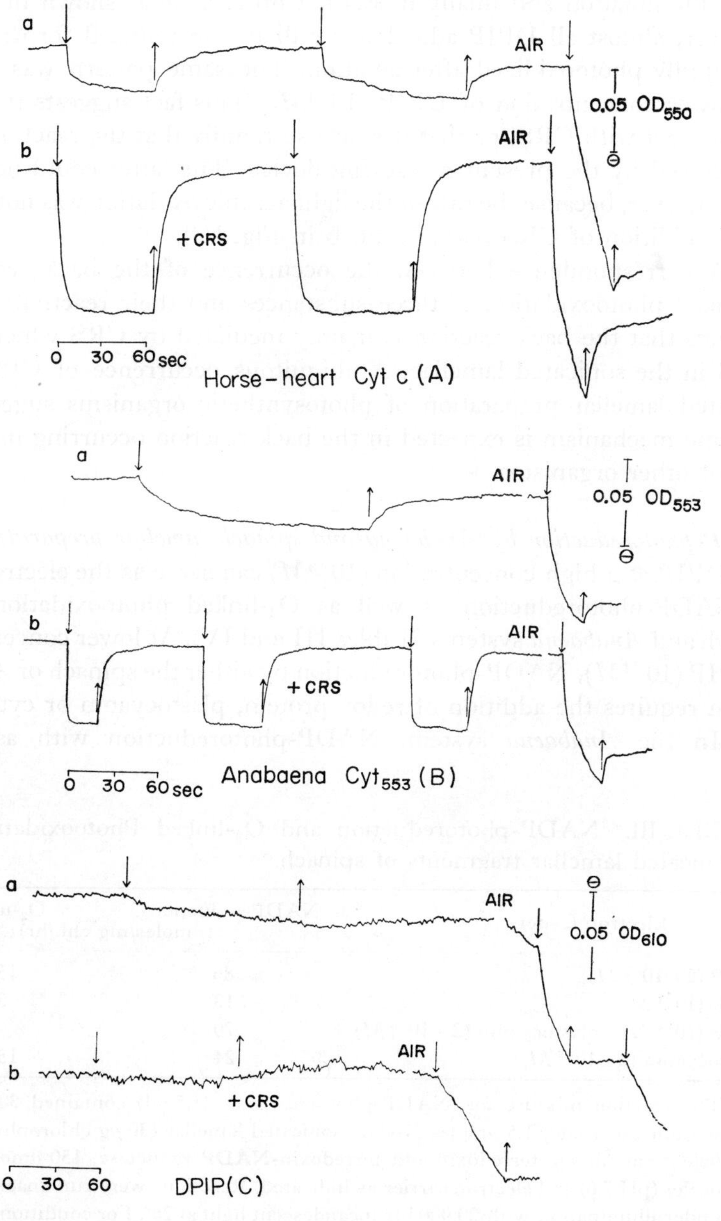
a
AIR
0.05 OD550
b
AIR
+CRS
0 30 60 sec
Horse-heart Cyt c (A)
a
AIR
0.05 OD553
b
AIR
+CRS
0 30 60 sec
Anabaena Cyt553 (B)
a
AIR
0.05 OD610
b
AIR
+CRS
0 30 60
DPIP (C)

Fig. 1

tion, DPIP was significantly oxidized, but not reduced in the following dark period; this light oxidation became less significant on further repetition of the illumination and finally it was not observable as shown in Fig. 1C. However, almost all DPIP added was still in the reduced form; the dye was rapidly photooxidized after aeration. The same pattern was obtained at a lower concentration of DPIP ($10^{-4}M$). This fact suggests that DPIP cannot react with CRS, or that it reacts so rapidly that the reaction cannot be followed by the present measuring device. The latter could not be the case, however, because the rate of the light aerobic oxidation was not lowered by the addition of CRS (curve a *vs.* b in Fig. 1C).

A correspondence between the occurrence of the back reaction in O_2-linked photooxidation of three substances and their reactivity to CRS indicates that the back reaction is mainly mediated by CRS which is contained in the sonicated lamellae. A ubiquitous occurrence of CRS in the sonicated lamellar preparation of photosynthetic organisms suggests that the same mechanism is expected in the back reaction occurring in the systems of other organisms.

NADP-photoreduction by Anabaena and spinach lamellar preparations

DPIP at a high concentration ($10^{-3}M$) can serve as the electron donor for NADP-photoreduction as well as O_2-linked photooxidation in the spinach and *Anabaena* systems (Tables III and IV). At lower concentrations of DPIP ($10^{-4}M$), NADP-photoreduction by either the spinach or *Anabaena* system requires the addition of redox protein, plastocyanin or cytochrome 553. In the *Anabaena* system, NADP-photoreduction with ascorbate-

TABLE III. NADP-photoreduction and O_2-linked Photooxidation by sonicated lamellar fragments of spinach.[a]

Electron feeder	NADP-reduction (μmoles/mg chl./hr)	O_2-uptake (μmoles/mg chl./hr)
DPIP (2×10^{-3} M)	86	152
DPIP (10^{-4} M)	13	38
DPIP (10^{-4} M)+plastocyanin (2×10^{-5} M)	70	—
Plastocyanin (2×10^{-5} M)	24	158

a. The reaction mixture for NADP-photoreduction (1.5 ml) contained 30 μmoles sodium ascorbate, 1.5 μmoles NADP, sonicated lamellae (30 μg chlorophyll), sufficient amounts of ferredoxin and ferredoxin-NADP reductase, 150 μmoles Tris buffer (pH 7.6) and electron carrier as indicated. Reactions were run anaerobically under illumination with 20,000 lux incandescent light at 26°. For conditions for the O_2-linked photooxidation reaction, see Table I.

TABLE IV. NADP-photoreduction and O_2-linked photooxidation by sonicated lamellar fragments of *Anabaena*.[a]

Electron feeder	NADP-reduction	O_2-uptake
	(μmoles/mg chl./hr)	
Expt. 1		
DPIP (2×10^{-3} *M*)	88	236
DPIP (10^{-4}*M*)	11	63
DPIP (2×10^{-5} *M*)	6	20
DPIP (10^{-4} *M*)+*Anabaena* cyt. 553 (5×10^{-6} *M*)	162	—
DPIP (2×10^{-5}*M*)+*Anabaena* cyt. 553 (5×10^{-6} *M*)	77	—
Anabaena cyt. 553 (5×10^{-6} *M*)	17	261
Expt. 2		
DPIP (2×10^{-3} *M*)	94	218
TMPD (2×10^{-3} *M*)	6	220
TMPD (10^{-4} *M*)	17	58
TMPD (10^{-4} *M*)+*Anabaena* cyt. 553 (5×10^{-6} *M*)	72	—
Anabaena cyt. 553 (5×10^{-5} *M*)	17	249

a. For experimental procedures, see Tables I and III.

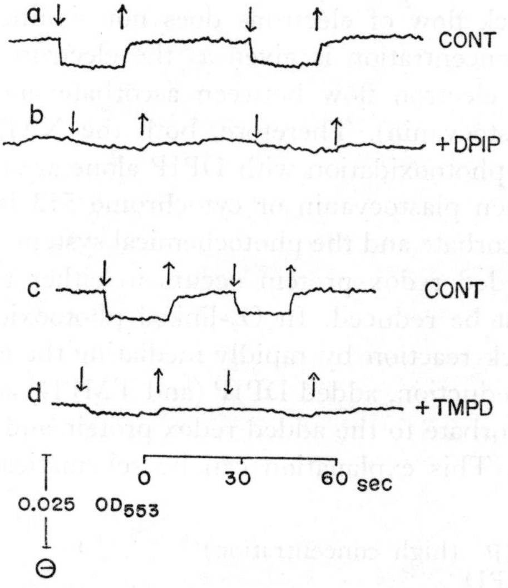

Fig. 2. A shift of redox state of *Anabaena* cytochrome 553 in the O_2-linked photooxidation reaction. The reaction mixture (1 ml) contained 20 μmoles sodium ascorbate, 50 mμmoles MV, 5 mμmoles *Anabaena* cytochrome 553, sonicated *Anabaena* lamellae (15 μg chlorophyll) and 100 μmoles Tris buffer (pH 7.6). The reaction was run aerobically at room temperature. The samples were illuminated with red light (20,000 lux).

TMPD was also supported by cytochrome 553, as was the case with the reaction by *Euglena* chloroplasts.[11] However, plastocyanin alone in the spinach system or cytochrome 553 alone in the *Anabaena* system did not serve as an efficient electron donor for NADP-photoreduction. The over-all reaction may be limited at the step of reduction of photooxidized plastocyanin or cytochrome 553 by ascorbate. DPIP or TMPD probably accelerates the electron flow from ascorbate to the redox protein. This is supported by the evidence that, on illumination, the reduced state of cytochrome 553 in the presence of excess ascorbate was shifted toward a slightly oxidized state (curves a and c in Fig. 2), while it remained unaltered in the presence of DPIP (curve b) or TMPD (curve d).

However, the reaction rate of the O_2-linked photooxidation of plastocyanin or cytochrome 553 was as high as (or even higher than) that of the DPIP-photooxidation, when MV is present. Thus, even if the plastocyanin- or cytochrome 553-reduction by ascorbate limits the rate of over-all reaction, the velocity should be high enough for supporting the NADP-reduction at least as high as the reduction supported by a high concentration of DPIP. These apparently contradicting phenomena can be explained by the occurrence of the back reaction which was observed in the O_2-linked photooxidation. The back flow of electrons does not significantly occur when DPIP at a high concentration is given as the electron donor, and DPIP can stimulate the electron flow between ascorbate and cytochrome 553 (and probably plastocyanin). Therefore, both the NADP-photoreduction and the O_2-linked photooxidation with DPIP alone are not affected by the back reaction. When plastocyanin or cytochrome 553 is the sole electron carrier between ascorbate and the photochemical system, however, the back reaction to the added redox protein occurs in either reaction. Thus the reaction rates must be reduced. In O_2-linked photooxidation, added MV suppresses the back reaction by rapidly mediating the electrons to O_2. In the NADP-photoreduction, added DPIP (and TMPD) accelerates the electron flow from ascorbate to the added redox protein and in turn suppresses the back reaction. This explanation can be schematically represented as follows:

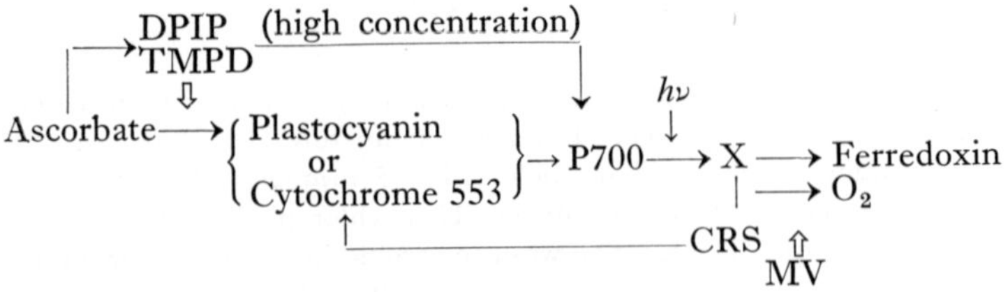

Ferredoxin-NADP reductase can reduce cytochrome f[12] and plastocyanin,[13] and has been assumed to mediate the back flow of electrons *in vivo*.[14] The back reaction in question, however, occurs in the absence of ferredoxin which is the electron donor for the reaction by this enzyme. Further, DPIP does not participate in the present back reaction, whereas the dye has been reported as an efficient electron acceptor for the diaphorase-action of the enzyme.[15] Thus we consider that the back reaction in question may not involve the action of this enzyme.

Acknowledgements

The authors wish to express their thanks to Dr. A. Hattori of the Institute of Applied Microbiology, University of Tokyo, for his help and stimulating discussion. *Porphyra* thalli were kindly supplied by Dr. T. Oohusa of Yamamoto Nori Institute, to whom the authors' thanks are due. This work was partly supported by a grant in aid for special project research on biophysics from the Ministry of Education.

REFERENCES

1. Rumberg, B., Schmid-Mende, P., Weikard, J., and Witt, H. T., in *Photosynthetic Mechanisms of Green Plants* (B. Kok and A. T. Jagendorf, eds.), p. 18, Nat. Acad. Sci.-Nat. Res. Council, 1963.
2. Vredenberg, W. J., and Amesz, J., in *Currents in Photosynthesis*, p. 349, Ad. Donker, Rotterdam, 1966.
3. Kok, B., Rurainski, H. J. and Harmon, A. E., *Plant Physiol.*, **39**, 513 (1964).
4. Fujita, Y., and Myers, J., *Arch. Biochem. Biophys.*, **119**, 8 (1967).
5. Fujita, Y., and Myers, J., *Arch. Biochem. Biophys.*, **113**, 730 (1966).
6. Fujita, Y., and Murano, F., *Plant and Cell Physiol.*, **8**, 269 (1967).
7. Murano, F., and Fujita, Y., *Plant and Cell Physiol.*, **8**, 673 (1967).
8. Losada, M., and Arnon, D. I., in *Modern Methods of Plant Analysis*, Vol. 7, p. 569, Springer, Heiderberg, 1964.
9. Katoh, S., Shiratori, I., and Takamiya, A., *J. Biochem.* (Tokyo), **51**, 32 (1962).
10. Fujita, Y., and Myers, J., *Plant and Cell Physiol.*, **7**, 599 (1966).
11. Katoh, S., and San Pietro, A., *Arch. Biochem. Biophys.*, **118**, 488 (1967).
12. Forti, G., Bertole, M. L., and Parisi, B., in *Photosynthetic Mechanisms of Green Plants* (B. Kok and A. T. Jagendorf, eds.), p. 284, Nat. Acad. Sci.-Nat. Res. Council, 1963.
13. Katoh, S., *Plant and Cell Physiol.*, **2**, 165 (1961).
14. Nishimura, M., in *Brookhaven Symposia in Biology*, No. 19, p. 132 (1966).
15. Avron, M., and Jagendorf, A. T., *Arch. Biochem. Biophys.*, **72**, 17 (1957).

A New Factor Required for Oxygen Evolution

HIROSI HUZISIGE, MASAO ISIMOTO AND HIROSI INOUE

Department of Biology, Faculty of Science, Okayama University, Okayama, Japan

INTRODUCTION

Recent studies on the fractionation of the photosynthetic system have brought forward considerable evidence supporting the view that chloroplasts can be fragmented into particles of different kinds by incubation with various detergents and by sonication, followed by differential centrifugation.[1–9] It is now well known that chloroplasts of green plants contain two kinds of particles of different photochemical activities and of different pigment compositions, although the question as to what structural and functional relationship may exist between them remains as yet largely unsettled.

The heavier particles pelleted by forces of 10,000 to 30,000$\times g$ have a lower ratio of chlorophyll *a* to *b* than that obtained for the original chloroplasts, whereas the lighter particles sedimented at 144,000$\times g$ have a higher ratio. The lighter particles have no activity of oxygen evolution, but they can photoreduce NADP when provided with an electron donor couple, e.g. sodium ascorbate and dichloro-phenolindophenol. On the other hand, the heavier particles show a measurable activity of oxygen evolution, but the rate of NADP reduction is small or negligible. The lighter particles, therefore, appear to be the particles of photosystem I (abbreviated to particle I), whereas the heavier particles so far obtained appear to be richer in the particles of photosystem II (particle II). For example, Anderson and Boardman mentioned in their recent report[9] that their preparation of the heavier particles contained both particles I and II, in an approximate ratio of 1:2.

Particle I seems to be rather easily isolated in a pure state free from particle II. On the other hand, the purification of particle II seems to be more difficult because of its medium particle weight, and there has been no paper which described a procedure suitable for removing the contaminating particle I from particle II. The separation and purification of par-

---

Abbreviations: OEF, oxygen-evolution factor; BQ, p-benzoquinone; FeCN, ferricyanide; PQ, plastoquinone.

ticle II was greatly improved by use of either of the following two combination techniques; the combination of Triton X-100 treatment, sonication and density-gradient centrifugation, or the combination of digitonin treatment, sonication, Triton X-100 treatment, differential and density-gradient centrifugation and Sephadex-gel filtration. The preparations of particle II thus obtained showed a low chlorophyll *a* to *b* ratio and a noticeable activity of oxygen evolution but no activity of NADP photoreduction. The experimental details of these combination procedures will be reported elsewhere.[10]

The activity of oxygen evolution of particle II thus purified was found in the present study to be easily abolished by various treatments. The activity was easily lost by aging or heat treatment but was not affected at all by detergent treatment or sonication. On the other hand, the treatment with a concentrated Tris-HCl buffer at a high pH value strongly inhibited the activity. It was found through the investigation of this treatment that a factor that accelerates the oxygen evolution activity of chloroplast fragments is present in the supernatant obtained by centrifugation of a suspension of chloroplasts that have been treated with a concentrated Tris-HCl buffer at high pH. This new factor is called the oxygen-evolution factor and is denoted by OEF in this paper. The present paper is concerned with the isolation and purification of this soluble factor, OEF, and its mode of action.

MATERIALS AND METHODS

Preparation of chloroplast fragments

Chloroplast fragments used in this study were prepared in the following manner. A suspension of spinach chloroplasts, which was prepared as described in a previous report,[11] was fragmented by incubation at 0° in 0.05% Triton X-100 and by sonication at 20 kc for 5 min. The resulting chlorophyll-containing particles were separated by differential centrifugation. The first centrifugation was made at $15{,}000\times g$ for 30 min. and this was followed by centrifugations at $30{,}000\times g$ for 30 min., $70{,}000\times g$ for 30 min. and $144{,}000\times g$ for 60 min. The pellet from each spin was resuspended in 0.017 *M* Tris-HCl buffer, pH 7.2. In this report, various fractions obtained by centrifugation are referred to by giving the centrifugal force used for sedimentation, and the forces are expressed in a unit of $10{,}000\times g$. For example, the fraction which sedimented between $15{,}000\times g$ and 30,000 *g* was called the 3 G fraction. The original granum preparation before centrifugation is referred to as the G_0 fraction. The 3 G fraction was used in most of the experiments as particle II, although the fraction was not completely free from particle I. This is because a highly purified preparation of particle II could not be obtained in large quantities even by the improved combination method.[10]

The chlorophyll content of each fraction was determined by the method of Ogawa and Shibata.[12]

Preparation of OEF

OEF was purified by the following procedure. Spinach leaves (*ca.* 3 kg) were homogenized in 1.5 *l* of cold 0.017 *M* Tris-HCl buffer, pH 7.2, in a Waring blendor at top speed for 1 min. The homogenate was filtered through cheesecloth to remove coarse materials and the filtrate was centrifuged at $2{,}000 \times g$ for 5 min. The supernatant was filtered through 4 layers of cheesecloth, and the filtrate was centrifuged at $20{,}000 \times g$ for 20 min. The sediment suspended in 0.8 *M* Tris-HCl buffer, pH 8.0, was filtered through 4 layers of cheesecloth. The suspension after having been sonicated at 20 kc for 5 min. and left standing for 12 hours in the dark at 0–4° was centrifuged at $50{,}000 \times g$ for 30 min. The yellow green supernatant (*ca.* 1.0 *l*) was fractionated by precipitation with cold acetone. The precipitate obtained between 50% and 80% saturation was resuspended in *ca.* 70 ml of 0.05 *M* Tris-HCl buffer, pH 7.2. After centrifugation at $20{,}000 \times g$ for 20 min., the yellow brown supernatant was diluted 5-fold with the buffer, and was applied on a DEAE-cellulose column which had been equilibrated with the same buffer. OEF was adsorbed onto the column as a dark brown layer, and a yellow solution passed through the column. After washing the column with 0.05 *M* Tris-HCl buffer, pH 7.2, and eluting contaminating minor components with the same buffer containing 0.1 *M* NaCl, OEF was eluted with the same buffer containing 0.2 *M* NaCl. The whole procedure of extraction and purification was carried out in a cold room (0–4°).

Assay for oxygen evolution

The oxygen-evolution activity was measured by a polarographic method in the presence of a suitable Hill oxidant. The basal reaction mixture contained, in a final volume of 2.2 ml, chloroplast fragments with 50 μg chlorophyll, 17 μmole Tris-HCl buffer (pH 7.2) and 1.0 μmole Hill oxidant. When necessary, 0.1 or 0.2 ml of an OEF solution was added to the basal reaction mixture. The OEF solution (0.2 ml) having an optical density of 0.70 at 265 mμ obtained after the 5-fold dilution was assumed to be a tentative unit of OEF.

RESULTS

Accelerating effect of OEF

OEF remarkably accelerated the oxygen evolution by a fresh 3 G fraction, but showed no accelerating effect on the oxygen evolution by the G_0 fraction, which may contain a large quantity of OEF. As shown in Table I, OEF showed no effect either on an aged 3 G fraction, which had been kept in the dark for 24 hours at 4° and had lost most of the activity

TABLE I. Effect of OEF on fresh and aged 3 G fractions.

	Fraction	OEF added (ml)	O_2 evolved (mμ moles/ 100 μg Chl/ 3 min)	Activity (%)
(A) Fresh granum fractions	G_0	0	146	100
		0.2	148	101
	3 G	0	41	100
		0.05	80	195
		0.1	110	270
		0.2	126	308
	14 G	0	0	0
		0.2	0	0
(B) Aged granum fractions (24 hrs, 4°)	G_0	0	25	100
		0.2	25	100
	3 G	0	0	0
		0.2	0	0
	14 G	0	0	0
		0	0	0
OEF alone		0.2	0	

TABLE II. Effect of OEF on variously treated 3 G fractions.

Treatment	OEF added (ml)	O_2 evolved (mμ moles/ 100 μg Chl/3 min)	Activity (%)
Control	0	72	100
	0.2	165	230
DCMU (10^{-7} *M*)	0	33	100
	0.2	50	150
Conc. Tris-HCl treatment (0.8 *M*, pH 8.0)	0	0	0
	0.2	0	0
Heat treatment (50°C, 3 min)	0	0	0
	0.2	0	0

of oxygen evolution, or on the 14 G fraction lacking the oxygen evolving system. OEF was also without effect on the 3 G fraction treated as shown in the first column of Table II. OEF alone showed no activity of oxygen evolution as shown in the bottom row of Table I.

Relation between the accelerating effect of OEF and heat treatment

The results summarized in Table III (A) show that OEF is rather heat-stable. The accelerating effect was still noticeable after treatment at 60° for 5 min. After treatment at 80° for 5 min, however, the accelerating

TABLE III. Relations between the accelerating effect of OEF and heat treatment.

(A) Heat treatment of OEF				
Fraction	Heat-treatment of OEF	OEF added (ml)	O_2 evolved (mμ moles/100 μg Chl/3 min)	Activity (%)
Fresh 3 G	Without heat-treatment	0	58	100
	" "	0.2	162	280
	40°, 5 min	0.2	140	240
	60°, 5 min	0.2	129	222
	80°, 5 min	0.2	58	100
	100°, 5 min	0.2	60	103
(B) Heat treatment of granum fractions				
Fraction	Heat treatment of granum fractions	OEF added (ml)	O_2 evolved (mμ moles/100 μg Chl/3 min)	Activity (%)
3 G	Without heat-treatment	0	47	100
	" "	0.1	77	164
	40°, 2 min	0	6	12
	" "	0.1	8	17
	50°, 2 min	0	0	0
	" "	0.1	0	0

effect was completely lost. On the contrary, such heat treatments of the 3 G fraction remarkably abolished its oxygen evolving activity, as shown in Table III (B). The treatment at 40° for 2 min. was enough to destroy almost all of the activity, and the treatment at 50° caused complete inactivation. OEF had no accelerating effect on the oxygen-evolution activity of the heat-treated 3 G fraction.

These data may be interpreted as indicating the presence of at least two factors involved in the oxygen evolution. One is OEF, which has a remarkable accelerating effect on the oxygen evolution by the fresh 3 G fraction, and the other is a factor present in the 3 G fraction which is easily destroyed by aging or heat treatment. For convenience, the latter heat-labile factor is designated the X-factor, although it has not yet been extracted.

Relation between the accelerating effect of OEF and light intensity

OEF showed a remarkable accelerating effect under strong light of 40,000 or 90,000 lux, but no effect under weak light of 7,000 lux, as can be seen from Table IV. The stronger the light intensity, the greater was the accelerating effect.

The effect of OEF as estimated from the rate of reduction of the Hill oxidant

As mentioned above, the oxygen-evolution activity was measured by a polarographic method using benzoquinone or ferricyanide as the Hill oxidant. Benzoquinone was more effective than ferricyanide as the Hill

TABLE IV. Relation between the accelerating effect of OEF and light intensity. (Material used: fresh 3 G fraction.)

Light intensity (lux)	OEF added (ml)	O_2 evolved (mμ moles/ 100 μg Chl/3 min)	Relative activity (%)
7,000	0	33	100
"	0.2	33	100
40,000	0	69	100
"	0.2	143	207
90,000	0	66	100
"	0.2	154	233

TABLE V. Accelerating effects of OEF as measured in terms of oxygen evolved and the degree of reduction of added Hill oxidant.

Fraction	OEF added (ml)	*p*-Benzoquinone (1 μmole)				Ferricyanide (1 μmole)			
		O_2 evolved		BQ reduced		O_2 evolved		FeCN reduced	
		mμ moles/ 30 μg Chl/ 3 min	Activity (%)	mμ moles/ 30 μg Chl/ 3 min	Activity (%)	mμ moles/ 30 μg Chl/ 3 min	Activity (%)	mμ moles/ 30 μg Chl/ 3 min	Activity (%)
G_0	0	187	100	130	100	72	100	270	100
	0.2	198	106	158	124	72	100	360	132
3 G	0	72	100	82	100	36	100	220	100
	0.2	165	230	109	133	99	274	300	136
OEF alone	0.2	0		0		0		0	

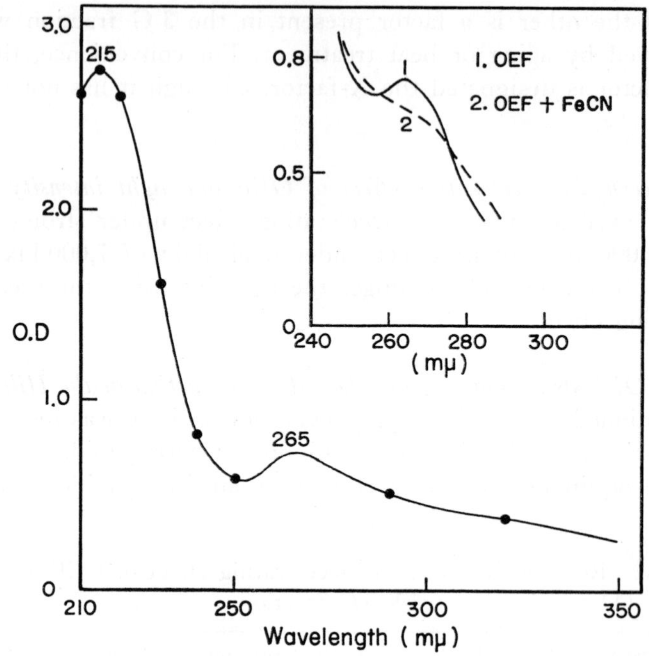

Fig. 1. Absorption spectrum of OEF.
Inset: Change in the absorption spectrum of OEF upon the addition of a small amount of ferricyanide. (For further explanation, see text.)

oxidant, and the increment of oxygen evolution by the addition of OEF was greater when benzoquinone was used as the Hill oxidant, as shown in Table V. Table V also shows the relation between the degree of the accelerating effect measured in terms of oxygen evolved and that measured in terms of the degree of reduction of the added Hill oxidant. As can be seen from the relation, the effect of OEF measured in terms of oxygen evolved was always greater than that estimated from the degree of reduction of the added Hill oxidant. For example, the degree of acceleration of oxygen evolution by OEF measured with the 3 G fraction was about 2.3 times, while the degree estimated from the rate of reduction of the Hill oxidants was not more than 1.4 times, irrespective of the kind of Hill oxidant used.

Some properties of OEF

OEF is water-soluble and is present in the supernatant obtained from the Tris-treated chloroplasts as described above. It is a rather heat-stable factor (Table III). OEF has two absorption peaks at 215 mμ and 265 mμ as shown in Fig. 1, and no absorption in the visible region. Upon the addition of a small amount of ferricyanide, the peak at 265 mμ decreased as shown in the upper right-hand corner of Fig. 1. This may indicate that OEF is a redox substance. OEF extracted from spinach chloroplasts was as effective with the 3 G fraction obtained from *Chlamydomonas* cells as with that obtained from spinach chloroplasts. Other properties of OEF are now under investigation.

DISCUSSION

All of the results reported here clearly indicate that OEF and the oxygen evolving system of photosynthesis have the following characteristics. OEF remarkably accelerates the oxygen evolution by a fresh 3 G fraction, but it shows no effect on both the G_0 fraction and the 14 G fraction. The 14 G fraction has no activity of oxygen evolution and the G_0 fraction seems to contain a large amount of OEF. OEF is also without effect on the 3 G fraction, which has lost the activity almost completely on various treatments (Tables I and II). It was deduced from these data that the presence of OEF alone is not sufficient for the occurrence of oxygen evolution, and that the 3 G fraction with added OEF alone is also incapable of evolving oxygen. There may be a certain additional factor besides OEF, which is easily destroyed by aging or heat treatment. This factor was tentatively designated the X-factor. The cooperation of OEF and this X-factor is therefore inferred to be necessary for the full activity of oxygen evolution (see Table III).

The accelerating effect of OEF is not observable under weak light, whereas the effect is remarkable under strong light (Table IV). This may be explained by assuming that the X-factor is inactivated by strong light illumination and that OEF functions to reactivate the inactive form of the X-factor. Regardless of their functions, attention should be paid in future studies of the oxygen evolving system to these labile factors involved in the oxygen evolution.

SUMMARY

It was found during the fragmentation of the particles of photosystem II that a new factor capable of accelerating the oxygen evolution of chloroplast fragments is present in the supernatant fluid obtained by centrifugation of the chloroplasts which have been incubated with a high concentration of Tris-HCl buffer at a high pH value. This factor was extracted and purified, and its mode of action was investigated and discussed. Instability of the separated oxygen evolving system, which made analysis of the system difficult, was also discussed.

REFERENCES

1. Wessels, J. S. C., *Proc. Roy. Soc. London*, **B157**, 345 (1964).
2. Boardman, N. K., and Anderson, J. M., *Nature*, **203**, 166 (1964).
3. Anderson, J. M., Boardman, N. K., and David, D. J., *Biochem. Biophys. Res. Comm.*, **17**, 685 (1964).
4. Ke, B., and Clenndenning, K. A., *Biochim. Biophys. Acta*, **19**, 74 (1965).
5. Becker, M. J., Shefner, A. M., and Gross, J. A., *Plant Physiol.*, **40**, 243 (1965).
6. Vernon, L. P., Shaw, E. R., and Ke, B., *J. Biol. Chem.*, **241**, 4101 (1966).
7. Briantais, J. M., *Photochem. and Photobiol.*, **5**, 135 (1966).
8. Anderson, J. M., Fork, D. C., and Amesz, J., *Biochem. Biophys. Res. Comm.*, **23**, 874 (1966).
9. Anderson, J. M., and Boardman, N. K., *Biochim. Biophys. Acta*, **112**, 403 (1966).
10. Huzisige, H., Ushiyama, H., and Aji, T., in preparation.
11. Huzisige, H., Satoh, K., Tanaka, K., and Hayashida, T., *Plant and Cell Physiol.*, **4**, 307 (1963).
12. Ogawa, T., and Shibata, K., *Photochem. Photobiol.*, **4**, 193 (1965).

Donation of Electrons to Photosystem 2 in Chloroplasts by *p*-Phenylene Diamine*

TAKASHI YAMASHITA AND WARREN L. BUTLER

Department of Biology, Revelle College, University of California, San Diego, La Jolla, California

INTRODUCTION

It was reported previously that washing chloroplasts with 0.8 *M* tris-HCl inhibited the Hill reaction with water as the electron donor.[1] The tris-washed chloroplast fragments, however, were capable of photoreducing NADP with artificial electron donor systems such as DCIP and ascorbate or TMPD and ascorbate. The photoreduction of NADP with DCIP or TMPD plus ascorbate as the donor system resulted in photophosphorylation which was partially (DCIP) or completely (TMPD) inhibited by DCMU. In the present paper we wish to report that the photoreduction of NADP with *p*-phenylene diamine (PD) plus ascorbate as the donor system is inhibited by DCMU and to present evidence that the electron donor system can supply electrons to the electron transport chain at a site on the water side of PSII.

EXPERIMENTAL PROCEDURE

Chloroplasts were prepared by grinding 50 g of spinach leaves in a Waring blendor for 20 seconds in 150 ml of solution containing 0.4 *M* sucrose, 0.05 *M* tris-HCl (pH 7.8) and 0.01 *M* NaCl. The supernatant from a 1 minute centrifugation at 300×*g* was centrifuged at 600×*g* for 7 minutes in four centrifuge tubes. The precipitate from two tubes was suspended in 20 ml of the 0.4 *M* sucrose, 0.05 *M* tris-HCl and 0.01 *M* NaCl medium (normal chloroplasts). The precipitate from the other two

* This research was supported by grant GM 15048 from the U.S. Public Health Service and grant GB-3776 from the U.S. National Science Foundation.

Abbreviations: DCIP, 2,6,-dichlorophenol indophenol; DCMU, 3-(3,4-dichlorophenyl)-1,1-dimethylurea; Fd, ferredoxin; NADP and NADPH, nicotinamide adenine dinucleotide phosphate and its reduced form; PD, *p*-phenylene diamine; TMPD, tetramethyl-*p*-phenylene diamine; Tris-HCl, Tris(hydroxymethyl) aminomethane-HCl buffer; PS I and PS II, Photosystem 1 and Photosystem 2.

tubes was suspended in 20 ml of 0.8 *M* tris-HCl (pH 8.0) (tris-washed chloroplasts). After standing for 10 minutes, heavy particles of both batches of chloroplasts were removed by centrifugation at $300 \times g$ for 1 minute. The remaining supernatants were centrifuged at $1500 \times g$ for 7 minutes and the precipitates resuspended in 2 ml of 0.4 *M* sucrose, 0.05 *M* tris-HCl (pH 7.8) and 0.01 *M* NaCl.

Chlorophyll concentration was determined by the method of MacKinney.[2] Ferredoxin was purified from spinach leaves. DCMU, donated by Dr. P. G. Heytler, was used as a methanol solution. NADP and *p*-phenylene diamine were obtained from Boehringer Manheim Company and Eastman Organic Chemistry Company, respectively.

NADP photoreduction was measured with an Aminco-Chance dual wavelength spectrophotometer (λ_1 : 340 mμ, λ_2: 374mμ). The light signal was measured with an EMI 9524 phototube, blocked with Corning 5840 and 5970 filters.

The fluorescence-yield measurements were made with an instrument similar to that used previously by Duysens and Sweers[3] and by Butler and Bishop.[4] Fluorescence of the chloroplast suspension was excited by a weak monochromatic beam (50 ergs/cm^2/sec, 650 mμ) chopped at a frequency of 360 cps. The phototube signal was measured with a PAR Lock-in Amplifier synchronized to the chopping frequency. A steady actinic beam (3×10^4 ergs/cm^2/sec, 645 mμ) was used to irradiate the sample in a 1 cm cuvette from the side. Corning filters 9830 and 2030 placed directly in front of the phototube blocked the measuring and actinic beams but passed fluorescence of wavelengths longer than 680 mμ. The lock-in amplifier responded only to the modulated fluorescence excited by the weak, chopped measuring beam but not to the steady fluorescence excited by the actinic beam. Changes in fluorescence yield due to the actinic beam were monitored by the measuring beam.

The standard reaction mixture consisted of 15 m*M* tris-HCl (pH 7.8), 4 m*M* KPO_4 (pH 7.8), 4 m*M* $MgCl_2$ and 20 m*M* NaCl. Reactions were measured in a 3 ml volume with a 1 cm pathlength at about 22°C.

RESULTS

Measurements of the photoreduction of NADP in the presence of added ferredoxin are shown in Fig. 1. The rate of reduction by red light with intact chloroplasts and water as the electron donor was 62 μmoles NADPH/mg Chl/hr(curve 6). Chloroplasts which had been washed with 0.8 *M* tris-HCl showed no photoreduction of NADP by water (curve 1).

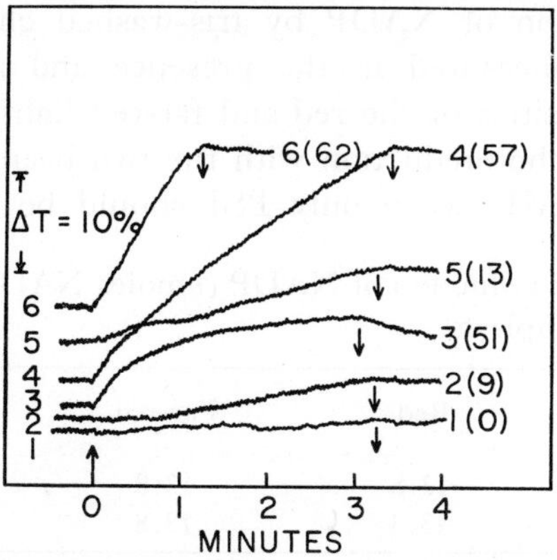

Fig. 1. Photoreduction of NADP by 0.8 *M* tris-HCl washed chloroplasts (curves 1–5) and by normal chloroplasts (curve 6). Standard reaction medium plus $1 \times 10^{-6} M$ ferredoxin, $3.3 \times 10^{-4} M$ NADP, 10 μg Chl/ml (curves 1–5) or 11 μg Chl/ml (curve 6). Curve 1, no further additions; curve 2, $3.3 \times 10^{-4} M$ ascorbate; curve 3, $3.3 \times 10^{-5} M$ PD; curve 4, $3.3 \times 10^{-5} M$ PD, $3.3 \times 10^{-4} M$ ascorbate; curve 5, $3.3 \times 10^{-5} M$ PD, $3.3 \times 10^{-4} M$ ascorbate, $3.3 \times 10^{-7} M$ DCMU; curve 6, no addition.

Numbers in parentheses are the initial rates of photoreduction of NADP (μmoles NADPH/mg Chl/hr) except for curve 2 where the number indicates the steady-state rate.

Red light (645 mμ, 7×10^{4} ergs/cm²/sec) on at upward arrow, off at downward arrow.

Addition of $3.3 \times 10^{-4} M$ ascorbate to the tris-washed chloroplasts resulted in a low rate of photoreduction (curve 2). Addition of $3.3 \times 10^{-5} M$ PD alone in the reduced state gave a relatively high initial rate of photoreduction of NADP which progressively decreased as the reaction proceeded (curve 3). Presumably, oxidized PD can accept electrons more readily than NADP so that a type of cyclic electron transport begins to predominate when some oxidized form of PD is allowed to accumulate. In the presence of ascorbate, however, PD supported a sustained photoreduction of NADP(curve 4). The initial rate (57 μmoles NADPH/mg Chl/hr) was comparable to the control rate by normal chloroplasts. The somewhat slower steady-state rate may be due to a limiting rate of electron transport from ascorbate to PD. The photoreduction of NADP by PD plus ascorbate was largely inhibited by $3.3 \times 10^{-7} M$ DCMU (curve 5). Thus, PD supplies electrons to the electron transport chain at a site prior to the DCMU block.

The relative effectiveness of red (645 mμ) and far-red (715 mμ) light for the photoreduction of NADP by tris-washed chloroplasts with PD and ascorbate was measured in the presence and absence of DCMU (Table I). The intensities of the red and far-red light used were adjusted so that the rates of photoreduction with the two beams were the same in the presence of DCMU when only PSI should be functioning. In the

TABLE I. Rate of reduction of NADP (μmoles NADPH/mg Chl/hr) by tris-washed chloroplasts.

DCMU (3.3×10^{-7} *M*)	Red	Far-red	R/FR
+	13.8	13.8	1.0
−	43.3	13.8	3.1

The standard reaction medium (3 ml) contained 11 μg Chl/ml, 3.3×10^{-5} *M* PD, 3.3×10^{-4} *M* ascorbate, 1×10^{-6} *M* ferredoxin, 3.3×10^{-4} *M* NADP. The intensity of both red (645 mμ) and far-red (715 mμ) was about 7×10^{4} ergs/cm^2/sec.

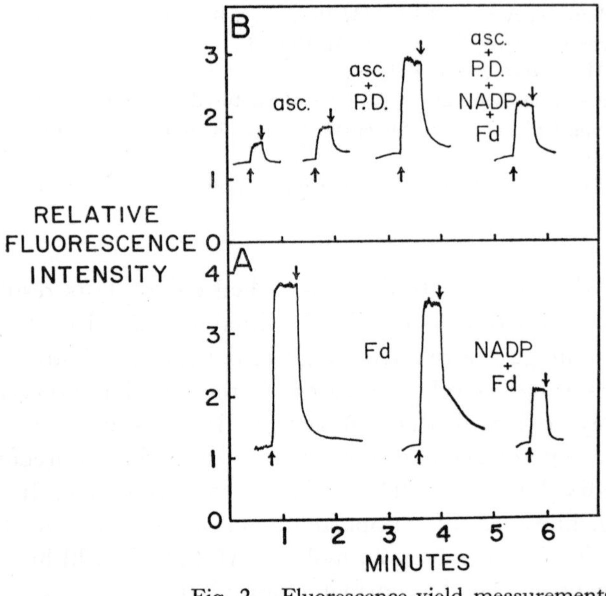

Fig. 2. Fluorescence-yield measurements.

A. Normal chloroplasts (11 μg Chl/ml) in standard reaction medium. 1×10^{-6}*M* ferredoxin and 3.3×10^{-4}*M* NADP added where indicated. Red light (645 mμ, 3×10^{4} ergs/cm^2/sec) on at upward arrow, off at downward arrow.

B. 0.8 *M* tris-HCl washed chloroplasts (12 μg Chl/ml) in standard reaction medium. 3.3×10^{-5}*M* PD, 3.3×10^{-4}*M* ascorbate, 1×10^{-6}*M* ferredoxin and 3.3×10^{-4}*M* NADP added where indicated. Actinic irradiation was the same as in A.

absence of DCMU red light was approximately three-fold more effective than the far-red light. Thus, a "red drop" phenomenon indicating the cooperative action of both PS II and PS I is demonstrated for the photoreduction of NADP from PD plus ascorbate.

Changes in fluorescence yield of chlorophyll in photosynthetic systems have been ascribed to changes in the oxidation state of the primary reductant of PS II, denoted "Q".[3] "Q" quenches chlorophyll fluorescence when in the oxidized state but not in the reduced state. In the dark (or in weak measuring light) "Q" is fully (or largely) oxidized so that the fluorescence yield is low. In the absence of an electron acceptor for the Hill reaction, irradiation with actinic light absorbed by PS II causes "Q" to go fully reduced and the fluorescence yield increases about three-fold (Fig. 2A). In the presence of an electron acceptor, the steady-state rate of electron flow through "Q" during irradiation sets "Q" at an intermediate state of reduction so that the fluorescence yield is less than maximal. Such effects of electron acceptors on fluorescence yield of chloroplasts have been noted previously.[5]

The tris-washed chloroplasts (Fig. 2B), however, do not show the large light-induced fluorescence-yield changes of normal chloroplasts, presumably because the source of electrons for the reduction of "Q" has been blocked by the washing treatment. Addition of ascorbate alone has little effect but addition of ascorbate plus PD restores most of the light-induced fluorescence-yield increase. Other electron donor systems such as TMPD or DCIP with ascorbate also restored the light-induced fluorescence-yield increase but not quite as effectively as PD and ascorbate. The fluorescence yield of the tris-washed chloroplasts without the electron donor system is also increased from 50 to 100% in both measuring and actinic light by the addition of 3.3×10^{-7} *M* DCMU.

DISCUSSION

The washing treatment with 0.8 *M* tris-HCl blocks the electron transport system of chloroplasts at a site prior to the DCMU block and thus permits an examination of the electron transport reactions in this region. The site of the DCMU block was placed just beyond the primary reductant of PS II, largely on the basis of fluorescence-yield measurements.[3] The site of inhibition due to the tris-wash treatment can be placed between water and PS II.

PD
$H_2O \rightarrow \vdots \xrightarrow{PD} Z\,(PS\ II)\ Q \vdots \rightarrow \rightarrow \rightarrow (PS\ I) \rightarrow Fd \rightarrow NADP$
tris wash (at the first block) DCMU (at the second block)

The sensitivity of the PD supported photoreduction of NADP to DCMU indicates that PD can donate electrons to the system between these two sites of inhibition. The demonstration of the "red drop" phenomenon for the PD supported NADP photoreduction further supports the contention that PD donates electrons on the water side of PS II.

The fluorescence-yield measurements give an independent conformation of the proposed sites for the tris-wash inhibition and the PD electron donation. The tris-wash treatment blocks the flow of electrons from water so that "Q" remains largely oxidized in the light and the fluorescence-yield remains low. (In the presence of DCMU which blocks electron transport beyond "Q" the endogenous electron donors are sufficient to reduce "Q" in the light.) Addition of the artificial electron donor system restores the PS II-mediated photoreduction of "Q" and the attendant light-induced fluorescence-yield increase.

The results of Katoh and San Pietro reported in this volume also show that electrons can be donated to PS II. In their work with heated *Euglena* chloroplasts, ascorbate supplied electrons to PS II. In our work PD was a much more efficient electron donor to PS II than ascorbate but ascorbate could have been responsible for a low rate of electron flow through PS II.

SUMMARY

Chloroplasts washed with 0.8 *M* tris-HCl showed no Hill reaction (photoreduction of NADP) with water as the electron donor but were capable of photoreducing NADP when *p*-phenylene diamine and ascorbate were added as an artificial electron donor system. The photoreduction of NADP by *p*-phenylene diamine with the tris-washed chloroplasts was largely inhibited by 3.3×10^{-7} *M* DCMU. DCMU primarily inhibited the action of red light rather than that of far-red light. These results indicate that the tris-wash treatment inhibits the photosynthetic electron transport system on the water side of the DCMU block and that *p*-phenylene diamine supplies electrons to the electron transport chain on the water side of photosystem 2.

Light-induced fluorescence-yield measurements support these conclusions. The tris-washed chloroplasts showed very little increase in fluorescence yield on irradiation with actinic light, in contrast to normal chloroplasts, indicating a limitation in reducing power generated by photosystem 2. On addition of *p*-phenylene diamine plus ascorbate most of the light-induced fluorescence-yield increase was restored. These results also indicate that the supply of electrons to photosystem 2 (from water) is blocked by washing the chloroplasts with 0.8 *M* tris-HCl but can be restored by addition of reduced *p*-phenylene diamine.

REFERENCES

1. Yamashita, T., and Horio, T., submitted to *Biochim. Biophys. Acta.*
2. MacKinney, G., *J. Biol. Chem.*, **140**, 315 (1941).
3. Duysens, L. N. M., and Sweers, H. E., in *Microalgae and Photosynthetic Bacteria*, p. 353, Japanese Society of Plant Physiologists, University of Tokyo Press, Tokyo (1963).
4. Butler, W. L., and Bishop, N. I., in *Photosynthetic Mechanisms of Green Plants* (B. Kok and A. T. Jagendorf, eds.), p. 91, Nat. Acad. Sci.-Nat. Res. Council, 1963.
5. Kok, B., in *Photosynthetic Mechanisms of Green Plants* (B. Kok and A. T. Jagendorf, eds.), p. 45, Nat. Acad. Sci.-Nat. Res. Council, 1963.

Light-induced Reactions in *Chromatium* Chromatophores under Controlled Redox Conditions

M. A. CUSANOVICH, R. G. BARTSCH AND J. M. OLSON*

Dept. of Chemistry, Univ. of California at San Diego, La Jolla, California
and
**Biology Dept., Brookhaven National Laboratory, Upton, New York*

INTRODUCTION

Information about the cellular components involved in bacterial photosynthesis is still scant. Our efforts to catalog and characterize soluble enzymes including the heme and non-heme iron oxidation-reduction proteins which may be closely involved in light-driven reactions are by no means complete. Here are summarized some of these efforts as related to the cytochrome content of *Chromatium* strain D.

As compared with most non-sulfur purple bacteria so far investigated, light-induced reactions of cytochromes are especially easy to detect in *Chromatium* cells[1] and in chromatophores.[2,3] This is primarily because of the relatively high concentration of cytochromes in *Chromatium* cells and chromatophores and also because of the relative lack of interference between the spectra of the cytochromes and the other reactive colored components of this organism. Olson and Chance[1] and later Morita, Edwards and Gibson[4] reported that with whole cells of *Chromatium* three or four distinguishable light-induced changes, probably involving cytochromes, could be detected spectrophotometrically. These components were labeled cytochrome $c_{423.5}$, c_{426}, c_{422}, and c_{430}, according to the wavelength maxima of the Soret peaks in light-minus-dark difference spectra. Based on the wavelength maxima of the alpha peaks in the difference spectra, $c_{423.5}$ may alternatively be designated c_{552} and c_{422} may be designated c_{555}.

Clayton[2] and Nishimura[3] studied light-induced reactions with more or less aerobic *Chromatium* chromatophore preparations and observed at least one reaction involving a cytochrome. Olson and Chance[1] and Vredenberg and Duysens[5] emphasized the distinction between two easily detected cytochromes. One cytochrome, designated c_{422}, underwent light-induced oxidation when the cell suspension was aerobic, and the second, designated

$c_{423.5}$, became reactive when anaerobic conditions were established. Because the rate of reduction of photo-oxidized $c_{423.5}$ was much lower than that of c_{422} in a succeeding dark interval, Duysens concluded that $c_{423.5}$ did not participate in the oxidation of hydrogen donors.[6] In the whole cell the light-induced oxidation of each of these two cytochromes proceeds with quantum efficiencies equal to 1.[5,7] Morita *et al.*[4] suggested that three *Chromatium* cytochromes may each react directly with BChl, thereby linking different substrate-level reactions with the photo-oxidizing system of the cell.

Further observations on the distinction between two of the chromatophore-bound cytochromes of *Chromatium* are reported here.

METHODS AND RESULTS

A portion of the work reported here is taken from unpublished experiments of Bartsch and Olson. Chromatophore preparations were made from photo-autotrophically grown cells by the method of Newton and Newton.[8] The particulate fraction which sedimented between 3×10^4 and $1 \times 10^5 \times g$ within three hours was washed twice with the extracting buffer (100 mM potassium phosphate, pH 7). In most instances the particles were then treated with approximately 10 mM potassium ferricyanide in 50 mM phosphate buffer, pH 7, in the cold for 30 min before the mixture was dialyzed vs. two changes of 5 mM phosphate buffer in the cold for 12–18 hours. The object of these treatments was to remove reducing material capable of rapidly interacting with particle-bound cytochromes and other redox constituents of the chromatophores. As will be seen later, this object was not fully realized. If the oxidized chromatophores were stored aerobically at ice temperature, the oxidation level of the particle-bound cytochromes remained constant for several days, as indicated by the minimal level of light-induced oxidation observed. The light-induced difference spectrum measured before appreciable re-reduction of cytochrome had occurred showed changes only in the regions associated with presumed BChl peaks at 390, 435, 500 and 620 mμ.

Light-induced cytochrome changes could be restored in the chromatophore suspension in several ways. These changes were measured with the dual wavelength spectrophotometer described by Morita *et al.*[9]

The simplest method used to prime the particles for a light reaction was to make the particle suspension anaerobic by bubbling with pre-purified argon (Matheson Co.) via two syringe needles inserted into a serum bottle stopper sealing the reaction cuvette. To facilitate the deaeration, the

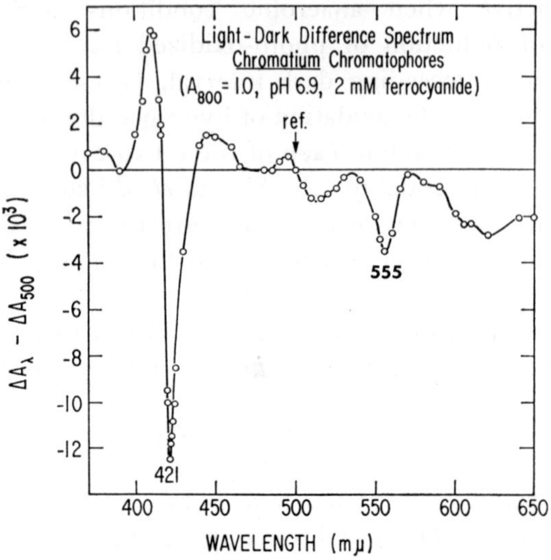

Fig. 1. Light-minus-dark difference spectrum of *Chromatium* chromatophores—High potential cytochrome c_{555}. ($A_{800}=1.0$, optical path$=0.3$ cm., 50 mM potassium phosphate buffer, pH 6.9, reduced with 2 mM potassium ferrocyanide, actinic light 890 mμ, $\sim 10^{-8}$ E cm^{-2} sec^{-1}).

cuvette was first evacuated. After 15 to 40 min bubbling, the sample became sufficiently reduced at the expense of an endogenous reducing system to permit nearly maximum observable light-induced oxidation of a cytochrome with a spectrum characterized by an alpha peak at 555 mμ and a Soret peak at 422 mμ. The anaerobic particle suspension could also be reduced by the addition of reduced *Chromatium* HiPIP[10] or more conveniently, by the addition of freshly prepared 2 mM potassium ferrocyanide solution.

Fig. 1 shows the steady state light-minus-dark difference spectrum of particles reduced with 2 mM ferrocyanide. The spectrum is essentially that of c_{422}.

The average quantum efficiency, ϕ, for oxidation of c_{422} (c_{555}) was 1/8, an efficiency significantly lower than the value of 1 observed by Vredenberg and Duysens[5] with whole cells. Nevertheless the reaction proceeded with reasonable efficiency considering the probable damage done to the particles during preparation.

As is illustrated in Fig. 2 (solid line), c_{422} (c_{555}) could be directly observed in an oxidized-minus-reduced difference spectrum. It was therefore feasible to measure the approximate oxidation-reduction potential of

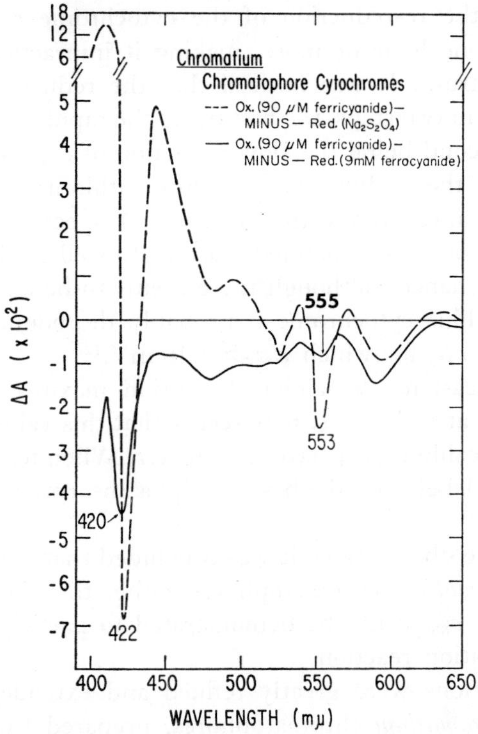

Fig. 2. Oxidized-minus-reduced difference spectra of *Chromatium* chromatophores. (A_{800}=3.0, optical path=1 cm, 50 mM phosphate buffer, pH 7.5.)

this particle-bound cytochrome by equilibrating the treated chromatophores with mixtures of ferro- and ferricyanide and measuring the change in cytochrome Soret and alpha absorption in a difference spectrum vs. an untreated sample. Then E_m equals the E_h of the redox buffer (defined by the ratio of ferrocyanide/ferricyanide) for which one-half the total spectral change occurred.[11] The sample was completely oxidized in the presence of 90 μM ferricyanide and completely reduced upon the further addition of 90 mM ferrocyanide. The redox potential of bound c_{555} was found to be $E_{m,7} \simeq 340$ mV, based on $E_{m,7} = 430$ mV for the ferro-ferricyanide reaction couple.

The light-induced oxidation of a second cytochrome became evident when a reducing agent such as 1 mM 2-mercaptoethanol was added to the oxidized particle suspension. The reduced α and γ absorption maxima of this cytochrome occurred at 552 and 422–423 mμ. A complete light-minus-dark difference spectrum of this cytochrome species was not traced with the dual wavelength apparatus. Even with the high concentration of reducing agent present, the re-reduction of the cytochrome c_{552} after photo-oxidation required one hour or more, making it impractical to trace the point-by-point spectrum. It was noticed that the reduction after photo-oxidation proceeded in two steps, the first being the rapid reduction of light-oxidized c_{555}, the second being the slow reduction of c_{552}. The latter cytochrome dominated the reduced (by sodium dithionite)-minus-oxidized difference spectrum measured with the Cary 14 spectrophotometer (see Fig. 2, dashed line), and corresponded reasonably well to the spectrum of $c_{423.5}$ of Olson and Chance,[1] although there seems to be a slight blue shift of the Soret peak. This cytochrome is probably the bound equivalent of purified cytochrome c_{552} for which $E_{m,7} \simeq +10$ mV.[12]

On various occasions the alpha absorption maxima of cytochrome c_{552} was stated to be at 553 mμ. It now seems that this value was obtained with an improperly calibrated spectrophotometer. When feasible the *c*-type cytochromes will be labeled on the basis of alpha absorption maxima rather than Soret maxima.

From the above observations it was concluded that dependent on the redox level of *Chromatium* chromatophores, either the high potential c_{555} or the low potential c_{552} could be demonstrated to participate in a light-induced photo-oxidation reaction.

These observations were greatly refined and extended by Cusanovich.[13,14] Crude *Chromatium* chromatophores, prepared from cells grown photoheterotrophically with succinate as the substrate,[15] were separated into two fractions according to the method of Worden and Sistrom,[16] and were extensively characterized.[13,17] The lower density portion, which amounted to approximately two-thirds of the total particle-bound BChl,

---

Fig. 3. Effect of redox potential on light-minus-dark difference spectra of low density *Chromatium* chromatophores. ($A_{800}=1.0$, optical path=1 cm., 35 μ*M* BChl, in 10% W/V sucrose in 100 m*M* phosphate buffer, pH 7.5, plus the following ingredients for each designated E_h which was attained by action of the endogenous reducing system for $E_h \geqq 200$ mV and by the addition of 1% sodium dithionite in 10 μ*M* NaOH solution for $E_h \leqq 200$ mV:

473 mV, 393 mV	100 μ*M* potassium ferricyanide (A)
378 mV, 200 mV	(A) plus 100 μ*M* 2,6-dibromobenzenoneindol-3-carboxyphenol
109 mV	300 μ*M* ferric chloride plus 10 m*M* disodium EDTA
−1 mV, −102 mV	300 μ*M* ferric chloride plus 10 m*M* dipotassium oxalate plus 100 μ*M* indigotetrasulfonic acid plus 100 μ*M* indigodisulfonic acid

Actinic light, 888 mμ, 10 mμ band width, $\sim 1.6 \times 10^{-8}$ E cm^{-2} sec^{-1}.

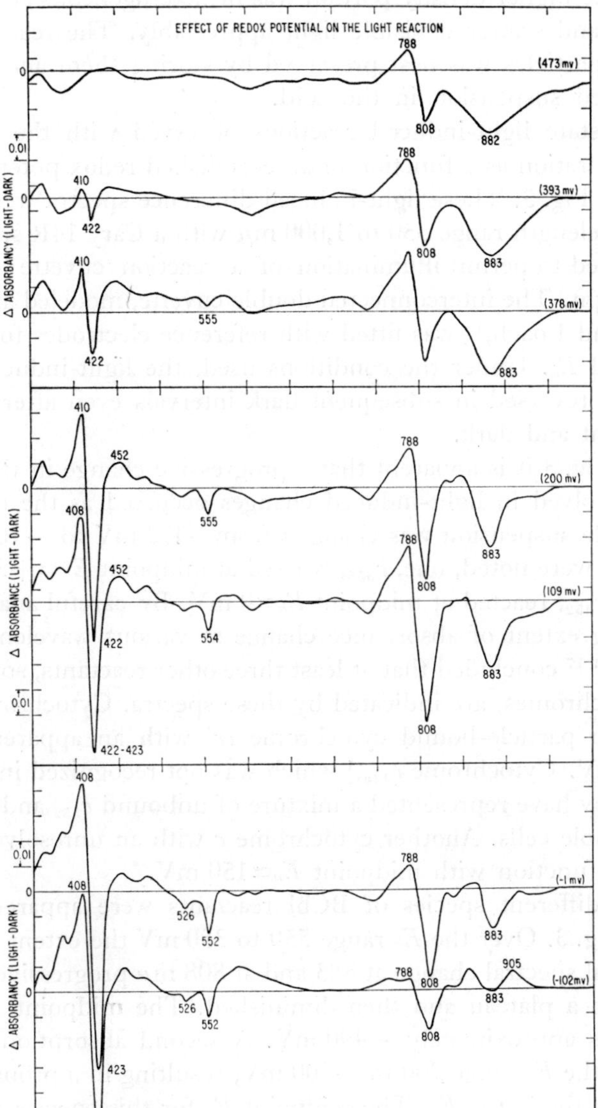
EFFECT OF REDOX POTENTIAL ON THE LIGHT REACTION
(473 mv)
788
808
882
0.01
410
422
788
(393 mv)
808
883
410
422
555
(378 mv)
883
410
452
788
(200 mv)
555
883
408
452
788
808
(109 mv)
422
554
883
0.01
808
422-423
408
0.01
788
408
(-1 mv)
526
552
883
788
808
905
(-102mv)
526
552
883
808
423
423
Δ ABSORBANCY (LIGHT-DARK)
400
450
500
550
600
650
700
750
800
850
900
950
WAVELENGTH (mμ)

Fig. 3

consisted of relatively homogeneous particles of approximately 500±100 Å diameter. The higher density portion was of heterogeneous size and shape distribution and scattered visible light appreciably. The reactivity of the low density particles was best preserved by storing them anaerobically as a concentrated suspension in the cold.

Steady-state light-induced reactions observed with the low density particle preparation as a function of an established redox potential, E_h, are illustrated in Fig. 3. These light-induced difference spectra were measured over the wavelength range 350 to 1,000 mμ with a Cary 14R spectrophotometer modified to permit illumination of a reaction cuvette with monochromatic light. The interconnected double cuvette, modeled after those of Harbury[18] and Loach,[19] was fitted with reference electrodes for measuring both pH and E_h. Under the conditions used, the light-induced reactions were rapidly reversed in subsequent dark intervals even after at least ten cycles of light and dark.

From Fig. 3 it is apparent that a progressive change in the spectra of reactants involved in light-induced changes occurred as the measured E_h of the particle suspension was changed from +473 mV to −102 mV. Two cytochromes were noted, one, c_{555}, reacted at midpoint $E_h \simeq +319$ mV, and the second, c_{552}, reacted at midpoint $E_h \simeq 0$ mV. By careful analysis of data plots relating extent of absorbance change at various wavelengths *vs.* E_h, Cusanovich[14,17] concluded that at least three other reactants, some of which may be cytochromes, are indicated by these spectra. Cytochrome c_{430}[1] was equated with particle-bound cytochrome *cc′* with an apparent midpoint $E_h \simeq +175$ mV. Cytochrome c_{426},[1] which was not recognized in the particle reactions, may have represented a mixture of unbound c_{552} and cytochrome *cc′* in the whole cells. Another cytochrome *c* with an unresolved spectrum appeared to function with midpoint $E_h \simeq 150$ mV.

Three different species of BChl reactants were apparent from the spectra of Fig. 3. Over the E_h range 550 to 350 mV the extent of the light-induced BChl spectral change at 883 and at 808 mμ progressively increased in parallel to a plateau and then diminished. The midpoint E_h for these reactions was approximately +490 mV. A second absorption change appeared over the E_h range 200 to −100 mV, resulting in a pronounced peak at 905 mμ at the lower E_h. The midpoint E_h for this species of BChl was less than +100 mV. The optimum pH for the BChl change at 808 mμ was pH 7.2 and for that at 883 mμ was pH 8, indicating that these two species reacted independently. The 883 mμ species of BChl, designated P890, and the pigment responsible for the absorption peak at 905 mμ, designated P905, perhaps represent two types of active center bacteriochlorophyll. The

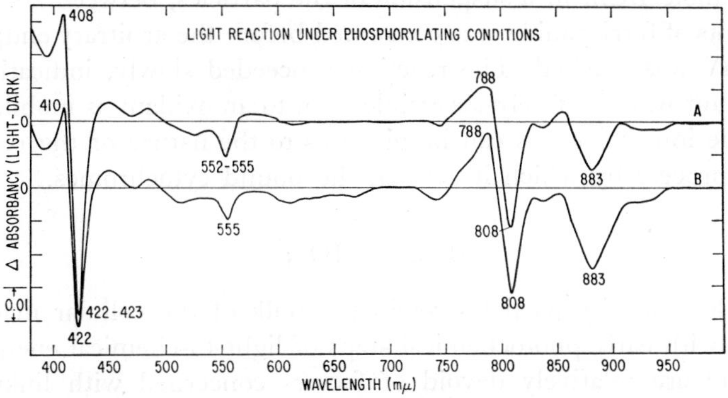

Fig. 4. Light-minus-dark difference spectra of *Chromatium* chromatophores under phosphorylating conditions. (A_{800}=1.0, optical path=1 cm, 35 μ*M* BChl, in 10% W/V sucrose in 100 m*M* phosphate buffer, pH 7.5, plus 500 μ*M* ADP, 300 μ*M* $MgCl_2$.) Curve A, anaerobic, self-reduced; Curve B, aerobic, self-reduced. Actinic light 888 mμ, $\sim 1.6 \times 10^{-8}$ E cm^{-2} sec^{-1}.

808 mμ BChl species may be an auxiliary type of BChl closely related to the active center.

In collateral experiments, Cusanovich[13,20] found that maximal rates of photophosphorylation were obtained with the low density particles poised in the E_h range 50 to 100 mV. Under saturating light (8.2×10^4 ergs cm^{-2} sec^{-1}, 880 mμ) 5.5 μmoles ATP/μmole BChl/hr. were produced without addition of soluble factors extracted from the cells.

The effect of components of the phosphorylation reaction mixture on the light-induced reactions was investigated. Upon addition of 500 μ*M* ADP plus 300 μ*M* $MgCl_2$ the light-induced difference spectrum of the low density particle suspension changed from that normally observed at $E_h \simeq$ 200–300 mV to one observed at $E_h \simeq 50$–100 mV. This phenomenon is illustrated in Fig. 4, where curve A is the light-minus-dark difference spectrum of the aerobic suspension, and curve B that of a comparable suspension under aerobic conditions. A possible explanation for this behavior would be that the reduction of the particles by the endogenous reducing system was facilitated by the presence of the added components.

It proved impractical to directly measure the E_h imposed by the endogenous reducing material, although as indicated by the character of the light-minus-dark difference spectra of anaerobic particles without added redox reagents, the endogenous E_h may have reached +200 mV in the absence of added reactants. An attempt was made to chemically titrate the reductant present in the particles. In several different experiments, for each μmole BChl in a suspension of the particles, between 20 and 100 μequivalents of ferricyanide was consumed before the arbitrary endpoint at $E_h \simeq 400$ mV was reached. The reaction proceeded slowly, indicating that the reductant was not freely accessible even to an oxidant as mobile as the ferricyanide ion. No guess can be given as to the nature of the reductant or to the process by which it acts on the bound cytochromes.

DISCUSSION

The low density particles retain the bulk of the cellular machinery concerned with early photochemical steps of light-to-chemical energy conversion and are relatively devoid of factors concerned with linking the photochemical system to the utilization of substrate. Application of the method of buffering the redox environment of the particles facilitates controlled studies of some of these photochemical reactions.

It is clear that with the experimental technique used here, many of the light-induced reactions of the *Chromatium* cell can be duplicated with chromatophores and that individual reaction components may be more easily studied than is feasible with the whole cell. The chromatophores retain as much as two-thirds of the several cytochromes of the cell, including c_{552} and cytochrome cc', and all of cytochrome c_{555}. No protoheme and therefore no cytochrome *b* was detected in cells or chromatophores. *Chromatium* offers the unusual example of an organism capable of photophosphorylation without a complement of cytochrome *b*. The absence of cytochrome *b* may be a factor in the inability of *Chromatium* to grow aerobically.

Some oxido-reductase enzyme activities, such as succinic dehydrogenase, remain bound to the particles whereas others, such as two NADH oxido-reductases, are quantitatively removed during preparation of the particles.[13] Therefore, it should be feasible to study the interaction between many of the electron transfer components and the reactants directly engaged in light-driven reactions.

The endogenous reducing system is responsible for the reduction of the high potential cytochrome c_{555} under aerobic conditions, even without added factors. Under anaerobic conditions the endogenous reducing system can reduce the low potential cytochrome c_{552} if ADP is added. It will be instructive to more completely reconstitute the reducing system and to identify the stored reductant in an effort to deduce at least one mechanism by which substrate interacts with the light-reactive electron transport system.

Acknowledgements

This research was supported in part by grants-in-aid from the National Institutes of Health (HD-01262) and the National Science Foundation (GB-2892) to Prof. M.D. Kamen. M.A.C. was supported by a training grant from the National Institutes of Health (5TO16MO1045). Part of the work was performed in the Brookhaven National Laboratory under the auspices of the U.S. Atomic Energy Commission.

REFERENCES

1. Olson, J. M., and Chance, B., *Arch. Biochem. Biophys.*, **88**, 26, 40 (1960).
2. Clayton, R. K., *Photochem. Photobiol.*, **1**, 313 (1962).
3. Nishimura, M., and Chance, B., *Biochim. Biophys. Acta*, **66**, 1 (1963).
4. Morita, S., Edwards, M., and Gibson, J., *Biochim. Biophys. Acta*, **108**, 45 (1965).
5. Vredenberg, W. J., and Duysens, L. N. M., *Biochim. Biophys. Acta*, **79**, 456 (1963).
6. Duysens, L. N. M., *Arch. Biol.* (Liege), **76**, 251 (1965).
7. Olson, J. M., *Science*, **135**, 101 (1962).
8. Newton, J. W., and Newton, G. A., *Arch. Biochem. Biophys.*, **71**, 250 (1957).
9. Morita, S., Olson, J. M., and Conti, S. F., *Arch. Biochem. Biophys.*, **104**, 346 (1964).
10. Bartsch, R. G., in *Bacterial Photosynthesis* (H. Gest, A. San Pietro and L. P. Vernon, eds.), p. 315, Antioch Press, Ohio, 1963.
11. Davenport, H. E., and Hill, R., *Proc. Roy. Soc.* (London), **B139**, 327 (1952).
12. Bartsch, R. G., and Kamen, M. D., *J. Biol. Chem.*, **235**, 825 (1960).
13. Cusanovich, M. A., *Ph. D. Thesis*, Univ. of California at San Diego, 1967.
14. Cusanovich, M. A., Bartsch, R. G., and Kamen, M. D., *Biochim. Biophys. Acta*, **153**, 397 (1967).
15. Bose, S. K., in *Bacterial Photosynthesis* (H. Gest, A. San Pietro and L. P. Vernon, eds.), p. 501, Antioch Press, Ohio, 1963.
16. Worden, P., and Sistrom, W. R., *J. Cell. Biol.*, **23**, 135 (1964).
17. Cusanovich, M. A., and Kamen, M. D., *Biochim. Biophys. Acta*, **153**, 376 (1967).
18. Harbury, H. A., *J. Biol. Chem.*, **225**, 1009 (1957).
19. Loach, P. A., *Biochemistry*, **5**, 592 (1966).
20. Cusanovich, M. A. and, Kamen, M. D., *Biochim. Biophys. Acta*, **153**, 418 (1967).

Cytochrome Reactions in *Porphyridium* and *Porphyra* —Reaction Times and Structural Rigidity

MITSUO NISHIMURA

Johnson Research Foundation, University of Pennsylvania, Philadelphia, Pennsylvania

INTRODUCTION

Duysens[1] first reported the light-induced oxidation of *f*-type cytochrome in a red alga, *Porphyridium*. The interaction of the two photochemical systems in the oxidation-reduction reactions of the cytochromes of *f*- and *b*-types was clearly demonstrated in *Porphyridium* and *Porphyra*.[2–4] We have studied the light-induced absorbance changes and delayed light emission in several species of *Porphyridium* and *Porphyra* (*Porphyridium cruentum*, *Porphyra umbilicalis*, *Porphyra yezoensis*, *Porphyra tenera* and *Porphyra suborbiculata*). The results of our analyses of the energy- and electron-transfer systems will be published elsewhere.[5] Here are presented the data on the reaction times of oxidation-reduction of cytochromes and the cytochrome reactions in glutaraldehyde-fixed cells.

RESULTS AND DISCUSSION

Reaction times of oxidation-reduction of cytochromes in Porphyridium and Porphyra

Fig. 1 shows the laser- or continuous illumination-activated absorbance changes in *Porphyra umbilicalis* and *Porphyridium cruentum*. Almost identical spectra were obtained in all 5 species studied. The light-minus-dark difference spectra were caused mainly by the cytochrome *f* (cytochrome-553) oxidation and smaller change of cytochrome *b*.

Table I shows the steady-state changes and "on" and "off" half times of the laser-induced absorbance changes in *Porphyra umbilicalis* and *Porphyridium cruentum*. The half-rise time for the rapid cytochrome oxidation was about 15 μsec in these two species. The half-reduction time was longer in *Porphyra* than in *Porphyridium*.

Examples of responses of cytochromes in the presence of *m*-Cl-CCP

Abbreviation: *m*-Cl-CCP, carbonyl cyanide *m*-chlorophenylhydrazone.

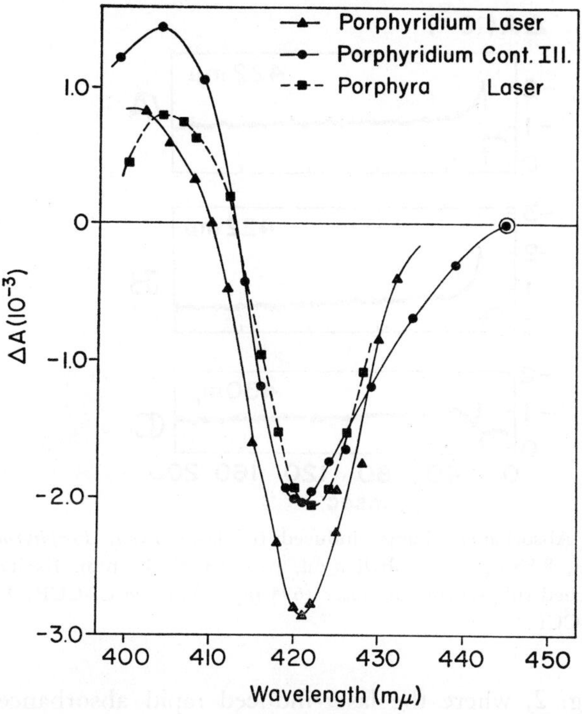

Fig. 1. Cytochrome oxidation by continuous illumination and laser flash. *Porphyridium cruentum:* continuous illumination, 690 mμ, 2.60 μg chlorophyll *a*/ml, path length 10 mm; laser activation, 694 mμ, 19.87 μg chlorophyll *a*/ml, path length 1.6 mm. *Porphyra umbilicalis:* laser activation, 694 mμ, thallus one layer.

TABLE I. Laser-induced cytochrome oxidation and reduction in *Porphyra* and *Porphyridium.*

		Position of maximum absorbance change (mμ)	ΔA (10^{-3})	$\tau_{1/2}$ on (μsec)	$\tau_{1/2}$ off (msec)
Porphyra umbilicalis		422	−1.04	15.7	86
Porphyridium cruentum	Untreated	421	−2.19	15.1	11.9
	5.0% glutaraldehyde fixed	421	−1.50	14.6	120

Porphyra umbilicalis, one layer; *Porphyridium cruentum,* 4.40 μg chlorophyll *a*/ml, 3.2 mm path length. Excitation light, Q-switched ruby-chromium laser (694 mμ).

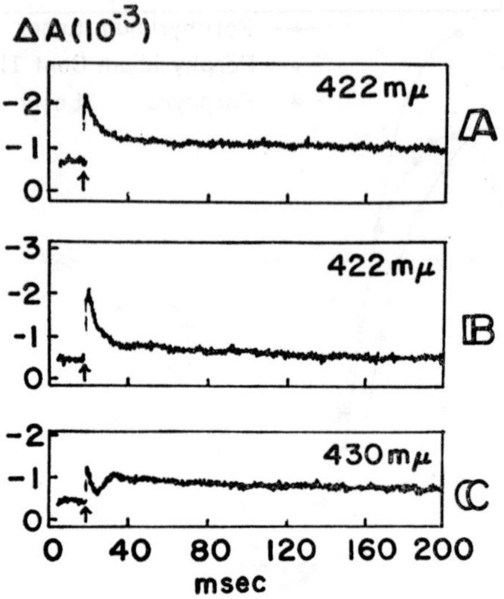

Fig. 2. Absorbance change induced by laser flash. *Porphyridium cruentum* cells, 8.88 μg chlorophyll *a*/ml, path length 1.6 mm. Excitation light, Q-switched ruby-chromium laser (694 mμ). A, no *m*-Cl-CCP; B, C, 20 μ*M* *m*-Cl-CCP.

are shown in Fig. 2, where the laser-induced rapid absorbance changes were recorded at 422 mμ and 430 mμ. The top trace represents the absorbance change at 422 mμ in the absence of *m*-Cl-CCP, the middle and the bottom traces show the changes in the presence of *m*-Cl-CCP at 422 mμ and 430 mμ, respectively. Clear rapid and slow oxidation-reduction cycles were observed. The rapid cycle had a maximum at 422 mμ, and the slow cycle had a maximum at 430 mμ. The maximum oxidation was attained in less than 50 μsec with a half-rise time of 14 μsec for the rapid phase. The slow cycle reached its maximum oxidation 20 msec after flash. This is interpreted as the very rapid oxidation of cytochrome *f* and its slower reduction and the oxidation of a *b*-type cytochrome. The reduction of cytochrome *f* and the oxidation of *b*-type cytochrome had half-change times of 9.7 msec and 7.0 msec. This indicates the transfer of electrons between these two cytochrome species in the millisecond range.

Light-induced electron transfer in glutaraldehyde-fixed cells

Glutaraldehyde has been widely used for fixing the microstructure of cells. This reagent was used in our study to see the effect of structural fixation or immobilization of molecules on the rate of electron transfer.

The degree of glutaraldehyde fixation was measured by the extractability of phycoerythrin from the treated cells. As seen in Fig. 3, the amount of phycoerythrin extracted decreased sharply on increasing the glutaraldehyde concentration. After thorough washing the light-induced cytochrome oxidation and reduction were measured with a double-beam spectrophotometer (continuous illumination) or with a rapid single-beam spectrophotometer (laser flash for activation of cells). Essentially no effect was observed on the "light-on" rate and the steady-state change of cytochrome oxidation in the cells treated with glutaraldehyde at concentrations of less

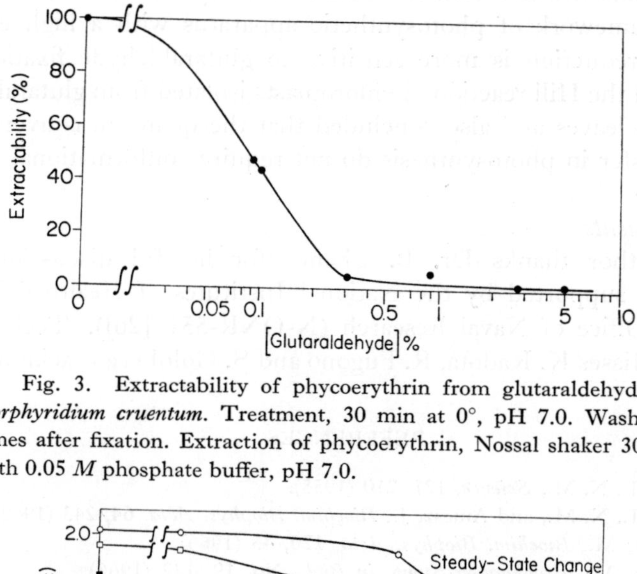

Fig. 3. Extractability of phycoerythrin from glutaraldehyde-treated *Porphyridium cruentum*. Treatment, 30 min at 0°, pH 7.0. Washed three times after fixation. Extraction of phycoerythrin, Nossal shaker 30 sec × 8, with 0.05 *M* phosphate buffer, pH 7.0.

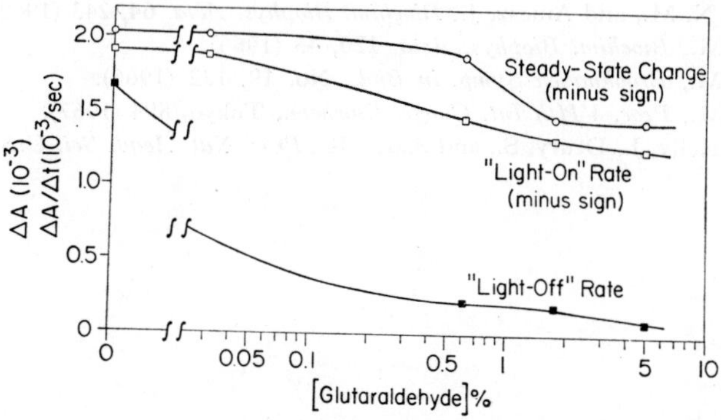

Fig. 4. Steady-state change and "on" and "off" rates of oxidation-reduction reactions of cytochrome *f* in glutaraldehyde-fixed *Porphyridium cruentum*. Glutaraldehyde treatment, 30 min at 0°, pH 7.0. Washed three times after fixation. 20 sec illumination—60 sec dark. Excitation wavelength, 690 mμ. Measuring wavelengths, 422 mμ–445 mμ.

than 1% (Fig. 4). In the same concentration range, the degree of phycoerythrin extraction decreased remarkably (or the degree of fixation was high). On the other hand, the "light-off" reduction of cytochrome *f* became very slow in the glutaraldehyde-treated cells.

The laser-activated rapid spectroscopic traces showed that the rapid cytochrome *f* oxidation was not affected by the glutaraldehyde treatment. The half-time of cytochrome oxidation, about 15 μsec, did not change by fixation (Table I). The "light-off" half-time of cytochrome reduction was increased ten-fold by the glutaraldehyde fixation.

These data indicate that the oxidation of cytochrome can take place in a rigid framework of photosynthetic apparatus with a high efficiency but that its reduction is more sensitive to glutaraldehyde fixation. Park *et al.*[6] studied the Hill reaction of chloroplasts isolated from glutaraldehyde-fixed spinach leaves and also concluded that the quantum conversion and electron transfer in photosynthesis do not require conformational changes.

Acknowledgements

The author thanks Dr. B. Chance for helpful discussions. This research was supported by the National Institutes of Health (PHS GM 12202) and Office of Naval Research (N-ONR-551 [26]). Technical assistance by Misses K. Kadota, R. Fugono and S. Goldberg is acknowledged.

REFERENCES

1. Duysens, L. N. M., *Science*, **121**, 210 (1955).
2. Duysens, L. N. M., and Amesz, J., *Biochim. Biophys. Acta*, **64**, 243 (1962).
3. Nishimura, M., *Biochim. Biophys. Acta*, **120**, 45 (1966).
4. Nishimura, M., *Brookhaven Symp. in Biol.*, No. 19, 132 (1966).
5. Nishimura, M., *Proc. VIIth Int. Congr. Biochem.*, Tokyo, 894 (1967).
6. Park, R. B., Kelly, J., Drury, S., and Sauer, K., *Proc. Nat. Acad. Sci. U.S.*, **55**, 1056 (1966).

Ferredoxin-dependent Photoreduction of Nitrate and Nitrite by Subcellular Preparations of *Anabaena Cylindrica**

AKIHIKO HATTORI† AND IZUMI UESUGI

Institute of Applied Microbiology, University of Tokyo, Tokyo

It has long been known that light accelerates the assimilation or reduction of nitrate in green plants. The enzymes, nitrate and nitrite reductases, were purified from various higher plants and their properties were extensively investigated. Although the pioneer work of Warburg on photoreduction of nitrate was carried out with *Chlorella*, only a few investigations have been done with algae on this subject at the enzyme level. The present work was undertaken with the purpose of elucidating the biochemical mechanism of photoreduction of nitrate and nitrite in the blue-green alga *Anabaena cylindrica* with special reference to the role of ferredoxin in the reactions.

About two week-old cells of *Anabaena cylindrica* Lemm (Fogg strain, IAM-M 1), which had been grown under continuous illumination in Allen and Arnon's medium[1] with nitrate as the sole source of nitrogen, were used as starting material. The algal cells harvested were suspended in 0.1 *M* phosphate buffer or 0.05 *M* Tris-HCl buffer at pH 7.4 and sonicated at 10 kc and 200 W for 5 minutes. The particulate fraction was collected by centrifugation at 20,000$\times g$ for 30 minutes and washed 4 to 5 time with the same buffers. The particulate fraction thus obtained showed a high activity for reduction of nitrate to nitrite (90—200 mμmoles NO_2^- formed/mg protein/hr) when reduced MV was given as an electron donor, but practically no activity for reduction of nitrite to ammonia.[2] This preparation was green in color and retained the activity for photochemical system I. In the presence of DPIP-ascorbate couple and a catalytic amount of MV, its photooxidation activity was of the order of 500 μmoles O_2/mg chlorophyll

* This work was supported by grants from the Ministry of Education (4090 and 955008).

† Present address: Ocean Research Institute, University of Tokyo, Nakano-ku, Tokyo.

Abbreviations: DPIP, 2,6-dichlorophenol indophenol; FAD, flavin adenine dinucleotide; FMN, flavin mononucleotide; MV, methyl viologen; NAD, nicotinamide adenine dinucleotide; NADH, reduced form of NAD; NADP, nicotinamide adenine dinucleotide phosphate; NADPH, reduced form of NADP.

a/hr. The nitrite reducing activity was recovered in the supernatant fraction obtained by centrifugation. A crude enzyme preparation could be obtained by passing the supernatant through a Sephadex G-75 column.[3] Further purification of the nitrite reductase was conducted by combined use of acetone precipitation and DEAE cellulose column chromatography.[4]

The activities for nitrate and nitrite reduction were estimated from the formation and consumption of nitrite, respectively. The methods used were essentially the same as those described in our previous papers.[2,3] The experiments were performed under N_2-atmosphere at 25°.

The ferredoxin requirement in photoreduction of nitrate by the washed particulate preparation is shown in Table I. At a high light intensity in the presence of DPIP-ascorbate couple ferredoxin supported the nitrate reduction to the same extent as did reduced MV at its optimum concentration in the dark. Reduction of nitrate was negligible when one of the components, ferredoxin, DPIP or ascorbate, was omitted or when the reaction was run in the dark. Ferredoxin could be successfully replaced

TABLE I. Ferredoxin-dependent photoreduction of nitrate by *Anabaena* particulate preparation.

System	Nitrite formed (mμmoles/10 min)
Complete*	129
Minus ferredoxin	2
Minus DPIP	5
Minus ascorbate	4
Complete with $Na_2S_2O_4$	7
Complete in dark	5
Complete with $Na_2S_2O_4$ in dark	7
Minus ferredoxin plus MV and $Na_2S_2O_4$ in dark	106
Complete†	76
Minus MV	0
Minus MV, plus FMN	4
Complete in dark	0
Complete with $Na_2S_2O_4$ in dark	127

* Complete system contained in 2.5 ml: 100 μmoles Tris-HCl (pH 7.4), 10 μmoles KNO_3, 1.2 mg ferredoxin, 2 μmole DPIP, 40 μmoles sodium ascorbate, and particulate preparation (2.7 mg protein, 53 μg chlorophyll *a*), and where indicated, 15 μmoles $Na_2S_2O_4$ and 3 μmoles MV.

† Ferredoxin was replaced with 3 μmoles MV. Particulate preparation (2.5 mg protein, 50 μg chlorophyll *a*). Where indicated, 1 μmole FMN was added.

The reaction was carried out for 10 minutes at 25° in the light (20,000 lux, white light), or in darkness, under N_2.

TABLE II. Effect of ferredoxin on NADPH-supported dark reduction of nitrate by *Anabaena* particulate preparation.

System	Nitrite formed (mμmoles/10 min)
NADPH	3
NADPH plus ferredoxin (3×10^{-6} *M*)	38
" (1×10^{-5} *M*)	56
" (3×10^{-5} *M*)	80
NADPH plus MV (2×10^{-4} *M*)	34
NADPH plus FMN (5×10^{-4} *M*)	3
Ferredoxin (3×10^{-5} *M*) with DPIP-ascorbate in light	79

Reaction mixture contained in 2.0 ml: 80 μmoles Tris-HCl (pH 7.4), 10 μmoles KNO_3, 3 μmoles NADPH, *Anabaena* particulate preparation (2.5 mg protein), and indicated amounts of ferredoxin, MV or FMN. Where indicated, 2 μmole DPIP and 40 μmoles sodium ascorbate were added. 25°, N_2-atmosphere. When necessary, illuminated with white light (20,000 lux).

with MV, but not with FMN which has been shown to be an active electron mediator in the spinach system.[5]

Ferredoxin which had been chemically reduced with dithionite was inactive either in the dark or in the light (Table I). NADPH alone was also inactive as an electron donor. However, if ferredoxin was added together with NADPH in darkness, nitrate reduction took place, depending upon its concentrations (Table II). NADP reductase was present in the particulate preparation. These results suggest a participation, between ferredoxin and nitrate reductase, of some cofactor which is very sensitive to dithionite. Inactivation of ferredoxin by dithionite was not the case since dithionite-reduced ferredoxin could support nitrite reduction under similar conditions (see below).

Ferredoxin-dependent nitrate reduction by spinach preparations was recently reported by Losada *et al.*[6] Ferredoxin which was reduced by illuminated grana or by *Clostridium* hydrogenase could not by itself serve as an electron donor in their system, however. Only in the presence of NAD could ferredoxin mediate the transfer of electrons to the nitrate reductase system. This is not the case with our *Anabaena* preparations. Our preparation does not require NAD in reducing nitrate in the light in the presence of ferredoxin, nor can it catalyze the reduction with NADH in the dark (Table III).

It has been shown in our previous study[2] that the particulate preparation obtained from acetone dry cells of *Anabaena* catalyzes the nitrate reduction either with NADPH or with NADH in darkness. Ferredoxin was required in the former case but not in the latter. The failure of the present preparation to reduce nitrate with NADH indicates a participation of another cofactor, presumably NAD-diaphorase, between NADH and nitrate reductase, which might be removed, or inactivated, by sonic treatment adopted to obtain this preparation. We conclude that in the *Anabaena* system nitrate reductase can accept electrons either from reduced ferredoxin or NADH through some unidentified cofactors. It is unlikely that the free form of FMN (and FAD) functions as an electron mediator between ferredoxin and nitrate reductase. We can also exclude the possibility that NADP reductase works as the cofactor in question, since the presence of

TABLE III. Effect of NAD on ferredoxin-dependent photoreduction of nitrate by *Anabaena* particulate preparation.

Addenda	Nitrite formed (mμmoles/10 min)
None	1
Ferredoxin	52
NAD	1
Ferredoxin plus NAD	49
NADH in dark	0

Reaction mixture contained in 2.5 ml : 100 μmoles Tris-HCl (pH 7.4), 10μmoles KNO_3, 2 μmole DPIP, 40 μmoles sodium ascorbate, particulate preparation (3.1 mg protein, 130 μg chlorophyll *a*), and where indicated, 1.2 mg ferredoxin, 1 μmole NAD and 3 μmoles NADH. 25°, 20,000 lux (or dark), N_2-atmosphere.

TABLE IV. Ferredoxin-dependent photoreduction of nitrite by *Anabaena* reconstructed system.

System	Nitrite consumed (mμmoles/10 min)
Complete	65
Minus ferredoxin	2
Minus DPIP	0
Minus nitrite reductase	0
Nitrite reductase heated at 70° for 5 min.	0
Complete in dark	0
Minus particulate preparation, plus $Na_2S_2O_4$ in dark	68
Minus particulate preparation and ferredoxin, plus MV and $Na_2S_2O_4$ in dark	158

Complete system contained in 2.5 ml : 100 μmoles Tris-HCl (pH 7.4), 0.4 μmole $NaNO_2$, 0.4 mg ferredoxin, 2 μmoles DPIP, 40 μmoles sodium ascorbate, 0.6 mg (chlorophyll *a*) particulate preparation, 1 mg nitrite reductase. Where indicated, 15 μmoles $Na_2S_2O_4$ and 3 μmoles MV were added. 25°, 20,000 lux (or dark), N_2-atmosphere.

ferredoxin (or MV) was indispensable for the reduction of nitrate with NADPH. Recently, Naik and Nicholas reported that a non-heme iron protein from *Azotobacter* could act as an electron mediator between NADH and *Micrococcus denitrificans* nitrate reductase.[7]

In Table IV are presented data showing the photoreduction of nitrite by a reconstructed system of *A. cylindrica*. In this experiment, purified nitrite reductase, together with other components, was added to the particulate preparation used in the preceding experiment. At a concentration of $1\times10^{-5}\,M$, chemically reduced ferredoxin supported nitrite reduction to the same extent as did photochemically reduced ferredoxin. Experimental evidence showing the participation of ferredoxin in the electron transport chain between NADPH and nitrite reductase has already been presented.[3] It appears that reduced ferredoxin can donate electrons directly to nitrite reductase. Similar results have been reported for the nitrite reducing systems of higher plants.[8–10]

The above results indicate that, in *Anabaena* cells, whole steps for reduction of nitrate to ammonia can be driven by continuous supply of reduced ferredoxin in the light without participation of pyridine nucleotides.

REFERENCES

1. Allen, M. B., and Arnon, D. I., *Plant Physiol.*, **30**, 366 (1955).
2. Hattori, A., and Myers, J., *Plant and Cell Physiol.*, **8**, 327 (1967).
3. Hattori, A., and Myers, J., *Plant Physiol.*, **41**, 1031 (1966).
4. Uesugi, I., and Hattori, A., unpublished data.
5. Paneque, A., Del Campo, F. F., Ramirez, J. M., and Losada, M., *Biochim. Biophys. Acta*, **109**, 79 (1965).
6. Losada, M., and Paneque, A., *Biochim. Biophys. Acta*, **126**, 578 (1966).
7. Naik, M. S., and Nicholas, D. J. D., *Biochim. Biophys. Acta*, **131**, 207 (1967).
8. Huzishige, H., Satoh, K., Tanaka, K., and Hayashida, T., *Plant and Cell Physiol.*, **4**, 307 (1963).
9. Joy, K. W., and Hageman, R. H., *Biochem. J.*, **100**, 263 (1966).
10. Ramirez, J. M., Del Campo, F. F., Paneque, A., and Losada, M., *Biochim. Biophys. Acta*, **118**, 58 (1966).

Cytochromes of Developing and Mutant Chloroplasts

N.K. BOARDMAN

C.S.I.R.O., Division of Plant Industry, Canberra, Australia

INTRODUCTION

Mature chloroplasts contain at least three cytochromes: cytochrome *f*(cyt *f*) and cytochrome b_6 (cyt b_6), with α-bands at 554 mμ and 563 mμ respectively,[1,2] and a cytochrome with an α-band at about 559 mμ at 20° and at 557 mμ at 77°K.[3,4] The latter cytochrome was named cytochrome 559 (chloroplasts)[4] to avoid confusion with the soluble cytochrome b_3 of Hill and Scarisbrick.[5] Cyt *f* and cyt 559 are in the reduced state in the isolated chloroplast, and cyt b_6 is oxidized. In the mature spinach chloroplast, the cytochromes are present in the approximate ratio of 1 mole of cyt *f* for 2 moles of cyt b_6 and 2 moles of cyt 559.[4] From cytochrome determinations on the fractions prepared by digitonin fragmentation of spinach chloroplasts, it was concluded that cyt *f* and cyt b_6 are associated with Photosystem 1, and cyt 559 with Photosystem 2.[4]

In the present study, the cytochromes of the developing plastid of the greening bean leaf were investigated. A comparison was made also between the cytochrome composition of the normal barley chloroplast, and a mutant chloroplast lacking chlorophyll *b*.

MATERIALS AND METHODS

Isolation of plastids

Bean plants (*Phaseolus vulgaris*) were grown in the dark for 10–12 days at 25°C, and subsequently illuminated with white light (400 f.c.). Plastids were isolated as described previously.[6] Normal bean chloroplasts were obtained from plants grown in a glasshouse. Barley chloroplasts were isolated by hand-grinding, both from normal plants and from plants which lacked chlorophyll *b*.[7]

Cytochrome determinations

The cytochrome contents of suspensions of developing or mature chloroplasts were determined without prior extraction of the chlorophyll. Reduced minus oxidized difference spectra were recorded at both 20°C and 77°K with a Cary Model 14R spectrophotometer, fitted with a scatter-transmission attachment. Full details are described elsewhere.[4] Chlorophyll or protochlorophyll was determined on acetone extracts, and molar ratios of chlorophyll/cytochrome calculated.[4]

RESULTS

Reduced minus oxidized difference spectra of mature bean chloroplasts at 20°C and 77°K are shown in Fig. 1 (g, h) and Fig. 2 (j, k, l) respectively. They resemble very closely the corresponding difference spectra of spinach chloroplasts, and the previous paper[4] should be consulted for a full explanation of these spectra. The ascorbate reduced minus ferricyanide oxidized (A-F) spectra show the cytochromes which are in the reduced state in the chloroplast, the dithionite reduced minus ferricyanide oxidized (D-F) spectra show all the cytochromes, and the dithionite reduced minus ascorbate reduced (D-A) spectra indicate the cytochromes which are oxidized. It is concluded that the mature bean chloroplast, like the spinach chloroplast contains at least 3 cytochromes: cyt *f*, cyt b_6, and cyt 559. The (D-A) spectra at 20°C are not shown in this paper, but they showed a single band at 562–563 mμ, as seen with spinach chloroplasts, and attributed to cyt b_6. The double peak in the (D-A) spectra at 77°K is thought tentatively to be due to the splitting of the α-band of cyt b_6,[4] although one cannot exclude the possibility that chloroplasts contain a second autoxidizable cytochrome in addition to cyt b_6. Attempts to purify cyt b_6 have not been successful so far.

A comparison of the difference spectra of normal bean chloroplasts with those of the proplastids from dark-grown plants (Figs. 1 and 2) shows that cyt 559 is not detectable in the proplastids. The (A-F) spectra of proplastids, both at 20°C and 77°K, are identical with those of purified cyt *f*.[8] The (D-F) spectrum of proplastids at 20°C shows peaks at 554 mμ and 562 mμ, suggesting the presence of cyt *f* and cyt b_6 only. At 77°K, the (D-F) spectrum shows a greatly reduced band at 557 mμ compared with the normal chloroplasts, in accord with the presence of cyt b_6, but the absence of cyt 559. The spectra of the proplastids resemble those obtained previously with the Photosystem 1 particle derived from spinach chloroplasts.[4]

If the difference spectra in Figs. 1 and 2 are scanned in a vertical direction, the gradual appearance of cyt 559 during greening is readily apparent. At room temperature (Fig. 1) the position of the maximum in the (A-F) spectrum moves to longer wavelengths, and the distinct peaks in the (D-F) spectrum of proplastids gradually merge into a single peak

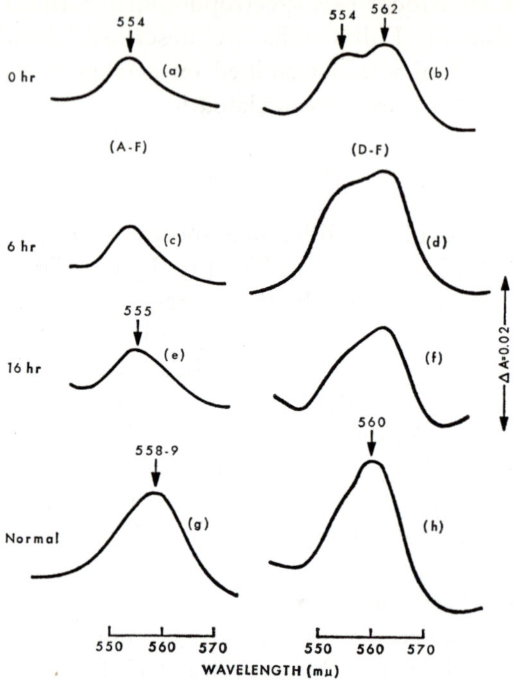

Fig. 1. Reduced minus oxidized difference spectra of developing and normal bean chloroplasts at 20°C. (a), (c) (e), and (g): Ascorbate reduced minus ferricyanide oxidized (A-F). (b), (d), (f) and (h): Dithionite reduced minus ferricyanide oxidized (D-F). Times of illumination of the bean plants are indicated on the left. ΔA=absorbancy change.

at 560 mμ with the mature chloroplast. The (A-F) and (D-F) spectra at 77°K show the gradual increase in the absorption at 557 mμ, due to the formation of cyt 559.

Chlorophyll-cytochrome ratios

The contents of cyt *f* and total cyt *b* in the proplastids and the 6-hr plastids were calculated from the room temperature difference spectra. Cyt *f* was estimated from the peak to trough height in the (A-F) spectrum, using a difference molar extinction coefficient ($\varepsilon_{M,\ 554\ m\mu} - \varepsilon_{M,\ 575\ m\mu}$) of 2.5×10^4. Total cyt *b* was obtained from the 562 mμ peak to trough height in the (D-F) spectrum, after correcting for the absorption of cyt *f*.[4] A difference molar extinction coefficient of 2.0×10^4 was used for cyt *b*. In the case of the proplastids the amount of cyt *b* calculated from the (D-F)

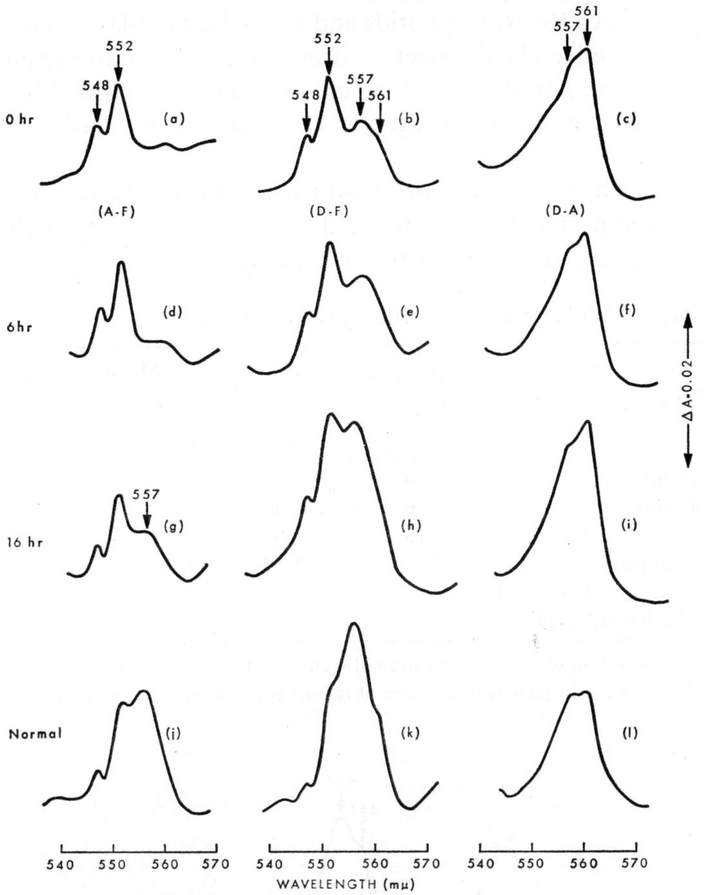

Fig. 2. Reduced minus oxidized difference spectra of developing and normal bean chloroplasts at 77°K. (a), (d), (g) and (j): (A-F); (b), (e), (h) and (k): (D-F); (c), (f), (i) and (l): Dithionite reduced minus ascorbate reduced (D-A). Times of illumination are indicated on the left.

spectrum was the same as the amount of cyt *b* calculated from the (D-A) spectrum. This confirms the conclusion that proplastids contain cyt b_6 and cyt *f*. Similar procedures were used to estimate the amounts of cyt *f*, and the total cyt *b* in the 6-hr plastids. The amount of cyt 559 was small enough not to contribute significantly to the absorption at 554 mμ in the (A-F) spectrum. Molar ratios of chl/cyt *f* and cyt *f*/cyt *b* could not be determined with the 16-hr plastids and the normal chloroplasts because of the presence of cyt 559. The figures in parentheses in Table I were obtained previously, using acetone powders of spinach chloroplasts. The chl/

cyt b ratios shown for 16-hr plastids and normal chloroplasts were calculated from the respective (D-F) spectra, after making small corrections for the absorption of cyt f at 560 mμ.[4] The chl/cyt b ratio for normal bean chloroplasts agree very well with the ratio of 118 obtained previously with spinach chloroplasts.

Table I shows that the proplastid has very low molar ratios of protochl/cyt f and protochl/cyt b compared with the corresponding chlorophyll/cytochrome ratios of the normal chloroplast.

TABLE I. Molar ratios of chlorophyll/cytochrome

Preparation	Time of greening (hr)	Molar ratio			
		chl a / chl b	chl / cyt f	chl / cyt b	cyt b / cyt f
Bean plastids	0	—	2.8	1.3	2.2
Bean plastids	6	3.20	76	29	2.6
Bean plastids	16	2.95	—	96	—
Bean chloroplasts		2.95	(430)	125	(3.6)
Normal barley chloroplasts		2.90	—	122	—
Mutant barley chloroplasts		—	—	64	—

See text for method of determination; chl=chl a+chl b or protochl. cyt b=cyt b_6+cyt 559. The figures in parentheses were obtained previously with normal spinach chloroplasts.[4]

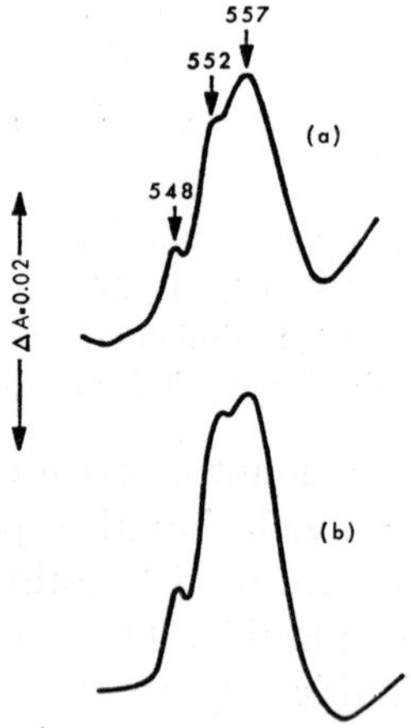

Fig. 3. Dithionite reduced minus ferricyanide oxidized difference spectra of (a) normal and (b) mutant barley chloroplasts at 77°K.

The relative proportion of cyt b_6 to cyt f in the proplastids agrees with our previous estimate of the cyt b_6/cyt f ratio in mature spinach chloroplasts.[4] The increase in the cyt b/cyt f ratio during greening is due to the formation of cyt 559.

Measurements of chlorophyll and cytochrome contents of greening bean leaves per mg dry weight, and per leaf are needed to establish whether there is any net synthesis of cyt b_6 and cyt f during greening. From a recent study, it was concluded that the number of 70S ribosomes per plastid increases only slightly during chloroplast development.[9]

A comparison of the (D-F) spectra of mature chloroplasts from normal and mutant barley plants shows that cyt 559 is present in the mutant chloroplasts (Fig. 3). Table I indicates, however, that the chl/cyt b molar ratio for mutant chloroplasts was only about one-half that of the normal chloroplasts.

DISCUSSION

The proplastid of the dark-grown bean leaf is well developed in size, approaching one-third to one-half the size of the mature bean chloroplast[10]. The grana and inter-grana lamellae normally observed in the higher plant chloroplast by electron microscopy are absent from the proplastid. The proplastids may contain one or more dense vesicular centres which have been termed prolamellar bodies.[11] Fluorescence microscopy indicates that the protochlorophyll of the proplastid is localized in discrete centres about 1 μ in diameter, which correspond in size to the prolamellar bodies.[10] The location of cytochromes f and b_6 in the proplastid is not known; perhaps they are also localized in the 1 μ fluorescing centres.

The fluorescing centres persist for several hours in the plastids during greening.[12] The majority of plastids isolated from plants illuminated for 6 hours showed centres, and it was observed that at least 12 hours of illumination were necessary for the disappearance of all the 1 μ centres.

These microscopic observations suggest one explanation for the present finding that the relative content of cyt 559, compared with cytochromes f and b_6, is low in the 6-hour plastid. Perhaps only some of the cyt f and cyt b_6 of the proplastid is incorporated into functional photosynthetic units during the first 6–8 hours of illumination, while a greater fraction of these two cytochromes remains associated with the proplastid-like fluorescent centres.

An alternative explanation is that most of the cyt f and cyt b_6 is incorporated into functional Photosystem 1 units at an early stage of greening, but the rate of synthesis of Photosystem 2 units and of cyt 559 is slower than the rate of formation of Photosystem 1 units. The observation by Bonner and Hill[13] of light-induced cyt *f* oxidation in etiolated mung bean leaves which had been greened for relatively short periods provides some support for the latter hypothesis.

A higher rate of formation of Photosystem 1, however, should be reflected in the chl *a*/chl *b* ratio, since Photosystem 1 has a higher chl *a*/chl *b* ratio (5.3) than Photosystem 2 (estimated as 1.7).[14] The chl *a*/chl *b* ratio of the 6-hr plastids was only slightly larger than that of normal chloroplasts (Table I) suggesting that the formation of Photosystem 2 does not lag very far behind the formation of Photosystem 1. The first explanation is therefore preferred.

In the barley mutant chloroplast, cyt 559 appears to be present to at least an equal extent as in the normal chloroplast. The lower chl/cyt *b* ratio observed with the mutant chloroplasts is apparently due to a decreased amount of chlorophyll in the light harvesting centres.[7,15] Hill-reaction measurements showed that the mutant chloroplasts were saturated at higher light intensities than the normal ones. The mutant chloroplasts were more active per mg of total chlorophyll at saturating intensities, but considerably less active at low intensities.[7] The quantum yields of fluorescence of the mutant chloroplasts were lower than those of the normal chloroplasts.[15] The mutant chloroplast apparently has a decreased amount of chlorophyll in Photosystem 2, compared with Photosystem 1, which results in a greater fraction of incident light being absorbed by Photosystem 1. A decreased amount of chlorophyll in the light-harvesting assemblies of Photosystem 2, but a normal complement of the components of the electron transport chain would account for the increased activity per mg. of total chlorophyll, which has been observed for the mutant chloroplasts at saturating intensities.

Acknowledgement

The technical assistance of Mrs. Sophie Sapiets is gratefully acknowledged.

SUMMARY

Cytochrome *f* and cytochrome b_6, but not cytochrome 559 are present in the plastids of the dark-grown bean plant. Cytochrome 559 is formed gradually during greening. Even after 16 hours of illumination, the proportion of cytochrome 559 relative to cytochromes *f* and b_6 is still below that observed for normal bean chloroplasts. These findings are discussed in relation to the structure of the developing bean chloroplast.

Cytochrome 559 is fully present in mutant barley chloroplasts lacking chlorophyll *b*, even though these chloroplasts apparently have a decreased amount of chlorophyll in Photosystem 2.

REFERENCES

1. Davenport, H. E., and Hill, R., *Proc. Roy. Soc.*, **B139**, 327 (1952).
2. Hill, R., *Nature*, **174**, 501 (1954).
3. Lundegardh, H., *Physiol. Plantarum*, **15**, 390 (1962).
4. Boardman, N. K., and Anderson, J. M., *Biochim. Biophys. Acta*, **143**, 187 (1967).
5. Hill, R., and Scarisbrick, R., *New Phytologist*, **50**, 98 (1951).
6. Anderson, J. M., and Boardman, N. K., *Australian J. Biol. Sci.*, **17**, 93 (1964).
7. Boardman, N. K., and Highkin, H. R., *Biochim. Biophys. Acta*, **126**, 189 (1966).
8. Hill, R., and Bonner, W. D., in *Light and Life* (W. D. McElroy and B. Glass, eds.), pp. 424–435, Johns Hopkins Press, Baltimore, 1961.
9. Boardman, N. K., *Exptl. Cell Res.*, **43**, 474 (1966).
10. Boardman, N. K., and Wildman, S. G., *Biochim. Biophys. Acta*, **59**, 222 (1962).
11. Hodge, A. J., McLean, J. D., and Mercher, F. V., *J. Biophys. Biochem. Cytol.*, **2**, 597 (1956).
12. Boardman, N. K., and Anderson, J. M., *Australian J. Biol. Sci.*, **17**, 86 (1964).
13. Bonner, W., and Hill, R., in *Photosynthetic Mechanisms of Green Plants* (B. Kok and A.T. Jagendorf, eds.), pp. 82–90, Nat. Acad. Sci.-Nat. Res. Council, 1963.
14. Anderson, J. M., and Boardman, N. K., *Biochim. Biophys. Acta*, **112**, 403 (1966).
15. Boardman, N. K., and Thorne, S. W., *submitted for publication.*

The Relevance of Light-induced Proton Uptake to the Mechanism of Energy Coupling in Photophosphorylation

STEVEN J. D. KARLISH AND MORDHAY AVRON

Biochemistry Section, Weizmann Institute of Science, Rehovoth, Israel

The relevance of energy-dependent proton movements to energy conservation mechanisms in oxidative and photophosphorylation has recently been extensively debated—with respect particularly to the "chemiosmotic"[1,2] and chemical[3,4] hypotheses. Among the few critical tests of the chemiosmotic theory are: (a) Ratio of protons taken up to electrons transported; the value of this H^+/e^- ratio should equal the number of sites of energy conservation in the electron transport path. (b) The effect of phosphorylating conditions; during phosphorylation one should observe a fall in the steady-state extent of the proton gradient, due to the stoichiometric utilization of one or two protons[2] per ATP synthesized.

Reversible light-induced pH changes in spinach chloroplasts have been described previously in detail.[5,6] This communication reports experiments designed to test the chemiosmotic theory by the above criteria. As will be shown the results cannot be reconciled with the theory as presently conceived.

Figs. 1a and b show the initial rates of proton and oxygen uptake in a suspension of chloroplasts, using diquat (N,N′-ethylene-2,2′-dipyridilium dibromide) as an electron acceptor. The most relevant experimental value of H^+/e^- (with respect to the chemiosmotic theory) is the maximal one, since lower values may represent degrees of uncoupling. Fig. la demonstrates that the H^+/e^- varied strongly with pH and had a maximum value of 6 at pH 6.0. The consequent prediction of 6 coupling sites in the electron transport path to diquat is hard to accept in the light of current knowledge. It is also difficult to reconcile the observed variation of the H^+/e^- ratio with a mechanism whereby translocation and electron transport are obligatorily coupled. We have observed a similar variation with pH, and a maximum H^+/e^- value of 4 using ferricyanide as an electron acceptor. Values of 5 H^+/e^- and 5 $H^+/h\nu$ have previously been reported using chloranil[7] and pyocyanine,[8] respectively.

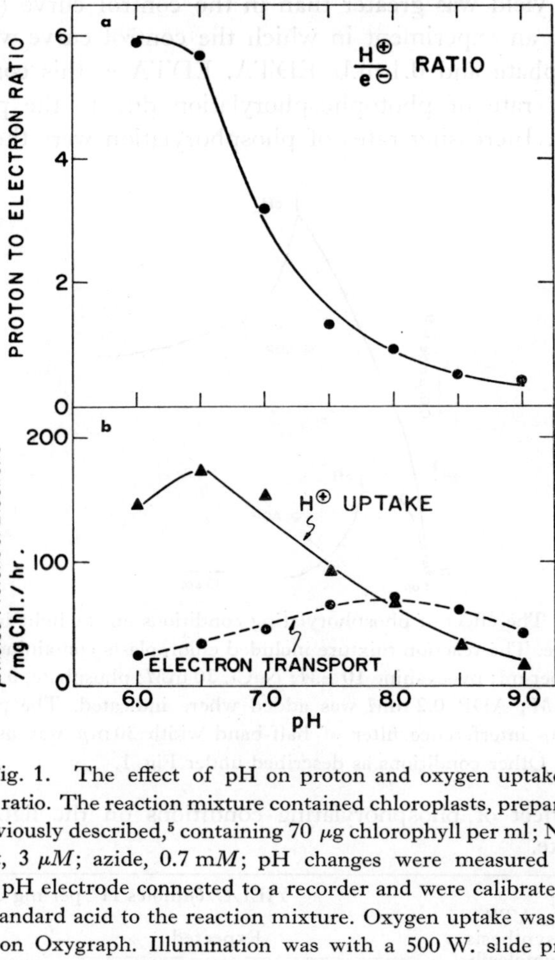

Fig. 1. The effect of pH on proton and oxygen uptake, and on the H^+/e^- ratio. The reaction mixture contained chloroplasts, prepared essentially as previously described,[5] containing 70 μg chlorophyll per ml; NaCl, 15 m*M*; diquat, 3 μ*M*; azide, 0.7 m*M*; pH changes were measured on a Radiometer pH electrode connected to a recorder and were calibrated by addition of a standard acid to the reaction mixture. Oxygen uptake was measured on a Gilson Oxygraph. Illumination was with a 500 W. slide projector fitted with a 580 mμ interference fitter, of 12 mμ half-band width. The intensity of illumination was about 1.0×10^5 ergs cm^{-2} sec^{-1}.

Fig. 2 shows the effect of the complete phosphorylating system on the pH changes. In the absence of ADP, the usual reversible light-induced proton uptake was observed, but in its presence two phases were seen in the light. The first represents the usual proton uptake and the second the irreversible proton consumption accompanying ATP synthesis at pH 8, $ADP^{3-} + HPO_4^{2-} + H^+ \longrightarrow ATP^{4-} + H_2O$.

On turning the light off the yield of reversible proton uptake which has accumulated during phosphorylation is recorded. Fig. 2 shows that, surprisingly, this yield was greater than in the control curve (-ADP). In Table 1 we show an experiment in which the control curve was attained with ADP, phosphate and 0.1 mM EDTA. EDTA at this concentration prevents the low rate of photophosphorylation due to the presence of endogeneous Mg. Increasing rates of phosphorylation were then initiated

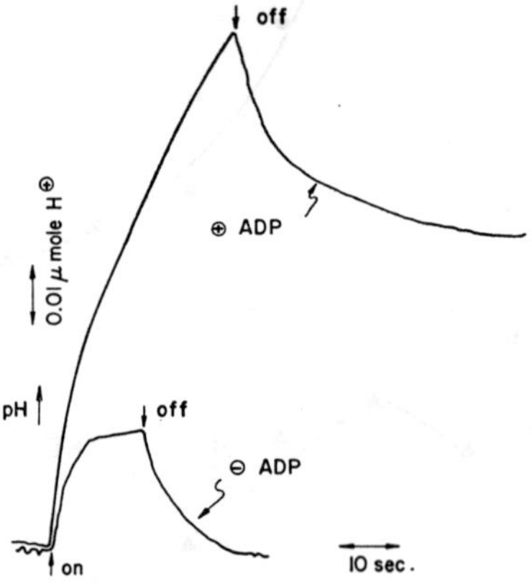

Fig. 2. The effect of phosphorylating conditions on the light-induced proton uptake. The reaction mixture included chloroplasts containing 50 μg chlorophyll per ml; pyocyanine 10 μM; NaCl, 30 mM; phosphate, 0.5 mM; $MgCl_2$ 0.5 mM; ADP, 0.2 mM was added where indicated. The pH was 7.8. A 700 mμ interference filter of half-band width 30 mμ was used for illumination. Other conditions as described under Fig. 1.

TABLE I. Effect of phosphorylating conditions on the light-induced proton uptake.

Mg^{++} concentration (mM)	ATP synthesis (μmoles/hr/ mg chl)	YIELD (μmoles H^+ per mg chl)		
		Expected		Observed
		Assuming $2H^+$/ATP	Assuming $1H^+$/ATP	
0	0	0	0	0
0.06	15	0.01	0.08	0.18
0.09	60	−0.01	0.05	0.27
0.16	120	−0.10	0.0	0.38
0.32	200	−0.34	−0.08	0.58
0.64	235	−0.39	−0.11	0.75

The system included chloroplasts containing 60 μg chlorophyll per ml; pyocyanine, 12 μM; EDTA, 0.1 mM; NaCl, 16 mM; ADP, 0.24mM; phosphate, 0.48 mM; and $MgCl_2$ at the concentrations indicated. The pH was 8.0. Other conditions as in Fig. 1, and illumination as in Fig. 2.

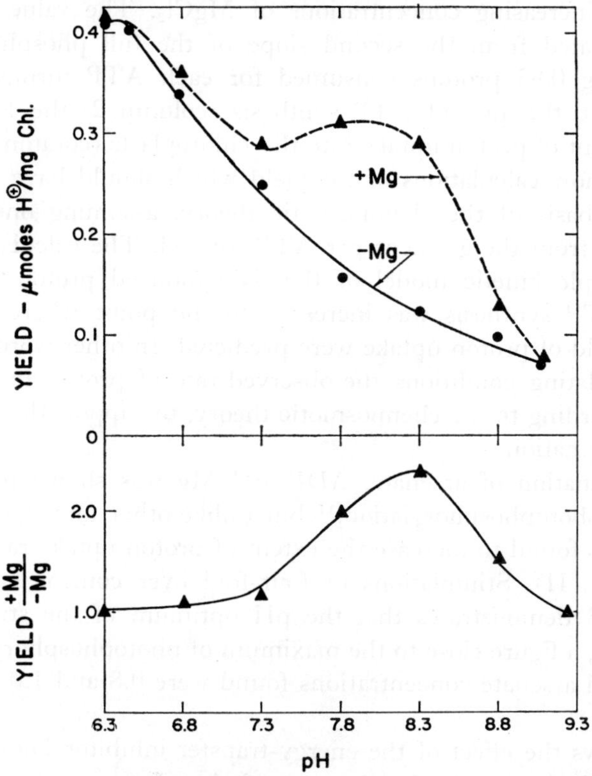

Fig. 3. The effect of pH on the arsenate, ADP and Mg stimulation of light-induced proton uptake. Conditions as in Table II except that arsenate was at 1.0 m*M* and illumination as in Fig. 2.

TABLE II. Effect of ADP, arsenate and Mg on light-induced proton uptake.

Reaction Mixture	YIELD (μmoles H^+/mg chl)
Control	0.12
plus Mg	0.12
plus Mg, Arsenate	0.13
plus ADP, Arsenate	0.11
plus Mg, ADP	0.14
plus Mg, ADP, Arsenate	0.32

The system included chloroplasts containing 50 μg of chlorophyll per ml; pyocyanine, 12 μM; EDTA, 0.1 m*M*; NaCl, 16 m*M*. ADP, 0.24 m*M*; arsenate, 0.25 m*M* and $MgCl_2$, 1.0 m*M* were included where indicated. The pH was 7.8. Other conditions as in Fig. 1, and illuminations as in Fig. 2.

by addition of increasing concentrations of $MgCl_2$. The value of these rates was calculated from the second slope of the full phosphorylation curves, assuming 0.93 protons consumed for each ATP formed at pH 7.8.[9] The higher the rate of ATP synthesis (column 2) the larger the steady-state extent of proton uptake into the chloroplasts (column 5). Columns 3 and 4 show calculations of the yield which should have been expected, on the basis of the chemiosmotic theory, assuming one or two protons utilized from the gradient per ATP formed. The calculations are based on a simple kinetic model of the light-induced proton uptake.[10] The rate of ATP synthesis was increased to the point where negative values of the yield of proton uptake were predicted. In other words, under such phosphorylating conditions the observed rate of proton uptake was insufficient, according to the chemiosmotic theory, to support the observed rate of ATP formation.

The combination of arsenate, ADP and Mg was shown previously to "uncouple" photophosphorylation,[11] but unlike other uncouplers,[6] this combination was found to increase the extent of proton uptake rather than inhibit it (Table II). Stimulations of four-fold over control have been observed. Fig. 3 demonstrates that the pH optimum of the stimulation was near pH 8.2, a figure close to the maximum of photophosphorylation.[12] Optimal Mg and arsenate concentrations found were 0.8 and 1.0 m*M*, respectively.

Fig. 4 shows the effect of the energy-transfer inhibitor Dio-9 on the light-induced pH changes, in the presence of phosphate or arsenate, ADP and Mg. Fig. 4a illustrates that the addition of Dio-9 to a complete phosphorylation system (+Mg), at a concentration which totally inhibited ATP synthesis (i.e. the irreversible proton uptake), did not eliminate the increase in the steady state proton uptake induced by the phosphorylation system. This striking observation is re-emphasized in Fig. 4b, where Dio-9 was included also in the control (-Mg). Subsequent addition of $MgCl_2$ still produced a large increase in the steady-state yield of proton uptake. It may be noted that under these conditions (pH 8) Dio-9 somewhat increases the usual (-Mg) proton uptake. In Figs. 4c and d, it can be seen that the same basic phenomena occurred when arsenate replaced phosphate. Thus Dio-9 at a concentration which completely inhibited ATP synthesis, did not markedly affect the stimulatory effect of arsenate, ADP, and Mg on the steady state proton uptake.

The finding that, in the presence of a complete phosphorylating system, the yield of light-induced proton uptake is higher than in its absence, is at odds both with the chemiosmotic *and* chemical theories. Both

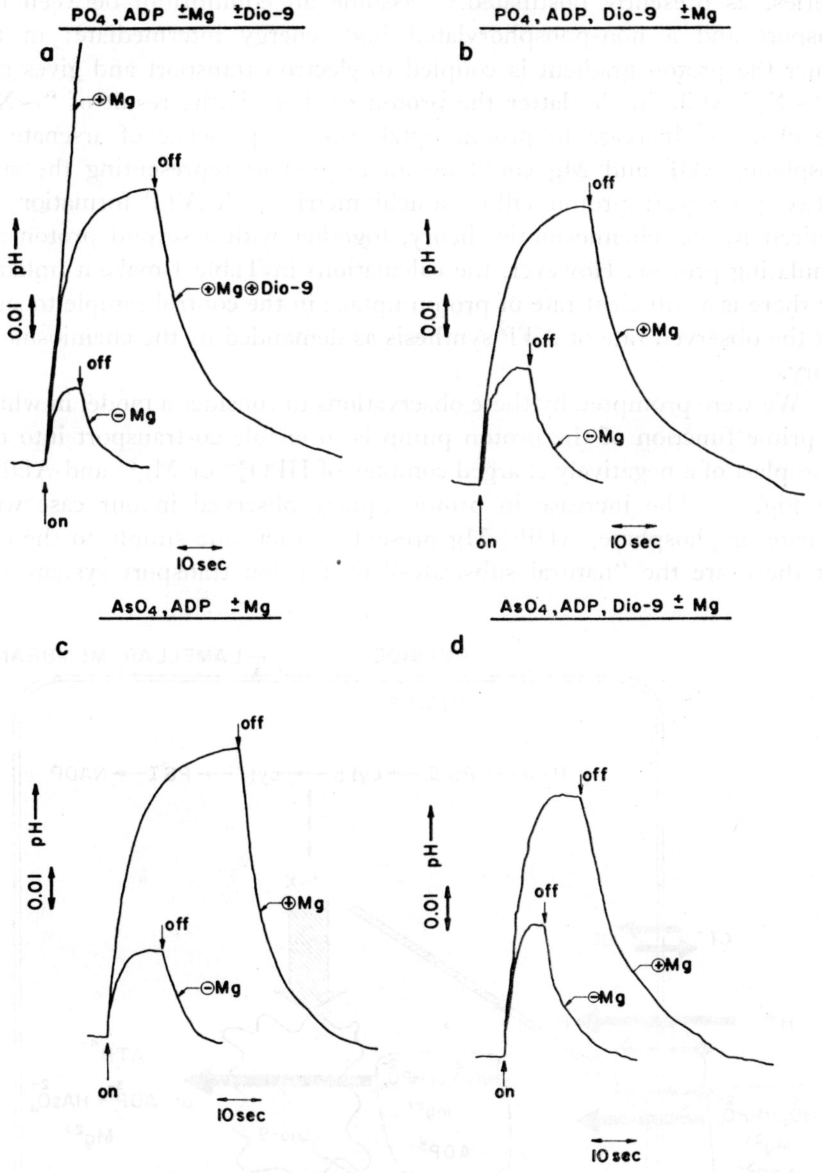

Fig. 4. The effect of Dio-9 on the phosphate or arsenate, ADP and Mg stimulation of light-induced proton uptake. Conditions as in Table I, except that arsenate or phosphate at 0.5 m*M*, $MgCl_2$ at 1.0 m*M* and Dio-9 at 4 μg/ml were added as indicated in the figures. Illumination as in Fig. 2.

theories, as presently postulated,[2,4] assume an equilibrium between ion transport and a non-phosphorylated high energy intermediate; in the former the proton gradient is coupled to electron transport and gives rise to "~X," while in the latter the proton gradient is the result of "~X." The observed increase in proton uptake in the presence of arsenate or phosphate, ADP and Mg could be interpreted as representing the sum of two processes: proton efflux stoichiometric with ATP formation, as required by the chemiosmotic theory, together with a second proton accumulating process. However, the calculations in Table I make it unlikely that there is a sufficient rate of proton uptake in the control sample to support the observed rate of ATP synthesis as demanded by the chemiosmotic theory.

We were prompted by these observations to consider a model in which the prime function of the proton pump is to enable co-transport into the chloroplast of a negatively charged complex of HPO_4^{2-} or Mg^{2+} and ADP^{3-} (see Fig. 5). The increase in proton uptake observed in our case with arsenate or phosphate, ADP+Mg present, is then due simply to the fact that these are the "natural substrates" of the ion transport system and

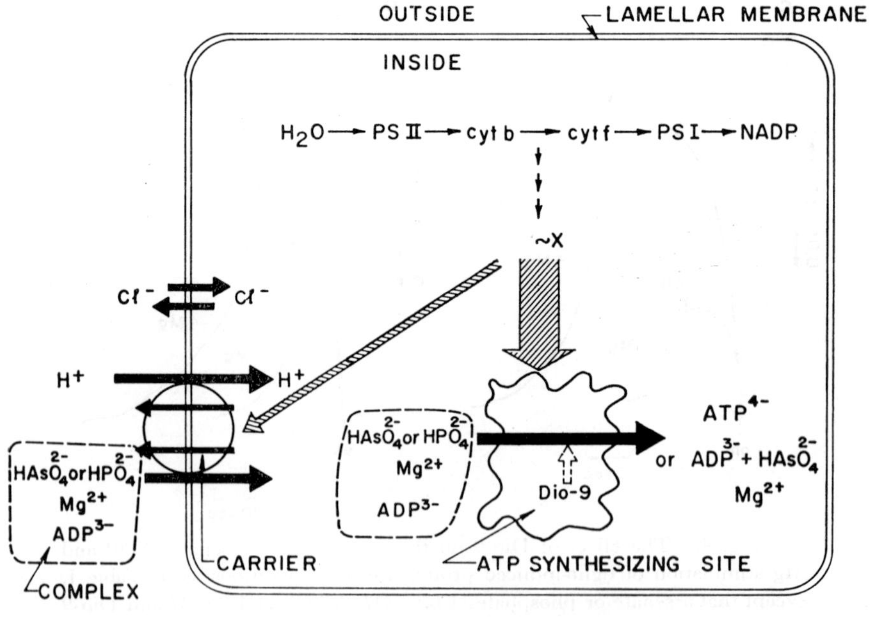

Fig. 5. A hypothetical model describing the proposed function of light-induced proton uptake in chloroplasts; see text for details.

thus permit a higher activity than other "substrates" with a lower affinity, such as chloride. This co-transport may be of an effectively uncharged complex of protons and HPO_4^{2-}, Mg^{2+} and ADP^{3-}. It has previously been shown that the uncharged species of several weak acids is that which best passively penetrates into the chloroplasts.[13]

The model envisaged in Fig. 5 allows that protons and a phosphate or arsenate, Mg, and ADP complex may be simultaneously transported on the same or different "carrier" proteins. ATP or ADP+AsO_4 will be formed subsequently on an ATP synthesizing enzyme complex. Dio-9 has been shown to inihibit the induced ATPase of the EDTA extracted coupling factor,[14] and is therefore assumed to act at the ATP synthesizing site. The experiments shown in Fig. 4 are consistent with the hypothesis presented, which allows the aresenate or phosphate, ADP, and Mg effects to lie before the site of action of Dio-9 (i.e. on a carrier in our hypothesis, Fig. 5).

Finally, it must be remarked that the above hypothesis requires that the actual energy expenditure in transporting the phoshpate or arsenate, ADP, Mg complex be a small fraction of that involved in the subsequent ATP synthesis reaction. Dio-9 has been shown to inhibit the uncoupling of photophosphorylation caused by arsenate ADP and Mg.[14] Hence no significant uncoupling of photophosphorylation (i.e. stimulation of electron transport) can be envisaged as occurring, due to the energy expenditure involved in the arsenate or phosphate, ADP+Mg uptake process.

REFERENCES

1. Mitchell, P., *Nature,* **191**, 144 (1961).
2. Mitchell, P., *Biol. Rev.*, **41**, 445 (1966).
3. Slater, E. C., *Nature,* **172**, 975 (1953).
4. Slater, E. C., *European J. Biochem.*, **1**, 317 (1967).
5. Neumann, J., and Jagendorf, A. T., *Arch. Biochem. Biophys.*, **107**, 1 (1964).
6. Jagendorf, A. T., and Neumann, J., *J. Biol. Chem.*, **240**, 3210 (1965).
7. Lynn, W., and Brown, R., *J. Biol. Chem.*, **242**, 426 (1967).
8. Dilley, R. A., and Vernon, L. P., *Proc. Natl. Acad. Sci. US*, **56**, 395 (1967).
9. Nishimura, N., Ito, T., and Chance, B., *Biochim. Biophys. Acta*, **59**, 177 (1962).
10. Karlish, S. J. D., and Avron, M., in preparation.
11. Avron, M., and Jagendorf, A. T., *J. Biol. Chem.*, **234**, 967 (1959).
12. Good, N., Izawa, S., and Hind, G., in *Current Topics in Bioenergetics* (D. R. Sanadi, ed.), Vol. I, p. 75 (1966).
13. Crofts, A. R., *Biochim. Biophys. Acta,* **131**, 97 (1967).
14. McCarty, R. E., Guillory, R. J., and Racker, E., *J. Biol. Chem.*, **240**, PC 4822 (1965).

Reconstitution of the Photo-sensitive Form of *Chenopodium* Chlorophyll Protein from Its Apoprotein

TERUYO MURATA, YASUHIRO ODAKA, KEIGO UCHINO AND EIJIRO YAKUSHIJI

Faculty of Science, Toho University, Narashino, Chiba, Japan

INTRODUCTION

A water-soluble chlorophyll protein (CP668) was found by Yakushiji *et al.* in the leaves of *Chenopodium album*, and was partially purified by column chromatography. The solution of this protein in a phosphate buffer has main absorption peaks at 277, 429 and 668 mμ. This protein is very sensitive to light and changes its absorption spectrum on several seconds illumination. The absorption peaks at 429 and 668 mμ diminish and new peaks appear at 362, 399, 567 and 743 mμ. Nearly complete photoconversion of CP668 (absorbancy at 668 mμ; 0.4 in 1-cm cell) to CP743 with 10,000 lux white light needs about 40 min.[2] Obata and Takamiya confirmed that the photoconversion of CP668 to CP743 did not occur in the absence of molecular oxygen and that ferricyanide, ferricytochrome *c* and DPIP could replace oxygen.[2] Nothing is known about the chemical nature of the prosthetic group of CP743 except that, in the acetone extract of CP743, a new pigment appears which can be separated from the residual chlorophylls by paper chromatography. Spectroscopic examination showed that this pigment is the chromophore group responsible for the 567 and 743 mμ absorption of CP743.

With the aim of elucidating the mechanism of photoconversion of CP668, we prepared the apoproteins of CP668 and 743 and compared the properties of the reconstituted chlorophyll proteins. The experimental procedures and the results obtained are presented herewith.

METHODS

Preparation of the chlorophyll protein

The method of preparation was as described in our previous paper.[1] The leaves of *Chenopodium album* were homogenized and extracted with a phosphate buffer ($M/100$, pH 7.2). The ammonium sulfate precipitate obtained between 0.3 and 0.7 saturation was dissolved in $M/20$ phosphate buffer, pH 7.2, and dialyzed against $M/100$ solution of the same buffer and purified on a column of Amberlite CG 50. All these procedures were carried out under dim green light.

Seasonal changes strongly influence the yield of the chlorophyll protein. The best season in Japan was found to be from the middle of July to the end of August.

Preparation of apoprotein

The solution of CP668 was dialyzed against $M/100$ phosphate buffer, pH 7.2, and the dialyzed solution was mixed with four parts of methyl ethyl ketone* and shaken. The temperature during the extraction with methyl ethyl ketone was kept below 5°C. The colorless water layer was separated from the organic solvent and the extraction with the solvent was repeated again to eliminate chorophyll. The apoprotein was precipitated from the solution with ammonium sulfate at 0.8 saturation. The ammonium sulfate precipitation was repeated several times to make the solution free from methyl ethyl ketone.

Preparation of chlorophylls and their derivatives

Chlorophyll $a+b$ was prepared by the method originally devised by Takamiya and Ogura.† An alcoholic solution of spinach chlorophyll was mixed with about one-seventh part of dioxane and water was added to the solution drop by drop until a slight turbidity was apparent. The solution was stored in an ice box for about one hour and the resulting microcrystals were gathered by centrifugation. The crude crystals thus obtained were dissolved in alcohol and recrystallized in the same way.

Chlorophylls *a* and *b* were separated by paper chromatography after the method of Jeffrey.

Phaeophytin was obtained by adding oxalic acid to an alcoholic solution of chlorophyll $a+b$. After the chlorophyll was completely converted to phaeophytin, the solution was mixed with petroleum ether, and the alcohol was washed out with water. The petroleum ether solution was air-dried and the phaeophytin was dissolved in alcohol for use.

---

* Dr. J. Fork (Camberra) recommended the use of methyl ethyl ketone for the preparation of apoprotein.

† Unpublished.

Preparation of chlorophyllide

Spinach leaves were homogenized in 50% acetone and stored at room temperature (about 20°C) for about 20 hours. The homogenate was filtered on a Buchner funnel and the filtrate was discarded. The material on the funnel was extracted with 80% acetone. Acetone extract was mixed slowly with water, under mild shaking, until the red fluorescence vanished. Talcum was then added to the solution and filtered on a Buchner funnel. The chlorophyllide adsorbed on the talcum was washed with petroleum ether and then dissolved in alcohol for use. The absence of chlorophyll in this preparation was confirmed by paper chromatography.

Reconstitution of chlorophyll proteins

An alcoholic solution of chlorophyll or its derivatives was added to the apoprotein solution under mild shaking. Sufficient amounts of chlorophyll were added as calculated from its content in the chlorophyll protein. The volume of the alcohol added was regulated not to exceed one-fifth of the apoprotein solution. After standing for about 20 minutes, the solution was saturated with ammonium sulfate and was filtered through a thin layer of talcum. The precipitate was dissolved in $M/20$ phosphate buffer, pH 7.2, whereby free chlorophyll remained adsorbed onto the cake of talcum. The ammonium sulfate precipitation was repeated several times. The reconstituted chlorophyll protein can be purified on a column of Dowex CG 50 as in the case of the original CP668. Phaeophytin- and chlorophyllide-proteins can be synthesized in the same way. The absorption spectra were measured with a Hitachi Recording Spectrophotometer EPS-2U.

RESULTS

Properties of the apoprotein

The solution of the apoprotein in $M/30$ phosphate buffer, pH 7.2, showed an absorption peak at 277 mμ with shoulders at 284 and 290 mμ (Fig. 1). The protein nature of the substance was confirmed by the Millon, Biuret and ninhydrin reactions.

Properties of the reconstituted chlorophyll protein

Apoprotein prepared from CP668 (prepared in the dark, photosensitive) combined with chlorophyll $a+b$ to form a water-soluble and light sensitive protein which had just the same absorption spectrum (Fig. 2) as that of the original CP668. Illumination of this reconstituted material in the presence of air caused a marked change in the absorption spectrum.

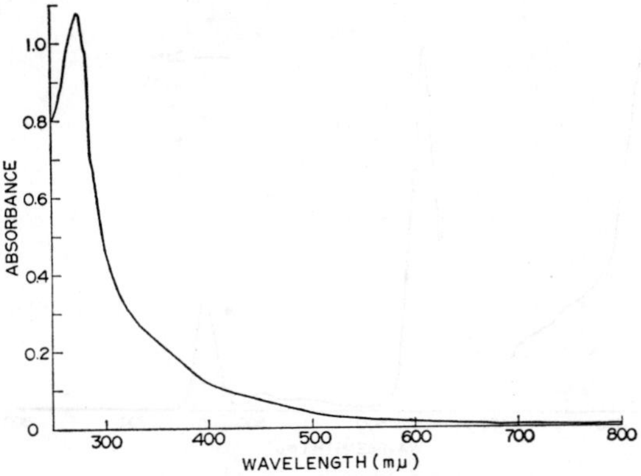

Fig. 1. Absorption spectrum of the apoprotein.

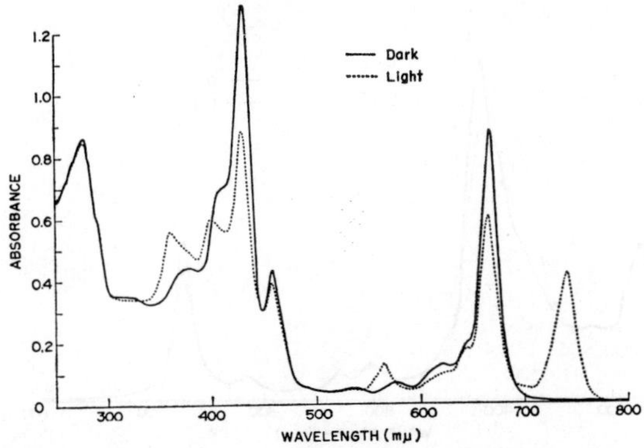

Fig. 2. Absorption spectrum of the reconstituted CP 668 (full line curve); Illuminated (10,000 lux, 1 min) form (dotted line curve).

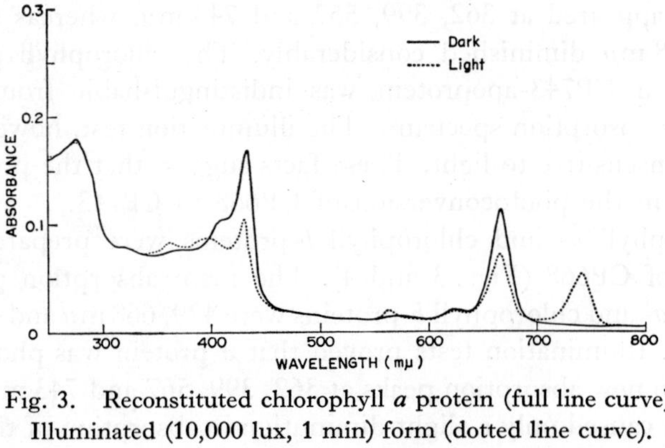

Fig. 3. Reconstituted chlorophyll *a* protein (full line curve); Illuminated (10,000 lux, 1 min) form (dotted line curve).

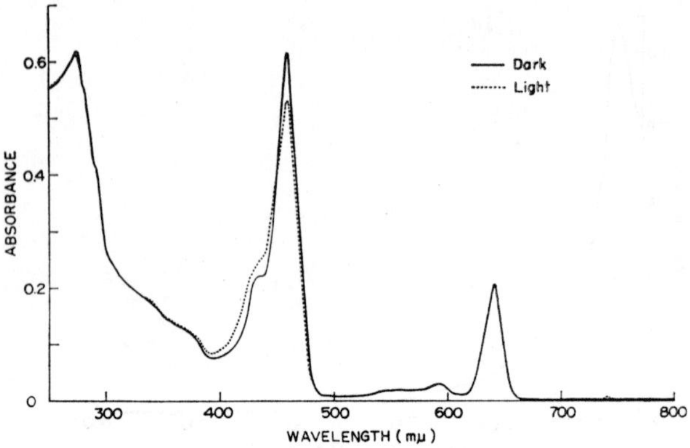

Fig. 4. Reconstituted chlorophyll *b* protein (full line curve); Illuminated (10,000 lux, 2 min) form (dotted line curve).

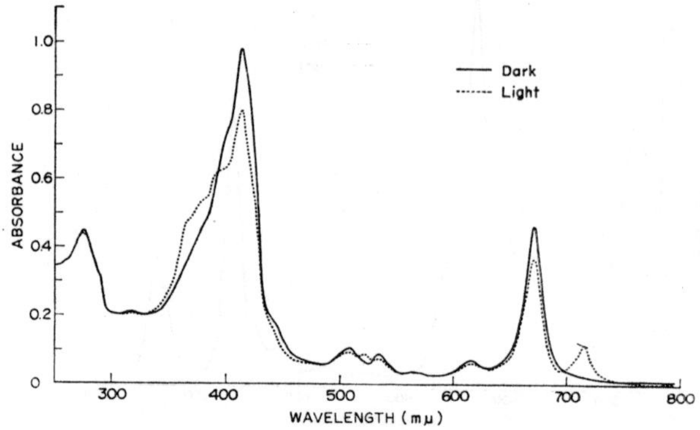

Fig. 5. Synthesized phaeophytin *a*+*b* protein (full line curve); Illuminated (10,000 lux, 1 min) form (dotted line curve).

New peaks appeared at 362, 399, 567 and 743 mμ, whereas the peaks at 429 and 668 mμ diminished considerably. The chlorophyll protein synthesized from CP743-apoprotein was indistinguishable from the native CP668 in its absorption spectrum. The illumination test, however, showed that it was insensitive to light. These facts suggest that the protein moiety participates in the photoconversion of CP668 to CP743.

Chlorophyll *a*- and chlorophyll *b*-proteins were prepared from the apoprotein of CP668 (Figs. 3 and 4). The main absorption peaks of the chlorophyll *a*- and chlorophyll *b*-proteins were 429, 668 mμ and 438, 645 mμ, respectively. Illumination tests proved that *a*-protein was photo-sensitive, giving rise to new absorption peaks at 362, 399, 567 and 743 mμ. However, illumination caused only a slight diminution in absorption of the *b*-protein.

These facts support our previous observation that during illumination of CP668, the *a* to *b* ratio underwent a gradual decrease.

Phaeophytin $a+b$ protein synthesized from the active apoprotein of CP668 showed absorption peaks at 414, 508, 537 and 675 mμ (Fig. 5). The photoconverted form of this protein showed new absorption maxima at 391, 522 and 716 mμ.

Chlorophyllide protein synthesized from the apoprotein of CP668 was found not to differ greatly from natural CP668 in its spectral features.

DISCUSSION

The photoconversion of CP668 to CP743 requires oxygen or some other oxidants such as ferricyanide, ferricytochrome *c* or DPIP.[2] The reverse reaction, a photoreduction to CP668, occurred when a CP743 solution was illuminated in the presence of sodium hydrosulfite or ascorbate under anaerobic conditions. The photoreduced product, however, could not be reoxidized to CP743 on illumination. The present study shows that the protein moiety is modified in the course of photoconversion of CP668 to CP743. On reconstitution, the modified apoprotein only gave rise to a photo-insensitive product. We tried to reactivate this photo-insensitive product with some reducing agents (ethyl mercaptane, ascorbate, etc.) or oxidizing agents (ferricyanide, plastocyanin+ferricyanide or high-potential cytochrome *c* of algae+ferricyanide); however, all efforts were unsuccessful. What kind of modification has occurred in the protein molecule remains open to future investigations.

The manner in which chlorophyll combines with the protein is still obscure. Phytol is not a requisite since chlorophyllide also is capable of combining with the protein. Although phaeophytin can also combine with the apoprotein, this does not necessarily exclude the possibility of Mg participating in the binding of chlorophyll with the apoprotein, since the mode of binding of phaeophytin to the protein might be different from that of chlorophyll. In addition, we have found that Ag-phaeophytin does not combine with the apoprotein, thus indicating the possible participation of the N-atoms of the chlorin in the binding mechanism. Details about this experiment will be reported elsewhere.

Acknowledgements

We would like to express our hearty thanks to Professor A. Takamiya of the University of Tokyo for his kind advice and help.

A part of this work was supported by a grant in aid for Special Project Research on Biophysics from the Ministry of Education, Japan.

SUMMARY

The apoprotein of the water-soluble *Chenopodium* chlorophyll protein (CP668)[1] was prepared by treating the protein solution with methyl ethyl ketone. Aqueous solution of apoprotein shows only one absorption peak at 277 mμ. The Biuret, ninhydrin and Millon reactions were positive.

The apoprotein of CP668 combines with chlorophyll $a+b$ to form photosensitive CP668 which can be converted by light to CP743. On addition of chlorophyll $a+b$ to the apoprotein prepared from CP743 (illuminated form of CP668), a chlorophyll protein is formed which shows the same absorption spectrum as that of the original, non-illuminated CP668 which, however, does not change its spectrum on illumination. This fact clearly demonstrates that protein plays an important role in the photo-conversion of CP668 to CP743.

Chlorophyll *a* protein and chlorophyll *b* protein were prepared from the apoprotein of CP668. Illumination tests of these two proteins proved that chlorophyll *a* protein was photo-sensitive, while chlorophyll *b* protein was insensitive to light.

The apoprotein of CP668 also combines with phaeophytin and chlorophyllide, to form the respective photo-sensitive and water-soluble chromoproteins.

REFERENCES

1. Yakushiji, E., Uchino, K., Sugimura, Y., Shiratori, I., and Takamiya, F., *Biochim. Biophys. Acta*, **75**, 293 (1963).
2. Takamiya, A., Obata, H., and Yakushiji, E., in *Photosynthetic Mechanisms of Green Plants* (B. Kok and A. T. Jagendorf, eds.), p. 479, Nat. Acad. Sci.-Nat. Res. Council, 1963.

A Transient Form of Chlorophyll Produced by Flash Photolysis of *Chenopodium* Chlorophyll Protein, CP668

ATUSI TAKAMIYA, YASUMARO KAMIMURA AND AKIRA KIRA

Department of Biophysics and Biochemistry, Faculty of Science, University of Tokyo, Tokyo

Department of Biology, Faculty of General Education, Toho University, Narashino, Chiba

Laboratory of Radiation Chemistry, Institute for Physical and Chemical Research, Tokyo

INTRODUCTION

In 1963 Yakushiji[1] discovered a water-soluble chlorophyll protein in the leaves of *Chenopodium album*. He has isolated the substance in pure state and investigated its properties. The most interesting feature of this new chlorophyll protein was that it underwent conversion when illuminated in the presence of molecular oxygen or some appropriate electron acceptors such as ferricyanide and dichlorophenolindophenol. The initial form having an absorption band at 668 mμ is converted to a far-red-absorbing form having an absorption band at 743 mμ (Fig. 1). This change could be reversed by illuminating the latter form in the presence of sodium dithionite as reducing agent.[2] The approximate value for the molecular weight of the chlorophyll protein was estimated by gel-filtration on a Sephadex G-100 column to be on the order on 78,000.[3] On this basis, about 6 molecules of chlorophyll *a* and about one molecule of chlorophyll *b* are calculated to be present in one molecule of the chlorophyll protein.

Being interested in this photochemical activity, as well as in the particular mode of occurrence of chlorophyll associated with the water-soluble protein, we made a study of its light-responses, using a technique of flash photolysis developed by Porter.[4] In the present report, the experimental results obtained are briefly described, with special reference to the possible mechanism of the photoconversion of the substance.

MATERIALS AND METHODS

The chlorophyll protein was isolated and purified from fresh leaves of *Chenopodium album* according to the original procedures of Yakushiji

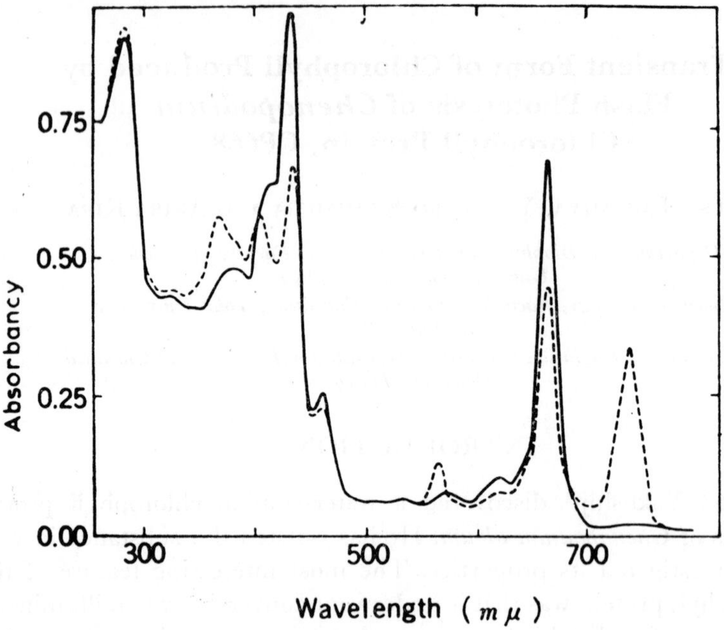

Fig. 1. Absorption spectra of *Chenopodium* chlorophyll protein CP 668. Solid line, CP668 (initial form); dotted line, CP743 (photo-converted form).

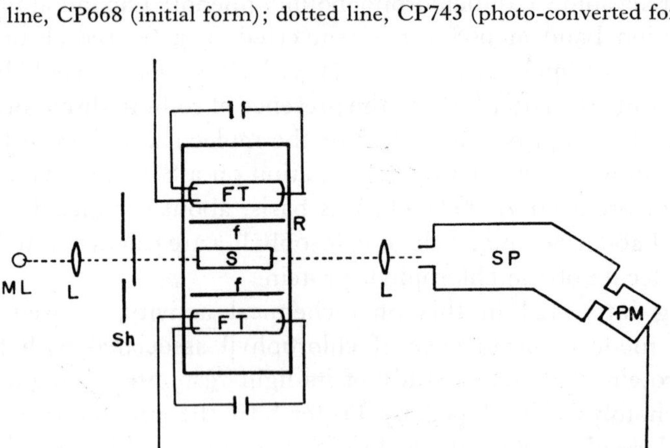

Fig. 2. Equipment for Flash Photolysis.
ML, measuring light lamp; L, lens; Sh, shutter; R, reflector; FT, flash light tube (Xenon lamp); f, filter; S, sample in cylindrical cuvette; SP, spectrometer; PM, photomultiplier; OSC, oscillograph; Tr, trigger; D, delay.

et al.[1] The initial red-absorbing form, CP668, prepared in the dark was used as test material. Purified samples showing values 0.9 to 1.5 for the purity index (i.e. the ratio of O.D. at 275 mμ to O.D. at 668 mμ) were used and essentially consistent results were obtained. The absorption spectra of the samples were measured in a Cary Spectrophotometer Model 14 or in a Hitachi Recording Spectrophotometer EPS-2U. The sample in the cylindrical cuvette for flash photolysis was exhaustively degassed by the usual procedure of repeated freezing and pumping. Molecular oxygen, when indicated, was introduced into the reaction mixture by equilibrating the evacuated sample solution with air at a given reduced pressure. The molar concentration of oxygen in the reaction mixture was calculated from the partial pressure of oxygen in the gas atmosphere, using a Bunsen coefficient of 0.031 at 25°. All the measurements were carried out at room temperature, about 25°.

The measurements of the absorption changes caused by illumination of flash light were carried out in Dr. Koidzumi's laboratory, Tohoku University, Sendai, and in Dr. Imamura's laboratory, Institute for Physical and Chemical Research, Tokyo, using the equipment for flash photolysis maintained and operated by one of the authors (Kira). Fig. 2 shows a diagram of the apparatus designed and constructed by Kira (details will be described elsewhere). The exciting flash light was furnished from a pair of Xenon lamps (Ushio Electric Co., Tokyo) placed on either side of the 10 cm long cylindrical cuvette containing chlorophyll protein solution. The light was filtered through one sheet of Toshiba glass filter VY–48 transmitting light of wavelengths longer than 460 mμ. The excitation lamp was operated at about 80–120 Jules, with an illuminating time of about 50 μ sec (corresponding to 10 μsec in half-life time). The intensity of exciting light was adjusted to suit the conditions of each set of experiments. The exciting flash was followed by a measuring flash with indicated time of delay. The measuring light source was another Xenon lamp operated at 10–20 Jules, with a half-life of 10 μsec. The spectrum of the measuring light flash passing through the sample solution in the cuvette was photographed with a grating spectrophotograph. The photographic record was analyzed with the use of a recording microphotometer (Narumi Co., Tokyo).

In kinetic studies, the measuring flash lamp was replaced by a Xenon arc lamp. The changes in light absorption were detected with the use of a photomultiplier attached to the spectrograph. The photocurrent was monitored by an oscillograph synchroscope (Iwasaki Co., Tokyo) and photographed and the records were examined at appropriate magnifications.

RESULTS

Fig. 3 shows the absorption spectrum of the transient species formed on flash illumination of CP668. This spectrum was obtained by subtracting the value of density of the photographic plate for the non-illuminated sample from that of the flash-illuminated sample of the chlorophyll protein. The chlorophyll protein has relatively low absorption at this range of wavelengths as compared with the markedly high absorption of the transient species. The difference, therefore, in first approximation may be regarded as the absorption spectrum of the transient species under investigation.

The absorption spectrum of the transient species is characterized by peaks at 480 mμ, 508 mμ, 550 mμ and 593 mμ and resembles in its general shape those reported as the absorption spectra of the metastable forms or the triplet excited states of chlorophylls (see references cited in Ref. 5).

The metastable nature of the transient species was visualized by varying the time lapse between the main and the recording flashes. A rapid decay was found to occur uniformly all over the range of wavelengths covered by the photographic method, indicating that the observed spectral change represents the spectral change caused by the formation of essentially one transient species from CP668.

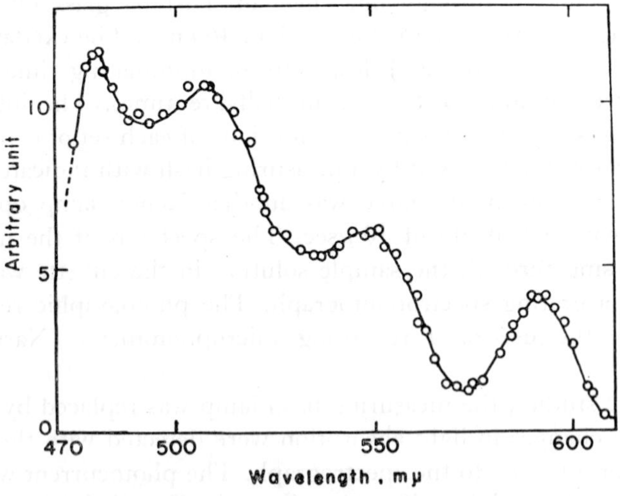

Fig. 3. Absorption spectrum of transient species of CP668. Time lapse between the main and the measuring flash in this particular experiment was 50 μsec.

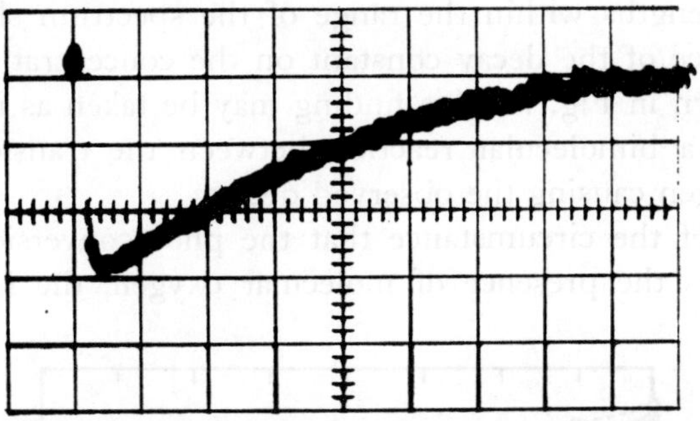

Fig. 4. Oscillograph record of time course of decay of transient species of CP668 (in vacuum) observed at 470 mμ. Time scale (abscissa): one msec per scale.

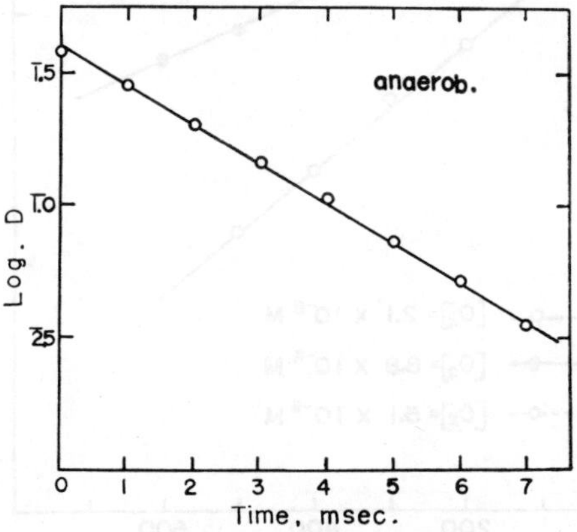

Fig. 5. Time course of decay of transient species of CP668 observed at 470 mμ in the absence of molecular oxygen. Data taken from Fig. 4 (half-life time was calculated to be 2 msec; $k_{obs.} = 3.3 \times 10^2$ sec^{-1}).

The kinetics of the decay was followed by electrophotometric measurements. Fig. 4 shows one of the oscillograph records of such decay obtained in the complete absence of molecular oxygen. The time course of decay is in accord with that of a unimolecular reaction through the entire course of the change. The unimolecular constant of decay can be calculated from the dip of the straight line plotting the logarithms of optical density against the time of decay (Fig. 5). The introduction of molecular oxygen into the reaction mixture markedly accelerated the decay. However, the unimolecular nature of the change was not altered (Fig. 6). The decay constants thus obtained for varied concentrations of oxygen at 473 mμ (measuring wavelength) are summarized in Table I. Similar results were also obtained at other wavelengths within the range of the spectrum shown in Fig. 3. The dependence of the decay constant on the concentration of molecular oxygen is shown in Fig. 7. This finding may be taken as indicative of the occurrence of a bimolecular reaction between the transient species and molecular oxygen causing the observed decay.

In view of the circumstance that the photoconversion of CP668 to CP743 requires the presence of molecular oxygen, the formation of the

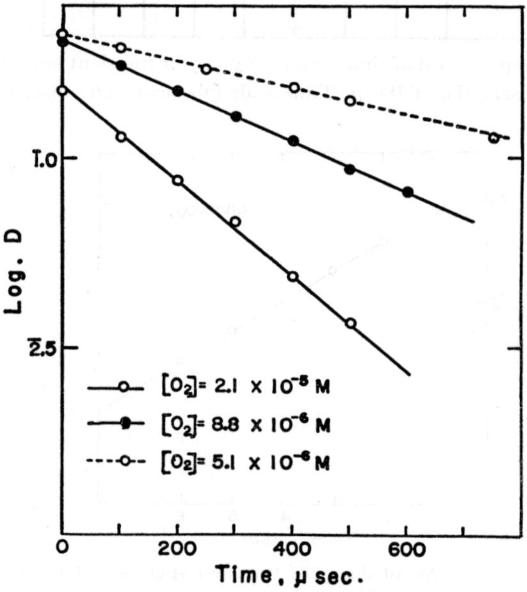

Fig. 6. Time course of decay of transient species of CP668 observed at 470 mμ in the presence of molecular oxygen.

TABLE I. Decay constants and half-life time of transient species of CP 668, observed at 470 mμ (measuring wavelength).

$[O_2]$ ($M \times 10^{-6}$)	$\tau_{1/2}$ (sec $\times 10^{-3}$)	$k_{obs.}$ (sec^{-1})
0	2.10	330
5.1	0.71	970
8.8	0.46	1490
12.0	0.31	2200
14.5	0.27	2500
(Preill. 3 min.	0.53	1300)
(6 min.	0.85	810)
21.0	0.24	2800

final product of photoconversion was also examined by following the time course of the changes at 740 mμ (Fig. 8). The first part of the time course of the change subsequent to the flash represents the formation of the transient species followed by its rapid decay in the presence of molecular oxygen (see above). The accurate value of the decay constant, however,

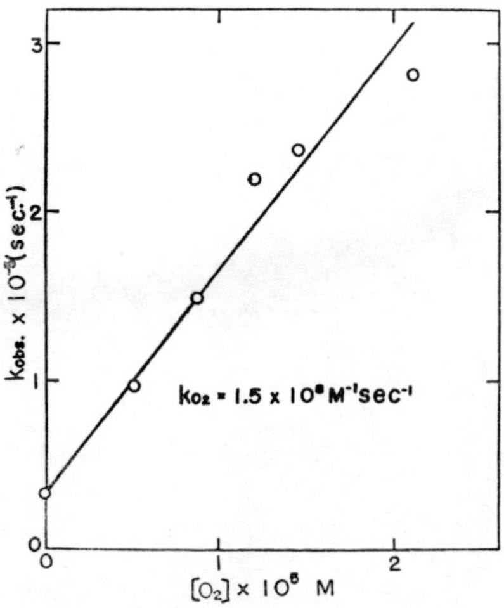

Fig. 7. O_2-dependency of the observed decay constants. Summarized data of observations at 470 mμ.

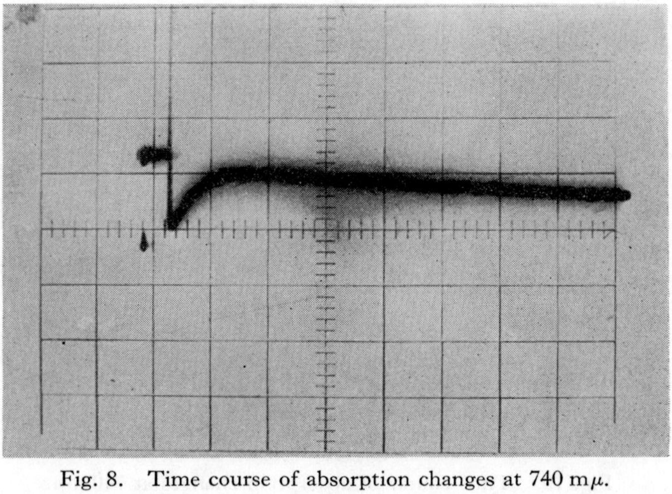

Fig. 8. Time course of absorption changes at 740 mμ. Time scale: 2 msec per scale (abscissa). Concentration of molecular oxygen in reaction mixture: $1.8 \times 10^{-5} M$.

cannot be calculated because of the overlapping of the subsequent change now under investigation. The change in this second part of the reaction is rather slow and reaches a steady level after a reaction time of a few seconds. On increasing the concentration of the molecular oxygen in the reaction mixture, the first part of the change becomes insignificant, leaving the second part of the change as shown in Figs. 9 and 10. The observed

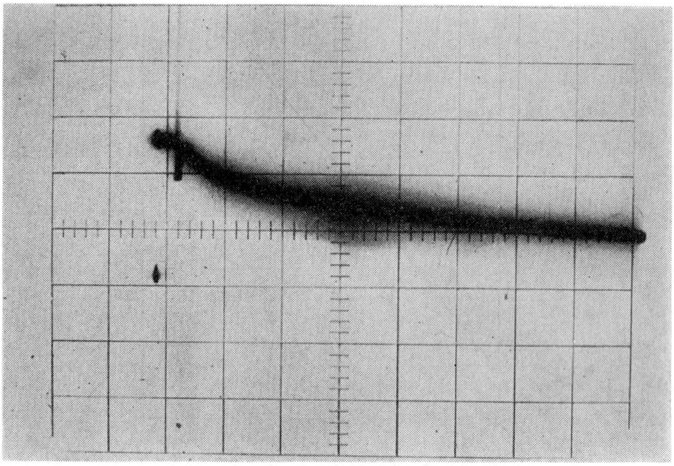

Fig. 9. Time course of absorption changes at 740 mμ.
Time scale: 5 msec per scale (abscissa). Concentration of molecular oxygen in reaction mixture: $5.5 \times 10^{-5} M$.

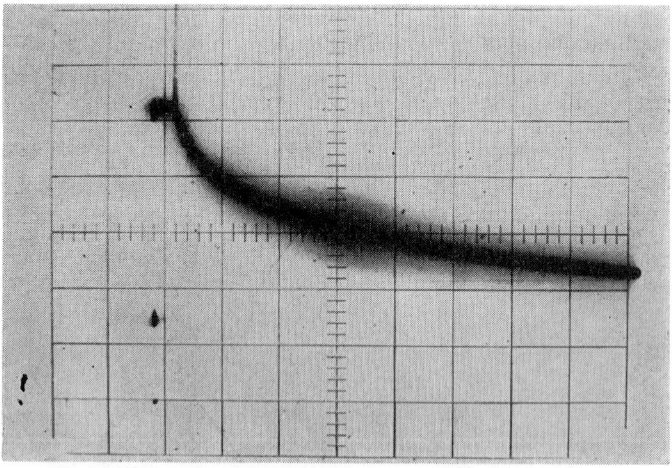

Fig. 10. Time course of absorption changes at 740 mμ.
Time scale: 5 msec per scale (abscissa). Concentration of molecular oxygen in reaction mixture: $2.8 \times 10^{-4} M$.

discrepancy between the rapid disappearance of the transient species, on the one hand, and the rather slow appearance of the 740 mμ-absorbing product (i.e. CP743) on the other, indicates that CP743 is not directly formed from the transient species under investigation (X). Detailed analysis of the records (Fig. 11) showed that the reaction rate of the second part of the change, i.e. the formation of CP743 from a still undetected intermediate or intermediates, is independent of the concentration of oxygen in the reaction mixture. The total yield of the final product, CP743, however, depends on the oxygen concentration as shown in Fig. 12.

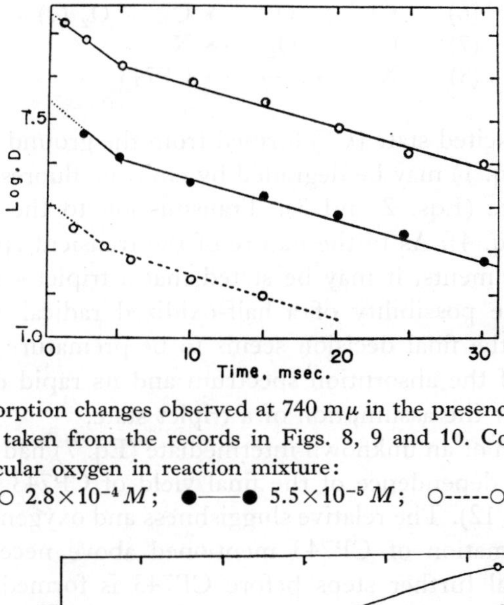

Fig. 11. Absorption changes observed at 740 mμ in the presence of molecular oxygen. Data taken from the records in Figs. 8, 9 and 10. Concentrations of molecular oxygen in reaction mixture:
○——○ 2.8×10^{-4} *M*; ●——● 5.5×10^{-5} *M*; ○-----○ 1.8×10^{-5} *M*

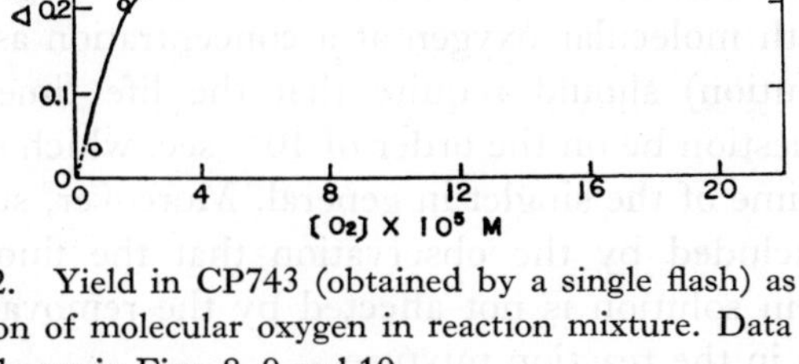

Fig. 12. Yield in CP743 (obtained by a single flash) as a function of concentration of molecular oxygen in reaction mixture. Data including experiments shown in Figs. 8, 9 and 10.

DISCUSSION

The following tentative reaction scheme is proposed to explain the experimental facts described above.

$$
\begin{array}{llll}
(1) & C + h\nu & \longrightarrow & C^* \\
(2) & C^* & \longrightarrow & C + h\nu \\
(3) & C^* & \longrightarrow & C + \text{heat} \\
(4) & C^* & \longrightarrow & C^T \\
(5) & C^T & \longrightarrow & C \\
(6) & C^T + O_2 & \longrightarrow & C + O_2 \ (?) \\
(7) & C^T + O_2 & \longrightarrow & X \\
(8) & X \rightarrow & \rightarrow & \rightarrow CP743
\end{array}
$$

The singlet excited state (C*) formed from the ground state of CP668 (C) by the light (Eq. 1) may be degraded by emitting fluorescence and heat to the ground state (Eqs. 2 and 3). Transmission to the transient state also takes place (Eq. 4). As to the nature of the transient species observed in the above experiments, it may be stated that a triplet state is the most probable case. The possibility of a half-oxidized radical also cannot be denied. Although the final decision seems to be premature, the similarity in general shape of the absorption spectrum and its rapid decay in a first order reaction favor the assumption of a triplet state.

The formation of an unknown intermediate (Eq. 7) had to be assumed to account for the dependence of the final yield of CP743 on the oxygen concentration (Fig. 12). The relative sluggishness and oxygen-independence of the rate of formation of CP743 mentioned above necessitate the assumption of several further steps before CP743 is formed from X (Fig. 11).

The involvement of the singlet excited state of CP668 (C*) as an intermediate of photoconversion is improbable when we consider the short life-time of the singlet excited state in general. If we assume a diffusion-controlled rate constant of 10^{10},[6] the reaction of the singlet excited state of chlorophyll with molecular oxygen at a concentration as low as 10^{-4} *M* (air-saturated solution) should require that the life time of the singlet excited state in question be on the order of 10^{-6} sec, which seems much too long for the life-time of the singlet in general. Moreover, such a possibility was definitely excluded by the observation that the fluorescence of the chlorophyll protein solution is not affected by the removal or addition of molecular oxygen in the reaction mixture.

Acknowledgements

The authors express their gratitude to Dr. M. Koidzumi, Tohoku University, and Dr. M. Imamura, the Institute for Physical and Chemical Research, for their generous support and the use of the equipment in their laboratories during the present work. Thanks are also due to Dr. N. Murata for advice and help in fluorescence measurements, and to Mrs. T. Murata for help in the preparation of the *Chenopodium* chlorophyll protein.

This work was supported in part by a grant from the Ministry of Education, Japan.

REFERENCES

1. Yakushiji, E., Uchino, K., Sugimura, Y., Shiratori, I., and Takamiya, F., *Biochim. Biophys. Acta*, **75**, 293 (1963).
2. Takamiya, A., Obata, H., and Yakushiji, E., in *Photosynthetic Mechanisms of Green Plants* (B. Kok and A. T. Jagendorf, eds.), pp. 479–485, Nat. Acad. Sci.-Nat. Res. Council, 1963.
3. Kamimura, Y., and Takamiya, A., unpublished data.
4. Porter, G., in *Techniques of Organic Chemistry* (A. Weissberger, ed.), Vol. 3, pp. 1056–1106, 1963.
5. Seely, G. R., in *The Chlorophylls* (L. P. Vernon and G. R. Seely, eds.), pp. 523–568, Academic Press, 1966.
6. Osborne, A. D., and Porter, G., *Pros. Roy. Soc.*, **A 284**, 9 (1965).

Induction Effects in EPR Signal I in Algae

Ellen C. Weaver*

Biophysics Laboratory, Stanford University, Stanford, California

INTRODUCTION

For more than ten years it has been known that one could detect a net change in the level of unpaired electrons when photosynthetic organisms were illuminated by means of electron paramagnetic resonance (EPR) spectroscopy. In plants there are two resonances, but the present communication is concerned solely with Signal I, characterized by a *g* value of 2.0025, a peak-to-peak width of 7.3 gauss, and a lack of detectable structure. The other resonance, Signal II, is involved in the oxygen evolving mechanism of plants and does not concern us here.

It now seems highly probable that Signal I is due to the oxidized form of the photoreactive center, P_{700}, and not to the bulk of chlorophyll. This hypothesis was first proposed by Beinert and Kok,[1,2] who noted that both the EPR signal and P_{700} bleaching were effected by red light and diminished by shorter wavelength light. They also demonstrated that preparations enriched for P_{700} by acetone extraction showed a corresponding increase in signal, and that the number of unpaired electrons, while always greater than the level of optically determined P_{700}, still corresponded to it within factors of 2 to 5. We do not know yet what accounts for the excess of spins over the amount of P_{700}. Oxidizing agents such as ferricyanide bleach P_{700} and also produce an EPR signal similar to that elicited by light. We have noted a light-induced increase in the EPR signal in ferricyanide-treated materials; Dr. Bacon Ke has also observed this effect in P_{700} absorbancy changes (personal communication).

The best evidence was provided by a mutant of *Scenedesmus* (Sc-8) which has as much chlorophyll as the wild type, yet lacks Signal I (Fig. 1). Kok[3] was unable to detect any P_{700} in this mutant, which lacked all but a trace of System I function.[4] A mutant of *Chlamydomonas* with the converse relationship, i.e. low chlorophyll but a high level of P_{700}, further reinforced

* Present Address: Exobiology Division, Ames Research Center, NASA, Moffett Field, California, U.S.A. 94035.

Abbreviations: DCMU, 3-(3,4-dichlorophenyl)-1,1-dimethylurea; PMS, phenazine methosulfate and reduced form; DPIP, 2,6-dichlorophenolindophenol and reduced form.

the conclusion that Signal I corresponded to the photoreactive center.[5]

Feeling confident that we were observing the photoreactive center I, we next chose to investigate the kinetics of signal formation in wild type

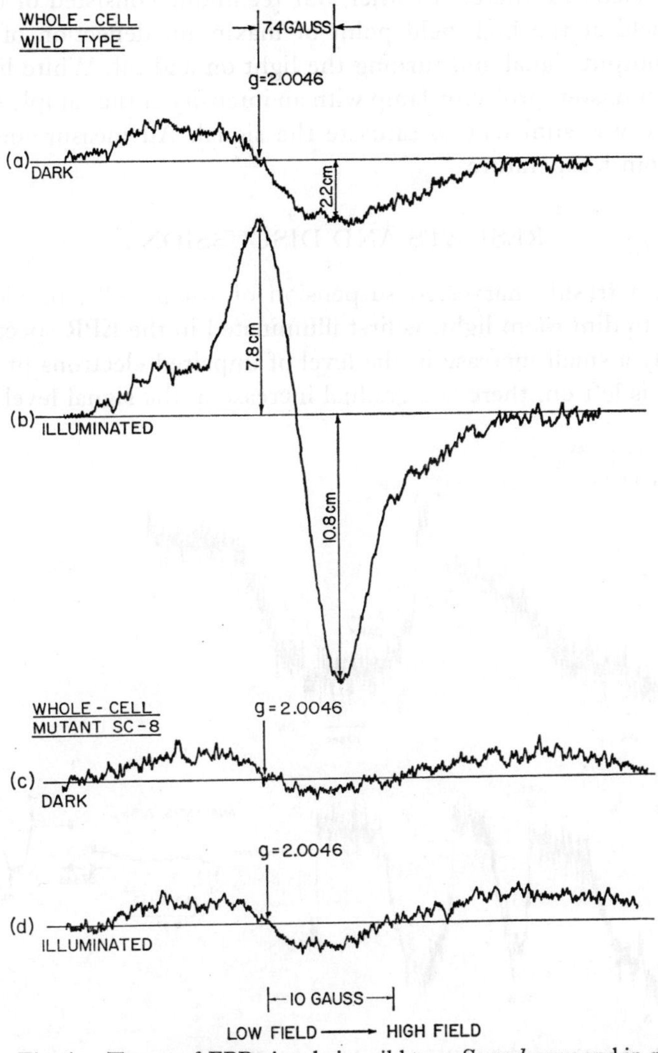

Fig. 1. Traces of EPR signals in wild type *Scenedesmus* and in mutant Sc-8. (a) and (c) are recorded without incident light; the observed resonance is Signal II, which decays in the absence of light over tens of minutes; (b) shows Signal I superimposed on Signal II; (d) is also recorded in the light but no Signal I is evident. Chlorophyll concentration is 1.2 mM for both samples.

algae (*Chlamydomonas*, *Scenedesmus* and *Anacystis*) and in mutant strains, all possessing P_{700}.

Growth conditions and details of our experimental methods are being published elsewhere.[6] In brief, our technique consisted of fixing the magnetic field at the high-field point of maximum deflection of the first derivative output signal and turning the light on and off. White light from a 500 watt tungsten projector lamp with an intensity at the sample of 5×10^4 ergs/cm²/sec was sufficient to saturate the signal. All measurements were made at room temperature.

RESULTS AND DISCUSSION

When a freshly harvested suspension of whole cells, previously exposed only to dim room light, is first illuminated in the EPR spectrometer, there is only a small increase in the level of unpaired electrons in Signal I. If the light is left on, there is a gradual increase in the signal level after the

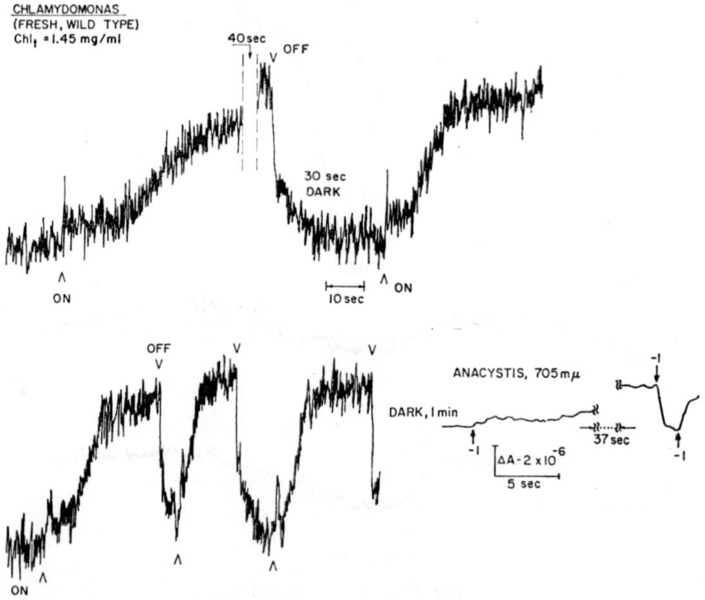

Fig. 2. Formation of Signal I with steady illumination. The initial spike is followed by a plateau, then a slow rise to saturation level. The length of the plateau and the rate of subsequent rise is related to the length of the preceding dark time. *Anacystis* behaves similarly. The P_{700} signal in *Anacystis* produced by actinic illumination of 430 nm is redrawn from Vredenberg and Amesz.[7] Note similarity to EPR signal.

initial plateau which may take considerably more than a minute to reach its maximum. Fig. 2 illustrates this phenomenon, and reproduces also a figure from Vredenberg and Amesz[7] for the oxidation of P_{700} which follows the same time course. After a limited dark period the time needed to reach the steady state signal is far shorter; the initial behavior can only be recovered after a dark period of several minutes or more. If one plots the relative number of electrons produced in the first ten seconds of saturating illumination versus the length of the preceding dark period, one observes an inverse relationship (Fig. 3). Again the similarity to data for P_{700} published by Vredenberg and his associates[7] is striking. These authors have considered this lag a case of induction effect, and the term seems to be equally applicable to EPR results.

From Fig. 3 it is seen that twice as many spins are produced after two seconds of darkness than after 14 seconds for this particular algal strain. If DCMU (10^{-5} *M*) is added, the dark period needed to halve the number of spins is lengthened by about four seconds, indicating only small effect from blocking photosystem II. If a mutant lacking photosystem II is used[8] there is an effect similar to that of DCMU on wild type. Therefore, the factors producing the induction effect seem to be mainly but not entirely part of photosystem I. If the dark and light periods are alternated, the signal produced in a given light period increases in size (Fig. 4). Therefore, some process takes place slowly in the dark whose effect is to prevent the rapid accumulation of spins.

In all cases where one observes a marked delay in growth of the signal, its decay with the cessation of light is rapid and instrument limited, i.e. faster than 0.1 second. Therefore, although the net rate of free radical accumulation appears to be slow, the actual rate must be rapid since a signal does build up.

In contrast to these observations on fresh, intact cells are those on cells which have been heated to 60° C for four minutes, on cells aged at 4°C several days, and on the small particles (PD-10) isolated from detergent-treated spinach chloroplasts by Dr. Leo Vernon (and kindly made available by him) which consist largely of photosystem I. Such materials have a maximal, rapidly produced EPR signal as soon as the light is turned on; there is absolutely no dependence on the length of the preceding dark period, i.e. no induction effect. The rate of signal decay is extremely slow, with half times of ten seconds or more. The addition of a reducing couple (PMS- or DPIP-ascorbate) to the particles speeds up the decay time.

Possible explanations for the induction effect in whole cells may be considered: Olson and Smillie[9] found that the cytochromes oxidized in

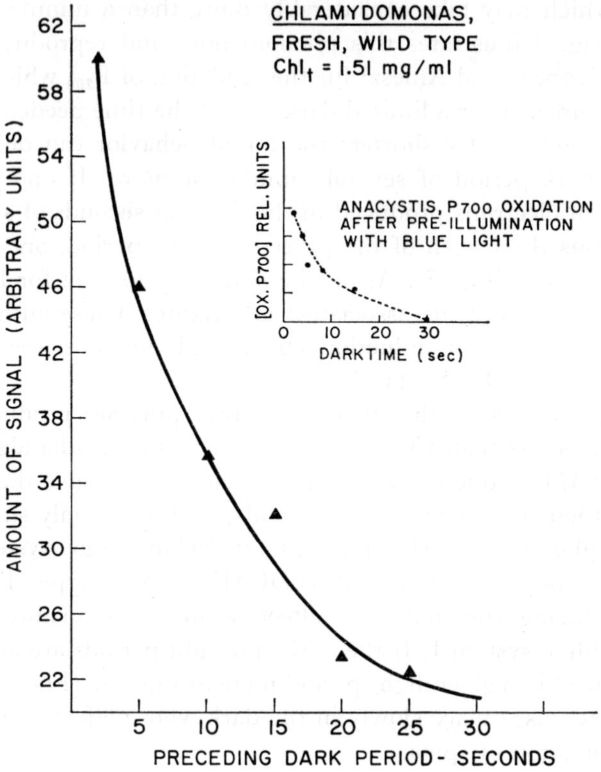

Fig. 3. Data taken from traces similar to those in Fig. 2. The relative number of unpaired electrons produced in the first ten seconds of illumination is plotted as a function of the preceding dark period. Insert: accumulation of P_{700} in blue light (1.1×10^{-9} einstein/cm^2/sec) as a function of darktime after pre-illumination in blue light. (Redrawn from Vredenberg and Amesz.)[9]

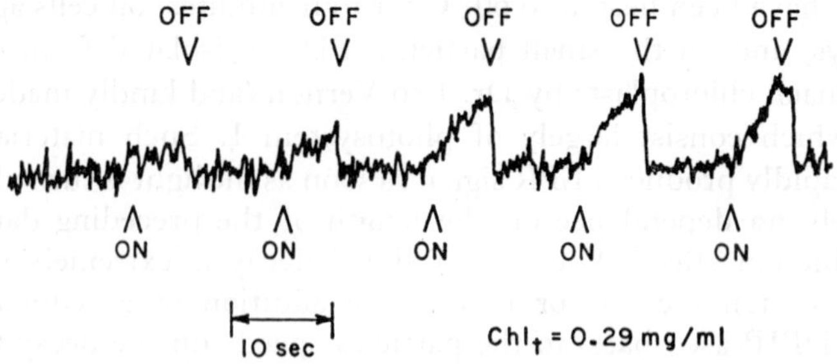

Fig. 4. Increase in light-induced signal with alternating six-second light and nine-second dark periods.

the light in whole *Anacystis* or *Euglena* returned to a reduced state in the dark even under conditions when photosystem II was totally blocked with DCMU. The lag in the appearance of the signal may be a measure of the extent of reduction of the cytochrome complement, or of the plastoquinone pool which also returns to the reduced state in the dark even in mutants lacking photosystem II.[10] Another factor which may be operative is the availability of the electron acceptor for P_{700}, which may be a conformational change during a period of illumination. However, the results with the small particles and heated cells would make this seem unlikely.

To summarize, the kinetics of the appearance of EPR Signal I in whole cells parallel those for the oxidation of P_{700} and cytochrome. EPR spectroscopy allows the study of the photoreactive center I without interference from overlapping pigments, since the detecting microwave radiation does not itself affect any photosynthetic processes. We plan to exploit this advantage for further studies on the factors affecting electron flow through the photoreactive center.

Acknowledgements

This work was supported by a grant from the Atomic Energy Commission, Contract AT(04–3)326–12.

REFERENCES

1. Kok, B., and Beinert, H., *Biochem. Biophys. Res. Comm.*, **9**, 349 (1962).
2. Beinert, H., and Kok, B., *Biochim. Biophys. Acta*, **88**, 278 (1964).
3. Kok, B., in *Photosynthetic Mechanisms of Green Plants* (B. Kok and A.T. Jagendorf, eds.), p. 45, Nat. Acad. Sci.-Nat. Res. Council, 1963.
4. Kok, B., and Datko, A., *Plant Physiol.*, **40**, 1171 (1965).
5. Weaver, E., *Abstr. WCl, Biophys. Soc.* (1966).
6. Weaver, E., *Photochem. Photobiol.*, in press.
7. Vredenberg, W. J., and Amesz, J., in *Currents in Photosynthesis, Proc. 2nd Western Europe Conf. on Photosynthesis* (J. B. Thomas and J. C. Goedheer, eds.), p. 349, A. D. Donker, Rotterdam, 1965.
8. Levine, R. P., and Volkmann, D., *Biochem. Biophys. Res. Comm.*, **6**, 264 (1961).
9. Olson, J. M., and Smillie, R. M., in *Photosynthetic Mechanisms of Green Plants* (B. Kok and A. T. Jagendorf, eds.), p. 56, Nat. Acad. Sci.-Nat. Res. Council, 1963.
10. Bishop, N. I., personal communication.

Aerobically Bound CO_2 in *Chlorella* Cells

SHIGETOH MIYACHI, RYUZI KANAI AND A.A. BENSON

Institute of Applied Microbiology, University of Tokyo, Tokyo
and
Scripps Institution of Oceanography, University of California, San Diego, California

INTRODUCTION

Warburg has suggested that photosynthetic oxygen comes from activated carbon dioxide ("photolyte"). According to this view, the mechanism of photosynthesis consists of the following cycle of reactions: By the absorption of a light quantum, a "photolyte" is reduced to carbohydrate and one molecule of oxygen is evolved, then two thirds of the carbohydrate formed by the light reaction is reoxidized by oxygen in a respiratory process (dark process), which provides the necessary energy for the resynthesis of "photolyte" from CO_2.[1] Existence of the photolyte in *Chlorella* cells was demonstrated by the following manometric observations: When HF was added to the cell suspension of *Chlorella* which had been kept with either 20% CO_2 in argon or 20% CO_2 in air under continued darkness, there occurred a significant evolution of CO_2. The amount of CO_2 evolved from the cells kept in 20% CO_2 in air greatly exceeded that of CO_2 evolved from the cells kept in 20% CO_2 in argon. According to Warburg, CO_2 released from the cells kept under anaerobic conditions results from enzymatic decarboxylation of glutamic acid as well as the CO_2 bound as bicarbonate within the cells; the extra CO_2 released from those kept under aerobic conditions is from the "aerobically bound CO_2". Evolution of this extra CO_2 was not observed when the cells kept in 20% CO_2 in air had been illuminated prior to the addition of HF, indicating that the aerobically bound CO_2 disappears on illumination. It was also shown that oxygen is developed when the aerobic CO_2 disappears in the light. The quantity of oxygen developed by illumination was equal to that of the aerobically bound CO_2 which disappeared. Based on these observations, he concluded that the aerobically bound CO_2 is the photolyte.[2]* His conclusion, however, has not yet been substantiated by experiments using $^{14}CO_2$ as tracer (see e.g., ref. 4). The "photolyte," if it exists, must therefore be a very labile compound.

Abbreviations: TCA, trichloroacetic acid; PCA, perchloric acid.

Recently we tried to confirm the existence of such a labile compound using a methylating reagent, diazomethane, to trap labile acids. It was found that when *Chlorella* cells[5] or chloroplast particles[6] which had been incubated in $^{14}CO_2$ were treated with diazomethane, more than 30 times as much radioactivity was fixed as after the usual hot ethanol treatment. The fact that the amount of ^{14}C remaining at the origin after two-dimensional paper chromatography of the stabilized products decreased when the chloroplasts were illuminated in the presence of ferricyanide seemed to indicate that this compound may function somewhere in the electron transport system of photosynthesis. However, it was also concluded that the ^{14}C-compound as observed by the diazomethane method is not related to the photolyte, because of the finding that oxygen is not required for the formation of the ^{14}C-compound.

The foregoing observations led us to re-investigate the existence of the photolyte in *Chlorella* cells. The HF-induced CO_2 evolution by algal cells has been studied by Bishop and Gaffron[7] as well as Vishniac.[8] However, their results are not pertinent in relation to the recent theory of Warburg, since the experiments of Bishop and Gaffron were done under anaerobic conditions and the essential experimental conditions were not fully described by Vishniac. We followed the experimental conditions adopted by Warburg as closely as possible. The results thus far obtained using manometry as well as $^{14}CO_2$ are briefly described in this paper.

RESULTS AND DISCUSSION

Effect of pre-illumination on hydrofluoric acid-induced CO_2 evolution

Chlorella ellipsoidea was grown in the S-medium of Warburg[1] at 20–25° with constant bubbling of air containing 5% CO_2. The light intensity was varied diurnally with 12-hour periods from 50 lux to 5,000 lux. The cells were harvested after several days of growth, centrifuged and then suspended in fresh S-medium (pH 3.8). Two ml of the suspension containing 0.1 ml (packed cell volume) of *Chlorella* cells were placed in three Warburg vessels, respectively. The gas phase in the first and the second vessel was 20% CO_2 in air. The third vessel was filled with argon containing 20% CO_2. The first and the third vessel were shaken in the dark for 30

* A recent experiment by Warburg *et al.*[3] showed that the nature of the aerobically bound CO_2 (extra CO_2) as determined by the HF method varies with the physiological state of the cells. In some cells, the extra CO_2 derives entirely from the photolyte but in other cells the extra CO_2 is due partly to the photolyte and partly to the precursor of photolyte, which does not evolve oxygen on illumination.

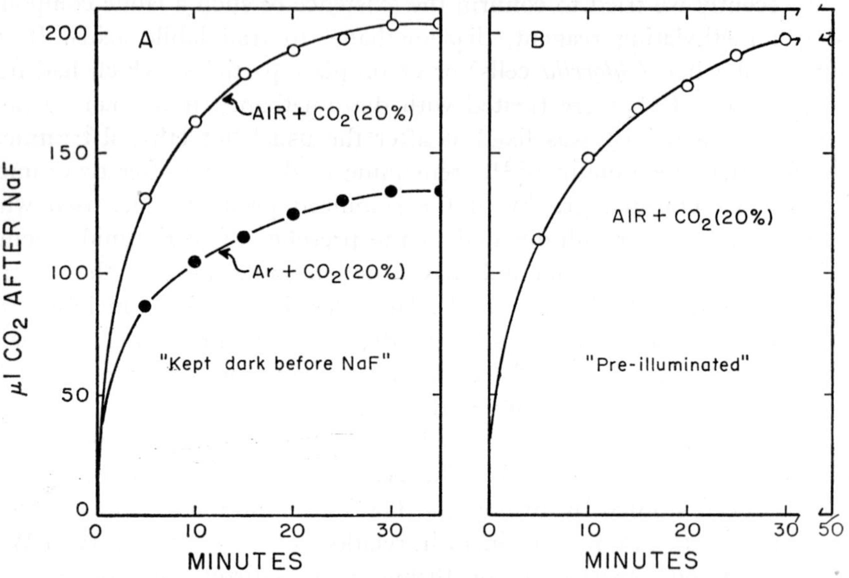

Fig. 1. Effect of pre-illumination on HF-induced CO_2 evolution.

minutes; then 0.1 ml of NaF solution (0.2 *M*, pH 3.8) was tipped in from the side arm. The second vessel was illuminated by a tungsten lamp (20,000 lux) for 20 minutes after which the NaF solution was tipped in immediately after turning off the light. The time courses of CO_2 evolution after the addition of fluoride are shown in Fig. 1. Fig. 1A shows that the amount of CO_2 released from the cells that had been equilibrated with 20% CO_2 in air greatly exceeded that released from the cells equilibrated with 20% CO_2 in argon, suggesting the formation of aerobically bound CO_2 in darkness. However, contrary to the reports of Warburg, the amount of CO_2 released on the addition of HF to the pre-illuminated cells was always equal to that released from the cells which had been kept in darkness (Fig. 1B), indicating that the aerobically bound CO_2 is not affected by illumination. Thus, we have been unable to confirm one of the experimental bases for Warburg's interpretation that "at least a part of the extra CO_2 is photolyte."

Effects of different acids on evolution of CO_2

The following experiments were performed using the cells of *Chlorella ellipsoidea* which had been cultured under continuous illumination in the medium routinely used in our laboratory.[9]

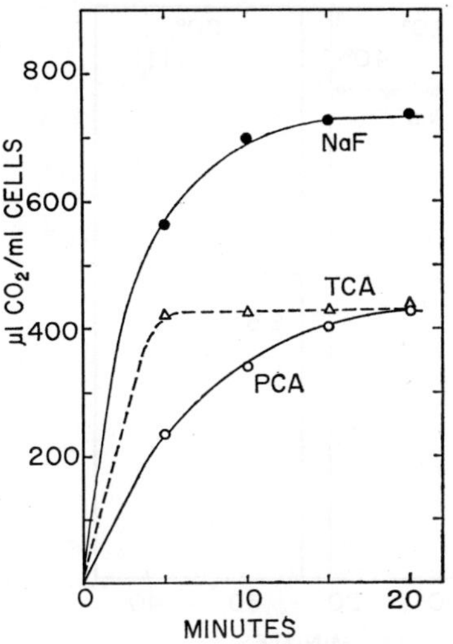

Fig. 2. Effects of different acids on evolution of CO_2.

Chlorella cells suspended in the fresh culture medium (pH value was adjusted to 4.0) were placed in Warburg vessels filled with air containing 20% CO_2. After shaking the vessels for 30 minutes in the dark, sulfuric acid, acetic acid, TCA, PCA or hydrofluoric acid was tipped in from the side arm and the respective time courses of the resulting gas evolution were followed. As may be seen in Fig. 2, the amounts of gas released after the addition of TCA (0.06–0.6 *N*) or PCA (0.6–1.0 *N*) were half or sometimes less than half of that released after the addition of HF. (When the concentrations of TCA or PCA added were lower than those indicated above, gradual evolution of gas ultimately attained the level observed when HF was added.) Although it is not shown in the figure, the amount of gas released after the addition of sulfuric acid (0.2 *N*) was practically equal to that observed after the addition of HF. The effect of acetic acid (0.2 *N*) was between those of HF and TCA or PCA. No additional gas evolution was observed when HF was added to the cell suspension which had been treated with PCA (0.6 *N*), indicating that HF-induced CO_2 evolution is due to enzymatic reactions.

The simplest explanation for the effect of TCA (0.06–0.6 *N*) or PCA (0.6–1.0 *N*) according to the theory of Warburg (see Introduction), would be that CO_2 was evolved from bicarbonate as well as from aerobically bound CO_2 within the cells, leaving glutamic acid intact. However, when the cells which had been treated with PCA at 20° were kept at 40° for 3

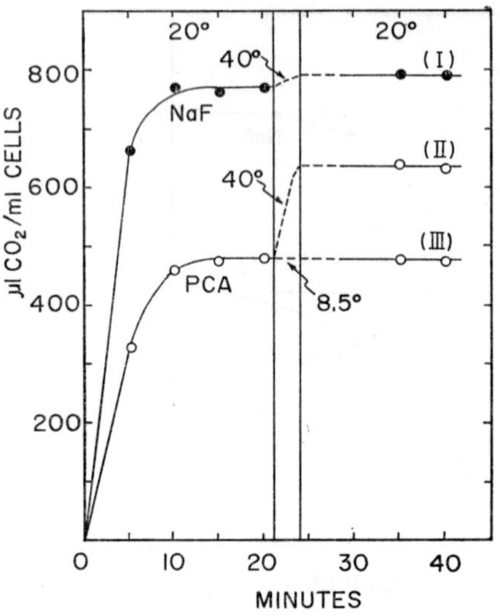

Fig. 3. Effect of temperature on acid induced CO_2 evolution. Warburg vessels containing suspensions of *Chlorella* cells (pH 4.0) were shaken for 30 minutes in the dark at 20°. Then, a solution of NaF (pH 4.0, I) or of PCA (II and III) was tipped into the vessel from the side arm at the time indicated as 0 minute in the figure. When the acid induced evolution of gas ceased (21 minutes after the addition of the acids), the vessel with HF-treated cells (I) as well as one of the vessels with PCA-treated cells (II) were transferred to a water bath kept at 40°; the remaining vessel with PCA-treated cells (III) was transferred to a bath kept at 8.5°. Each vessel was taken out after 3 minutes and transferred back to the manometric bath (20°) where the gas evolution was determined.

minutes, there occurred an additional evolution of gas (Fig. 3). No evolution of gas was observed when HF-treated cells were kept at 40° or when PCA treated cells were kept at 8.5° for 3 minutes. These results indicate that the cells of *Chlorella* form some heat labile C-compound in aerobic darkness. This compound is also decomposed by the effect of HF but is stable to TCA or PCA. It is difficult to imagine that glutamic acid could be decarboxylated by warming to 40° for 3 minutes. Probably the compound in question is at least a part of the aerobically bound CO_2.

$^{14}CO_2$ *Fixation Experiments*

Isolation of the aerobically bound CO_2 using ^{14}C as tracer has not been successful. The following experiments showed that glutamic acid and aspartic acid are decomposed by the effect of HF.

Cell suspensions of *Chlorella* were kept in three syringes with rubber caps attached to each tip, and air containing 20% $^{14}CO_2$ was injected in the dark through the rubber cap. The syringes were shaken in the dark for 30 minutes, then the rubber cap of one syringe was taken off and the cell suspension injected into 5% PCA solution kept at 0°. Sodium fluoride (final concentration, 0.02 *M*, pH 4.0) was injected into the other two syringes through the rubber cap and the syringes were shaken under continued darkness. After the scheduled time of contact with HF the cell suspension in each syringe was injected into cold PCA solution. The PCA-treated cell suspensions were centrifuged and washed twice with water. The supernatants were combined, neutralized with KOH and the resulting precipitate (potassium perchlorate) was discarded. (It has been confirmed that the precipitate does not include any radioactive compound.) A part of the clear solution and the washed cell residue were placed on planchets, dried *in vacuo*, and the radioactivity was determined. All the above mentioned treatments were done in the cold. Another part of the neutralized supernatant was put on a Whatman No. 1 filter paper and chromatographed two-dimensionally (solvent system: phenol-water (72:28 w/w) and butanol-propionic acid-water (142:71: 100 v/v)). After chromatography, radioautograms were prepared and the radioactivity of each radioactive spot was determined.

Table I shows that the radioactivities in glutamic acid and aspartic acid were decreased by the addition of HF. The HF-induced decomposition of glutamic acid was previously reported by Warburg[1] and Vishniac.[8] Warburg proposed that this is due to an enzymatic decarboxylation of glutamic acid to γ-aminobutyric acid and CO_2. Decarboxylation of aspartic acid induced by HF has not yet been reported.

Neither glutamic acid nor aspartic acid contain unstable aerobically bound CO_2, although no other compound showed such systematic decrease in radioactivity. Most probably the unstable radioactive compounds (derived

TABLE I. Effect of HF on dark $^{14}CO_2$-fixation.

	cpm $\times 10^{-3}$		
	0 min*	4 min*	8 min*
PCA-Extract (after neutralization)	19.9	16.3	12.7
PCA-Insoluble Fraction	1.5	1.5	1.6
Glutamic Acid	7.4	5.0	2.6
Aspartic Acid	8.0	6.5	4.7
Other Compounds	3.1	4.7	3.2

* Time after the addition of HF.

from these amino acids?) were decomposed during the treatment described. Indeed our preliminary experiments showed an appreciable decrease in the radioactivity of the PCA-extract after neutralization with KOH. The experiment is being extended towards the separation of the unstable compound.

Acknowledgements

This work was supported by a grant from the National Science Foundation, U.S.A. and a grant from the Ministry of Education, Japan. The authors wish to express their gratitude to Prof. H. Tamiya and Prof. E. Hase for their helpful discussion. Thanks are also due to Mr. T.H.Ji for his kind collaboration.

REFERENCES

1. Warburg, O., *New Methods of Cell Physiology,* Interscience Publishers, New York, 1962.
2. Warburg, O., in *La Photosynthèse,* p. 221, C.N.R.S., Paris, 1963.
3. Warburg, O., and Krippahl, G., *Biochem. Z.,* **346**, 418 (1967).
4. Vernon, L. P., and Avron, M., *Ann. Rev. Biochem.,* **34**, 269 (1965).
5. Miyachi, S., Ji, T. H., and Benson, A. A., in preparation.
6. Miyachi, S., *Brookhaven Symp. in Biol.,* No. 19, p. 491, 1966.
7. Bishop, N. I., and Gaffron, H., *Biochim. Biophys. Acta,* **28**, 35 (1958).
8. Vishniac, W., in *Comparative Biochemistry of Photoreactive Systems* (M. B. Allen, ed.), p. 377, Academic Press, New York, 1960.
9. Tamiya, H., Iwamura, T., Shibata, K., Hase, E., and Nihei, T., *Biochim. Biophys. Acta,* **12**, 23 (1953).

Ion Translocation in Isolated Chloroplasts: K^+ and H^+ Ion Movements Induced by Nigericin and Their Relation to Photophosphorylation*

N. SHAVIT†, R. A. DILLEY AND A. SAN PIETRO

Charles F. Kettering Research Laboratory, Yellow Springs, Ohio

INTRODUCTION

Nigericin is an antibiotic produced by a streptomycete isolated originally by Harned *et al.*[1] The inhibition of the growth of a variety of microorganisms by this antibiotic is unique insofar as it is reversed by the addition of K^+ to the assay medium.

This antibiotic has been shown to exhibit biphasic properties in mitochondria.[2,3] At low concentrations (below 1 μg/ml) it inhibits the oxidation of many NAD-linked substrates concurrent with the loss of alkali metal cations accumulated previously in mitochondria. At higher concentrations and in the presence of K^+ ions it induces an ATPase activity, uncouples oxidative phosphorylation and prevents the accumulation of K^+ and Pi.

Recently we reported[4] some of the effects of this antibiotic on photoreactions in chloroplasts. The intent of this communication is to present evidence in harmony with the suggestion that the action of nigericin in chloroplasts, as in mitochondria, must be at the level of the ion-translocating mechanism of the membrane. Nigericin might interact specifically with a component of the chloroplast membrane which then promotes a $K^+ \rightleftarrows H^+$ exchange resulting in the dissipation of the high energy intermediate or state derived from electron transport.

EXPERIMENTAL PROCEDURES

Chloroplast isolation from spinach leaves, ATP formation, ferricyanide

* Contribution No. 279 of the Charles F. Kettering Research Laboratory. This research was supported in part by Research Grants from the National Institutes of Health, United States Public Health Service (GM-10129 to A.S.P.) and the National Science Foundation (GB-5138 to R.A.D.).

† On leave of absence from the Biochemistry Section, The Weizmann Institute of Science, Rehovoth, Israel.

reduction and ATP hydrolysis were assayed as described elsewhere.[4] The H^+ and K^+ ion fluxes and turbidity changes were measured as described previously.[5] Nigericin was a gift from Dr. R. L. Harned, Commercial Solvents Corporation, Indiana.

RESULTS

K^+ -dependent uncoupling by nigericin

The relationship between ferricyanide photoreduction and accompanying phosphorylation and nigericin concentration in a medium containing K^+ ions is shown in Fig. 1. It is apparent that nigericin is a potent uncoupler of photophosphorylation and the consequence of this uncoupling action is an enhancement (5–6 fold) of ferricyanide photoreduction. A 50 per cent inhibition of ATP formation was attained with a concentration of nigericin of 2×10^{-8} *M*. It should be noted that the experiment described in Fig. 1 was carried out with a phosphate acceptor system present.

The requirement for K^+ ions in the assay medium is illustrated in Fig. 2. In this experiment the rate of electron transport was measured in the absence of the phosphate acceptor system. Maximal stimulation of electron transport was observed at a concentration of K^+ of about 50 m*M*. The apparent Km for K^+ under these conditions was about 15 m*M*. KCl

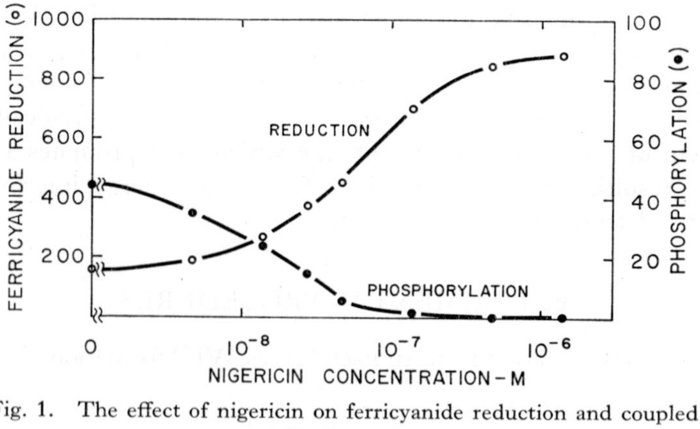

Fig. 1. The effect of nigericin on ferricyanide reduction and coupled ATP formation.

Each reaction mixture contained in a total volume of 3 ml at pH 7.8: 2 m*M* Tris-Cl, 2 m*M* $MgCl_2$, 0.33 m*M* ADP-Tris, 0.6 m*M* Pi-Tris, 0.25 m*M* $K_3Fe(CN)_6$, 50 m*M* KCl and 50–60 μg of chlorophyll as chloroplasts. Temperature, 22°; light intensity, 80,000 lux for 30 sec. Ferricyanide reduction and ATP formation are given in μmoles per mg chlorophyll per hour.

itself did not have any significant effect. Nigericin also inhibits the ATP formation with PMS and this is shown in Fig. 3. In the presence of 50 m*M* KCl the inhibitory pattern is comparable to that of the ferricyanide-supported phosphorylation. An inhibition of 50 per cent was obtained at a concentration of nigericin of about $5 \times 10^{-8} M$. In the absence of KCl, and with concentrations of nigericin of up to $10^{-6} M$, no significant inhibition

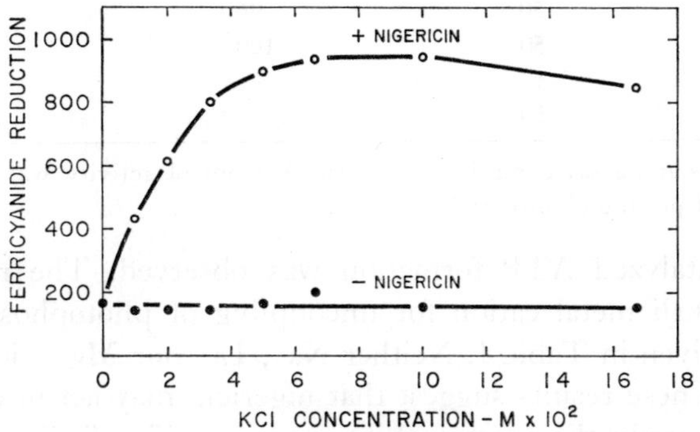

Fig. 2. The dependence on K^+ of the stimulation of electron transport by nigericin.

Reaction mixture and assay as described in Fig. 1, except that ADP, Pi and Mg^{++} were omitted. Nigericin concentration was $5 \times 10^{-7} M$. The rate of ferricyanide reduction is given in μmoles reduced per mg chlorophyll per hour.

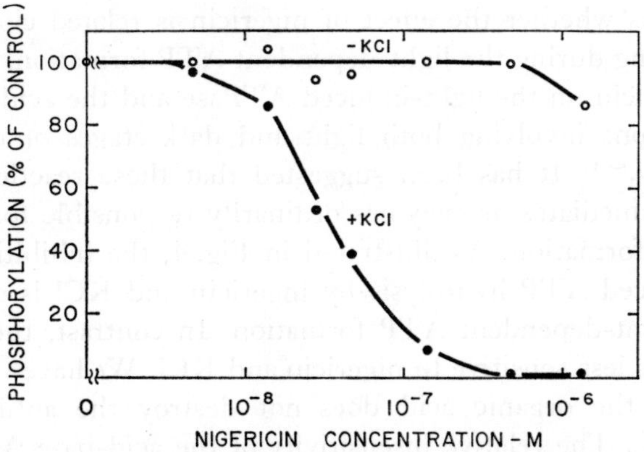

Fig. 3. The effect of nigericin on ATP formation with PMS in the presence and absence of KCl.

Reaction mixtures as described in Fig. 1, except for the presence of 0.03 m*M* PMS and, where indicated, 50 m*M* KCl. Control rates were 290 and 450 μmoles of ATP formed per mg chlorophyll per hour in the absence and presence of KCl, respectively.

TABLE I. K^+ specificity of nigericin uncoupling action.

Cation added		Nigericin concentration	
		$3\times10^{-8}\,M$	$5\times10^{-8}\,M$
	mM	per cent of control activity	
None	—	100	117
K^+	50	70	48
Na^+	50	96	92
Li^+	50	100	94
Mg^{++}	17	—	98
Mg^{++}	50	—	111

Reaction mixtures and assay conditions as in Fig. 3. Control activity was 270 μmoles of ATP formed per mg chlorophyll per hour.

of the PMS-catalyzed ATP formation was observed. The requirement for a specific alkali metal cation for uncoupling of photophosphorylation by nigericin is given in Table I. Neither Na^+, Li^+ nor Mg^{++} ions replaces K^+ efficiently. These results suggest that nigericin may act in chloroplasts by affecting primarily their permeability to a specific alkali metal cation.

Effect of nigericin on ATP formation and hydrolysis reactions

The light-dependent efflux of K^+ ions[6] and uptake of H^+ ions[7] observed in chloroplasts has been suggested to result in the formation of a high energy intermediate or state which energizes Pi and ADP to form ATP. In order to assess whether the effect of nigericin is related closely to the processes occurring during the light-dependent ATP formation, we studied the effect of nigericin on the light-induced ATPase and the acid-base ATP formation, reactions involving both light and dark stages or dark stages only, respectively.[8–10] It has been suggested that these reactions involve some of the intermediates or enzymes ordinarily responsible for the light-dependent ATP formation. As illustrated in Fig. 4, the inhibition pattern of the light-induced ATP hydrolysis by nigericin and KCl is comparable to that of the light-dependent ATP formation. In contrast, the acid-base ATP formation is less sensitive to nigericin and KCl. We have verified that incubation with the organic acid does not destroy the antibiotic (unpublished results). The relative insensitivity of the acid-base ATP formation to inhibition by nigericin and KCl could be a consequence of: (a) a change in the membrane permeability; (b) an alteration in the ion translocation mechanism; or (c) the inherent kinetics of the reaction. In the light-driven ATP formation, the energetic state is continually formed and dissipated in a cyclic manner. By comparison, the acid-base ATP formation results from the single establishment of an energetic state by an alteration of the pH external to the chloroplasts; the energetic state is postulated to

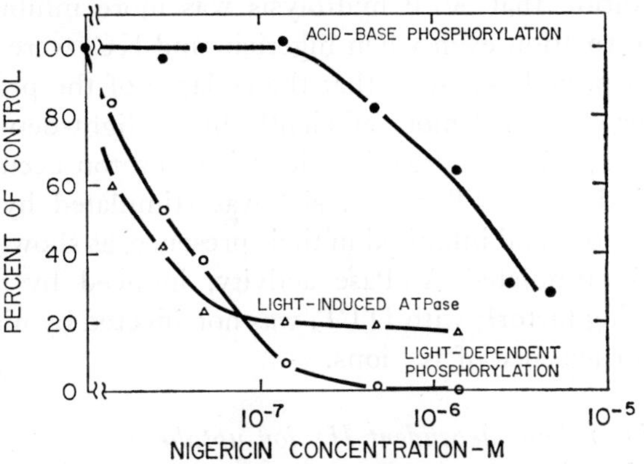

Fig. 4. The inhibition by nigericin of ATP formation and ATP hydrolysis reactions.

Reaction mixtures and assays were as described elsewhere.[4,10] Nigericin and 50 m*M* KCl were present during illumination in the light-dependent ATP formation. In the acid-base induced ATP formation and the light-induced ATP hydrolysis assays, nigericin and 50 m*M* KCl were added in the basic stage and the post-illumination stage, respectively. Control rates were: light-dependent ATP formation, 500 μmoles per mg chlorophyll per hour; acid-base induced ATP formation, 0.210 μmole per mg chlorophyll per minute; light-induced ATPase, 102 μmoles of Pi released per mg chlorophyll per hour.

TABLE II. Effect of nigericin on light-induced and coupling factor ATPase.

Reaction	Cation added		Nigericin concentration	
			$5\times10^{-8}\,M$	$5\times10^{-7}\,M$
		mM	per cent of control activity	
Light-induced ATPase	None	—	145	—
	K^+	50	66	—
Coupling factor ATPase*	None	—	108	100
	K^+	50	109	108

Light-induced ATPase assayed as described elsewhere.[4] Control activities were 120 and 107 μmoles of Pi released per mg chlorophyll per hour with and without KCl, respectively. The latent Ca^{++} -stimulated ATPase activity of a crude preparation of coupling factor was induced by incubation with DTT. (R.E. McCarty, personal communication). Control activities were 116 and 176 μmoles of Pi released per mg protein per hour with and without KCl, respectively.

* A gift from Drs. E. Racker and A. Bennun, Cornell University, Ithaca, New York.

be maintained by protons supplied by the internal organic acid. It is reasonable to expect that the kinetics of these two processes would be different.

The observation that ATP hydrolysis was more inhibited than the acid-base ATP formation even when nigericin and KCl were added in the post-illumination period, suggests that the collapse of the proton gradient (see below) is accomplished more efficiently in the light-dependent or induced processes than in the organic acid-driven proton accumulation.

The light-induced ATP hydrolysis[8] was stimulated by nigericin in the absence of K^+ ions and inhibited in their presence, as shown in Table II. The latent Ca^{++}-stimulated ATPase activity, induced by incubation of chloroplast coupling factor[11] with DTT, was not affected by nigericin either in the presence or absence of K^+ ions.

Effect of nigericin on light-dependent H^+ ion uptake

The foregoing data demonstrate that nigericin and K^+ interact with the energy transfer process in a unique manner. Other alkali metal cations or divalent cations cannot fulfill the observed K^+ requirement. This is further evident from the data presented in Fig. 5. The light-dependent H^+ ion gradient established in a medium containing KCl is collapsed by the addition of nigericin in the light; in contrast, the H^+ ion gradient established in a medium containing LiCl was not affected by the addition of nigericin. The antibiotic did not affect the H^+ ion gradient generated in a medium of either CsCl, NaCl or $MgCl_2$. Addition of nigericin prior

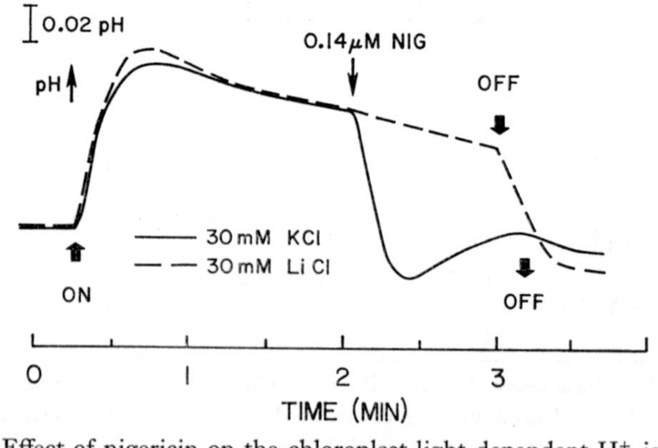

Fig. 5. Effect of nigericin on the chloroplast light-dependent H^+ ion gradient. The reaction contained 0.1 M sucrose, either 30 mM KCl (——) or 30 mM LiCl (– – – –), 200 μg of chlorophyll as chloroplasts and 15 μM pyocyanine in 10 ml total volume. Initial pH was 6.0–6.1. Illumination with red light as previously described.[5]

to illumination severely inhibited the generation of the H^+ ion gradient in KCl but not in media containing either CsCl, LiCl or NaCl. These results are consistent with an uncoupling effect of nigericin in the presence of K^+ ions and the effect of known uncouplers on H^+ ion uptake;[12] i.e. any compound tested to date which inhibits the H^+ ion uptake, also inhibits photophosphorylation. Although not shown, gramicidin inhibited the H^+ ion uptake in the presence of either CsCl, KCl, LiCl or NaCl, whereas valinomycin had no appreciable effect in any salt, in agreement with its lack of effect on photophosphorylation.[13]

Specific $K^+ \rightleftarrows H^+$ ion exchange induced by nigericin

Studies with mitochondria have revealed that a large number of antibiotic substances act as uncouplers of oxidative phosphorylation and increase mitochondrial permeability to monovalent cations; the specificity for K^+ varied with the uncoupler.[14] The effect of nigericin and K^+ ions in chloroplasts suggested that this antibiotic acts in a manner analogous to the transport-inducing antibiotics in mitochondria.[15,16] Some typical potassium and hydrogen electrode responses of chloroplast suspensions when nigericin or gramicidin was added are shown in Fig. 6. Traces a, b and c are K^+ electrode signals due to: (a) light-dependent K^+ efflux in the presence of pyocyanine; (b) nigericin-induced K^+ efflux in the dark; and (c) gramicidin-induced K^+ efflux in the dark, respectively. These signals were obtained in a medium containing 0.1 *M* tris-acetate in order to assure that the electrode responses obtained were not a consequence of a change of the pH of the medium. Under these conditions it is apparent that either nigericin or gramicidin causes a K^+ efflux in the dark which is equivalent to the light-dependent K^+ efflux.[6] However, this dark K^+ efflux is not unique to nigericin since gramicidin (Fig. 6, trace c) or valinomycin, Triton X-100 and chlorpromazine gave similar results (unpublished experiments). When the K^+ efflux response reached its maximal value either by illumination or in the dark by addition of nigericin or gramicidin, no further transport of K^+ could be observed by any combined addition of light or antibiotics. These observations suggest that both the dark and the light efflux originate from the same pool of internal K^+. The initial efflux of K^+ induced by nigericin and other uncouplers probably results from the existent concentration gradient, since these experiments were performed in the absence of externally added K^+ ions and the internal K^+ concentration was probably in the range of 10 to 40 m*M*.[5]

The response observed with the hydrogen electrode is given in Fig. 6, trace d. It is evident that the addition of nigericin or gramicidin to a chloro-

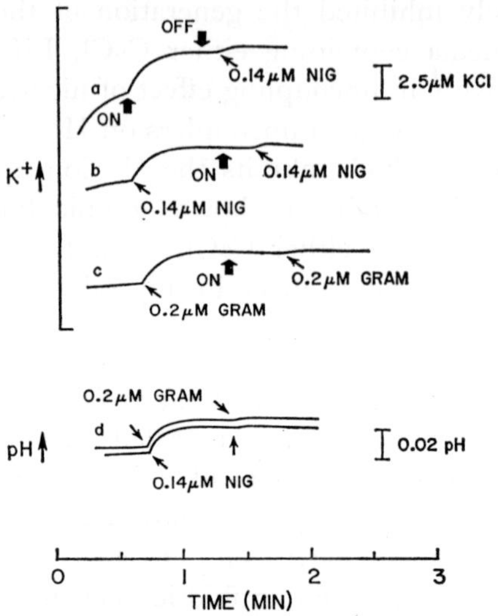

Fig. 6. Nigericin- and gramicidin-induced $K^+ \rightleftarrows H^+$ dark exchange in low salt medium.

Potassium efflux was measured in chloroplast suspensions with a cation electrode.[5] The reaction mixtures for K^+ measurement contained the following: 0.1 *M* tris-acetate pH 6.0, 155 μg of chlorophyll as chloroplasts and 15 μ*M* pyocyanine in a total volume of 10 ml. An increase in the K^+ concentration in the suspension is given by an upward deflection of the trace. (↑, ↓) Light on, light off. (↖) Addition of an antibiotic. (a) Light-induced K^+ efflux; (b and c) K^+ efflux caused by addition of 0.14 μ*M* nigericin or 0.2 μ*M* gramicidin, respectively, in the dark.

H^+ ion uptake was measured[5] in chloroplast suspensions containing the following: 0.2 *M* sucrose, 155 μg of chlorophyll as chloroplasts and 15 μ*M* pyocyanine in 10 ml total volume. The pH was adjusted to 6.0 with standard KOH. An increase in the pH of the suspension is given by an upward deflection. (d) H^+ uptake caused by addition of 0.14 μ*M* nigericin or 0.2 μ*M* gramicidin (indicated by the arrows) in the dark.

plast suspension, in the absence of externally added salts, causes an uptake of H^+ ions in the dark. This dark H^+ ion uptake appears to have similar kinetics to the dark K^+ efflux observed under the same conditions (Fig. 6, traces b and c). These results demonstrate the existence of a dark $K^+ \rightleftarrows H^+$ ion exchange induced by nigericin and gramicidin. The K^+ efflux induced by nigericin was verified using atomic absorption spectrophotometry. The stoichiometry of this exchange reaction was in the range of 1.2–1.5 K^+ effluxed per H^+ ion taken up.

The $K^+ \rightleftarrows H^+$ exchange reaction, observed in the absence of externally

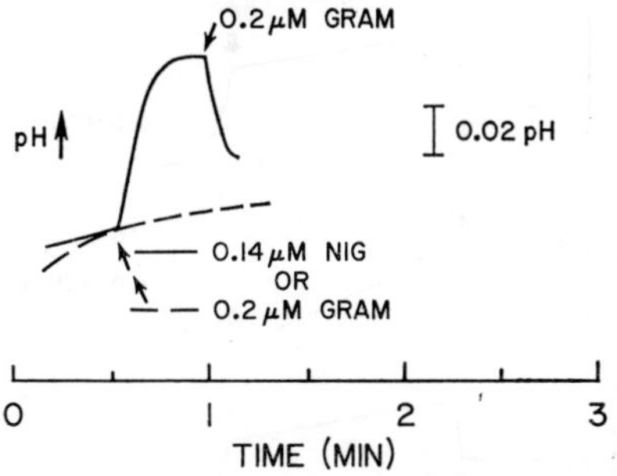

Fig. 7. Nigericin-induced H^+ uptake and gramicidin-induced H^+ efflux in the dark and in high salt.

The reaction mixture contained the following: 0.1 *M* LiCl, 0.2 *M* sucrose, and 200 μg of chlorophyll as chloroplasts in a total volume of 10 ml at pH 6.0. (——) 0.14 μ*M* nigericin added at bottom arrow and 0.2 μ*M* gramicidin added at top arrow; (– – – –) 0.2 μ*M* gramicidin added at bottom arrow.

added salts, is not unique to nigericin since it is induced by other compounds as well. It was of interest therefore to study the nigericin- and gramicidin-induced H^+ ion uptake in the presence of different externally added salts. As shown in Fig. 7, when the external salt present was LiCl, addition of nigericin induced an H^+ ion uptake (solid line). This H^+ uptake in the dark was partially reversed by the subsequent addition of gramicidin. Moreover, similar H^+ ion gradients could be induced by nigericin in either CsCl or NaCl but not in KCl. Gramicidin added initially did not cause an uptake of H^+ in 0.1 *M* LiCl (Fig. 7, dashed line), or in CsCl, NaCl, and KCl. This is as expected, since Chappell and Crofts[17] have shown that in mitochondria, gramicidin induces greater permeability to all the alkali metal cations. These results are in accord with the postulate that nigericin increases the permeability of the chloroplast membrane specifically to K^+, whereas gramicidin-induced permeability is not cation specific.[17] Valinomycin gave a response similar to nigericin in LiCl or NaCl media, including the gramicidin induced collapse of the pre-established H^+ gradient. However, with valinomycin the requirement for the cation is less specific than with nigericin, since Cs^+ can replace K^+. No H^+ ion uptake induced by valinomycin was observed in media containing either CsCl or KCl (unpublished results).

Turbidity changes induced by nigericin upon illumination

Nigericin caused a decrease in the turbidity of chloroplast suspensions

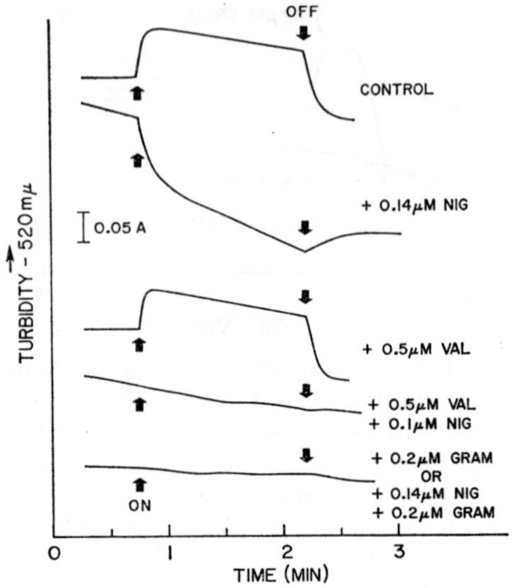

Fig. 8. Effect of nigericin on light-induced turbidity changes of chloroplast suspensions.

The reaction mixtures contained 0.1 *M* KCl, 20 μg of chlorophyll as chloroplasts and 15 μ*M* pyocyanine in a volume of 1 ml. Various additions were made as indicated.

upon illumination. This decrease in turbidity is rapid and dependent upon the presence of KCl in the assay medium. As shown in the uppermost trace of Fig. 8, chloroplasts suspended in KCl show, upon illumination, an increase in turbidity. In contrast, the presence of nigericin plus KCl reverses the direction of the light-induced turbidity change (Fig. 8, second trace). In the presence of CsCl, NaCl, or LiCl, nigericin gave results essentially identical to the control reaction, i.e. a light-induced turbidity increase occurred.

The light-induced increase in turbidity in the control experiments was found to be inhibited by gramicidin but not by valinomycin (Fig. 8, third and fifth traces). The light-induced decrease in turbidity induced by nigericin in the presence of KCl was inhibited both by gramicidin and valinomycin (Fig. 8, fourth and fifth traces).

DISCUSSION

The results presented above confirm and extend our prior observations[4] on the effect of nigericin on the energy transfer steps in photophosphorylation. Nigericin is the most potent uncoupler of photophosphorylation known to date, and is unique in that K^+ is specifically required for the uncoupling to occur. Proposed mechanisms of uncoupling of photophosphorylation assume that either the uncoupler acts by catalyzing the hydrolysis of a high-energy intermediate[18] or that the uncoupler discharges the proton potential across the membrane.[19] Uncoupling by nigericin in the presence of K^+ ions could be explained by assuming that it promotes specifically the uptake of K^+ supported by a high-energy intermediate which otherwise is utilized to form ATP from ADP and Pi. Whether nigericin induces the utilization of a high-energy intermediate or the dissipation of a H^+ ion gradient by increased permeability to K^+ is not distinguished in these experiments. The data presented indicate clearly that nigericin has a profound effect on the permeability of the chloroplast membrane to K^+ and H^+ ions.

Most of the uncouplers of photophosphorylation have been shown to inhibit the production and/or maintenance of the light-dependent H^+ ion gradient in chloroplasts.[7,12] The establishment of such an H^+ ion gradient has been postulated to be a necessary step in the synthesis of ATP by chloroplasts.[7,19,20] However, nigericin does not inhibit the H^+ ion uptake or ATP synthesis unless K^+ ions are present in the assay medium. In contrast, the inhibition of the proton uptake and ATP formation by other uncouplers is not dependent upon the presence of any specific cations.

In the present study three transport-inducing antibiotics were compared with respect to their effect on light-dependent reactions and the induction of ion exchanges in the dark. Gramicidin, valinomycin, and nigericin were equally effective in causing an exchange *in the dark* of external H^+ for internal K^+ when the chloroplasts were suspended in a medium without added salts (Fig. 6). The fact that the stoichiometry of the exchanges was close to unity supports the claim that we are dealing with an exchange reaction resulting probably from the collapse of the K^+ permeability barrier. There is diffusion of K^+ outward driven by the chemical potential counterbalanced by the inward flow of H^+. These agents appear to increase the permeability of the membrane toward alkali metal cations, without appreciably affecting the permeability to anions; otherwise anion transport could accompany the cation and exclude the cation exchange phenomena observed. Moreover, the effect of these three agents on the H^+-cation exchange when the concentration gradient of the cations is reversed (high salt, Fig. 7), indicates that gramicidin is indeed capable of exchanging H^+ for Li^+, Cs^+, K^+, or Na^+; valinomycin exchanges K^+ or Cs^+ for H^+; and nigericin only exchanges K^+ for H^+. These data agree with the known effects of the three agents on mitochondrial ion movements.[15,17]

The induction of a $K^+ \rightleftarrows H^+$ exchange in the dark by these three agents does not seem to us sufficient to explain their activity on chloroplast photo-reactions. Thus, nigericin uncouples photophosphorylation, inhibits the generation of a H^+ ion gradient and other processes (Fig. 4; Table II) in the presence of about 50 m*M* KCl; gramicidin affects them in the same manner in the presence or absence of any one of the alkali metal cations; and valinomycin is without effect on these reactions under similar conditions. The exchange reactions induced by these antibiotics *in the dark* clearly demonstrate their capacity to enhance the membrane permeability to cations (with significant differences in their specificity for alkali metal cations). However, the relationship between these dark exchanges and the effect of the antibiotics on light-induced chloroplast reactions is not ob-obvious, since the vectorial movement of alkali metal cations in the presence of the antibiotics under illumination should be opposite to that observed in the dark.

It appears that nigericin interacts with the ion translocation system of the chloroplast membrane in a manner akin to the action of some transport-inducing antibiotics in mitochondrial systems. This interaction could cause the dissipation of the H^+ ion gradient by affecting the exchange diffusion system, presumably by transport of K^+ inwards in exchange for protons. We may propose the following series of exchange reactions which occur subsequent to the light-driven H^+ ion uptake: (*a*) In the presence of nigericin, an exchange of internal H^+ for external K^+ resulting in the uptake of K^+. An exchange of H^+ for K^+ would not by itself lead to anion uptake since there would be no change in internal cation concentration. However, the H^+ uptake could continuously support H^+-K^+ exchange, leading to K^+ accumulation in an amount greater than the observed maximum H^+ gradient. This additional K^+ uptake could result in an electrochemical potential sufficient to drive Cl^- uptake with an osmotic equivalent of water, resulting in the observed swelling (Fig. 8). The observation that the rate of turbidity decrease was less in the presence of nigericin plus K_2SO_4 (i.e., a less permanent anion) compared to KCl is consistent with this mechanism. (*b*) In the presence of valinomycin alone, there is an increased permeability to K^+ but not to H^+ ions. Thus, valinomycin does not affect either the light-induced shrinkage or the H^+ ion gradient, nor does it induce a swelling as does nigericin (Fig. 8). It appears therefore that the mechanism of action of valinomycin is different from that of nigericin. The inhibition by valinomycin of the light-induced swelling observed with nigericin (Fig. 8) is consistent with the view that valinomycin causes an increase in K^+ ion permeability. Hence, no K^+ ion accumulation would occur because of the rapid equilibration of external and internal K^+ ions. In the absence of any accumulation of K^+ ions, there could be no uptake of anions (and water) and, therefore, no swelling would be observed. (*c*) For the case of gramicidin the mechanism of action must be different from that of nigericin and valinomycin, since gramicidin inhibits the H^+ ion uptake in the absence of any added alkali metal cation and does not induce swelling upon illumination of chloroplastsin the presence of salts. If gramicidin enhances the permeability of the membrane to alkali metal cations and H^+ ion, its uncoupling action does not involve the exchange reactions postulated to participate in the case of nigericin.

It remains to be demonstrated that nigericin induces the accumulation of K^+ under conditions in which nigericin and K^+ have been shown to inhibit ATP synthesis.

Acknowledgments

We thank Drs. L. P. Vernon and D. Fleischman for their helpful discussions. The excellent technical assistance of Mrs. Bonnie Grow and Mrs. Joan Platt is gratefully acknowledged.

REFERENCES

1. Harned, R. L., Hidy, P. H., Corum, C. J., and Jones, K. L., *Antibiot. Chemotherap.*, **1**, 594 (1951).
2. Graven, S. N., Estrada-O. S., and Lardy, H. A. *Biochemistry*, (Wash.), **6**, 365 (1967).
3. Estrada-O. S., Graven, S. N., and Lardy, H. A., *J. Biol. Chem.*, **242**, 2925 (1967).
4. Shavit, N., and San Pietro, A., *Biochem. Biophys. Res. Comm.*, **28**, 277 (1967).
5. Dilley, R. A., and Vernon, L. P., *Arch. Biochem. Biophys.*, **111**, 365 (1965).
6. Dilley, R. A., *Biochem. Biophys. Res. Comm.*, **17**, 716 (1964).
7. Neumann, J., and Jagendorf, A. T., *Arch. Biochem. Biophys.*, **107**, 109 (1964).
8. Carmeli, C., and Avron, M., *Biochem. Biophys. Res. Comm.*, **24**, 923 (1966).
9. Jagendorf, A. T., and Uribe, E., *Proc. Natl. Acad. Sci.* (Wash.), **55**, 170 (1966).
10. Shavit, N., and Boyer, P. D., *J. Biol. Chem.*, **241**, 5738 (1966).
11. Vambutas, V. K., and Racker, E., *J. Biol. Chem.*, **240**, 2660 (1965).
12. Jagendorf, A. T., and Neumann, J., *J. Biol. Chem.*, **240**, 3210 (1965).
13. Avron, M., and Shavit, N., *Biochim. Biophys. Acta*, **109**, 317 (1965).
14. Pressman, B. C., *Proc. Natl. Acad. Sci.* (Wash.), **53**, 1076 (1965).
15. Moore, C., and Pressman, B. C., *Biochem. Biophys. Res. Comm.*, **15**, 562 (1964).
16. Estrada-O. S., Graven, S. N., and Lardy, H. A., *Fed. Proc.*, **26**, 610 (1967).
17. Chappell, J. B., and Crofts, A. R., *Biochem. J.*, **95**, 393 (1965).
18. Slater, E. C., *European J. Biochem.*, **1**, 317 (1967).
19. Mitchell, P., *Biol. Rev.*, **41**, 445 (1966).
20. Dilley, R. A., *Brookhaven Symp. Biol.*, No. 19, p. 258 (1966).

Comparative Studies on Phosphorylation and Apparent Proton Changes in Illuminated Chromatophores*

Konstantine Cost and Albert W. Frenkel

Department of Botany, College of Biological Sciences, University of Minnesota, Minneapolis, Minnesota

INTRODUCTION

A light-induced interaction between chromatophores from *Rhodospirillum rubrum* and the acid base indicator dye bromthymol blue (BTB) has been described recently.[1,2] This interaction has been ascribed to internal light-induced pH changes which involve a proton flux,[2] or to apparent changes in pH caused by light-induced conformational changes in some protein component(s) of the chromatophore.[1] Antonini *et al.*[3] several years ago pointed out that BTB is a useful indicator of conformational changes in the Hemoglobin-Globin system; their theoretical analysis is an important starting point in the interpretation of other BTB interactions with complex protein systems.

The light-induced BTB-chromatophore absorbancy changes reported earlier have been shown to be sensitive to Antimycin A and to a number of other inhibitors of electron transport pointing to a requirement of electron transport for this phenomenon.[1] More recently, an attempt has been made to correlate such light-induced changes with light-induced phosphorylation activity.[4] It is apparent from the experimental evidence presented below, that the reaction system which gives rise to the light-induced BTB reaction also is required for light-induced phosphorylation; an active BTB reaction, however, is not a sufficient criterion for the operation of an intact phosphorylation system.

* Supported in part by grants from the National Institute of Allergy and Infectious Diseases of the National Institutes of Health, U. S. Public Health Service (AI-02218) and from the Graduate School of the University of Minnesota.

Abbreviations: BTB, bromthymol blue or 3,3′-dibromothymolsulfonphthalein; PMS, phenazine methosulfate or 5-methylphenazinium methyl sulfate; ADP, adenosine diphosphate; P_i, orthophosphate; PSP, photosynthetic phosphorylation; (Photophosphorylation; light-induced phosphorylation).

Light-induced absorbancy changes of BTB at 6080 Å in the presence of chromatophores illuminated with red light will be referred to in the text as "the light-induced BTB (chromatophore) reaction."

MATERIALS AND METHODS

Actively growing cells from 24–36-hour 0.5 *l* cultures were washed twice with water; the resulting cell pellet was then suspended in 20 ml of 0.2 *M* potassium glycylglycine buffer (pH 7.75) and this suspension was frozen at −20° C until such time when chromatophores were prepared from these cells. The chromatophores were prepared as described previously,[5] with the exception that they were washed three times with distilled water, and then were suspended either in an appropriate amount of distilled water, or in 0.1 *M* potassium glycylglycine buffer (pH 7.75), or in 0.1 *M* potassium phosphate buffer (pH 7.7). Except where indicated, chromatophores were stored at 2° C for not longer than 36 hours after their preparation. Reactions were carried out in modified Thunberg cuvettes which were evacuated exhaustively; this procedure compared favorably with argon flushing. Chromatophores used in the aging experiments for determining PSP activity were sparged with argon and then stored under argon at 22° C for the duration of the experiment.

Absorbancy changes of BTB were measured with a Cary Model 14 spectrophotometer. Illumination of the cuvette was through a CS7-69 deep red Corning filter at right angle to the measuring beam. The sensitive pH measurements were carried out with a Radiometer TTT1 meter in conjunction with a Texas servoriter as described elsewhere.[1,6] A glass electrode obtained from Arthur H. Thomas Co. (No. 4858-L15) proved most satisfactory for these measurements.

RESULTS

Stability of Light-induced BTB absorbancy changes

Chromatophores were incubated for the duration of the experiment at 22° C. At the start of the experiment and then approximately every 24 hours the cuvettes were illuminated with red light and optical density changes were monitored at 6080 Å until illumination produced no further change in optical density; the results of a sample experiment are shown in Figs. 1 and 2. The buffered reaction mixtures showed higher rates and greater total color changes than the unbuffered mixtures. Over the entire period of time, each reaction mixture showed an almost constant total absorbancy change at 6080 Å (Fig. 2); the rate of the absorbancy change, however, increased throughout the duration of the experiment (Fig. 1). Bacterial growth was not detected in the course of these experiments.

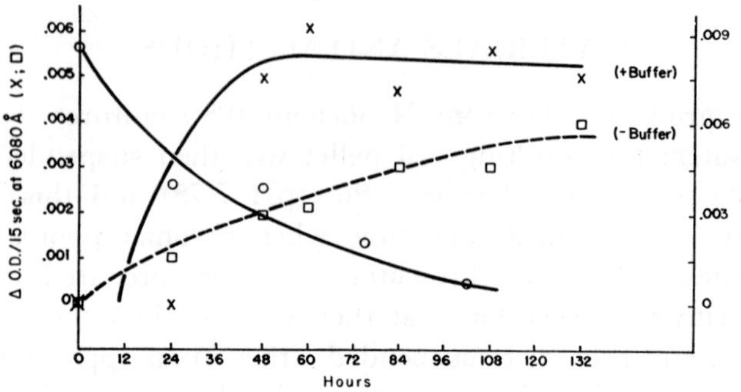

Fig. 1. The effect of aging on rates of light-induced absorbancy changes of BTB and of pH changes monitored with a glass electrode in the presence of chromatophores (washed 3×). Rates represent initial slopes (for 15 seconds); these were measured once a day up to 6 days. Chromatophore preparations were incubated at 22°C and rate measurements were carried out at the same temperature.

All concentrations represent final concentrations. Bacteriochlorophyll concentration in all preparations: $3\times10^{-5}M$.

×—× Additions: $3\times10^{-5}M$ BTB; $2\times10^{-3}M$ Pi. Final volume 5.0 ml. pH 7.7.

□—□ Additions: $3\times10^{-5}M$ BTB; (no buffer added). Final volume 5.0 ml. pH approximately 6.7.

○—○ Unbuffered chromatophore suspension incubated at 22°C in the dark. To aliquots of this preparation at each point the following additions were made: $2\times10^{-3}M$ Pi; $2\times10^{-4}M$ ADP (adjusted to pH 7.7 prior to addition). Final volume 2.5 ml. pH 7.7.

pH changes

A series of reaction mixtures identical with those described above except for the omission of BTB were monitored for pH changes during and after illumination; no pH changes could be observed under these conditions within the sensitivity range of the instrument (limit of detection, 6×10^{-4} of a pH unit).

Marked pH changes, whose initial rates are plotted in Fig. 1, were observed upon addition of ADP and orthophosphate to cuvettes containing pre-measured amounts of chromatophores in distilled water (incubated at 22° C). It is assumed that these pH changes are a measure of the light-induced phosphorylation activity of the chromatophores.[6] It is evident from Fig. 1 that the phosphorylating activity deteriorated by the end of 108 hours (22° C), while the light induced BTB reaction continued undiminished.

PMS, which is known to restore photosynthetic phosphorylation in

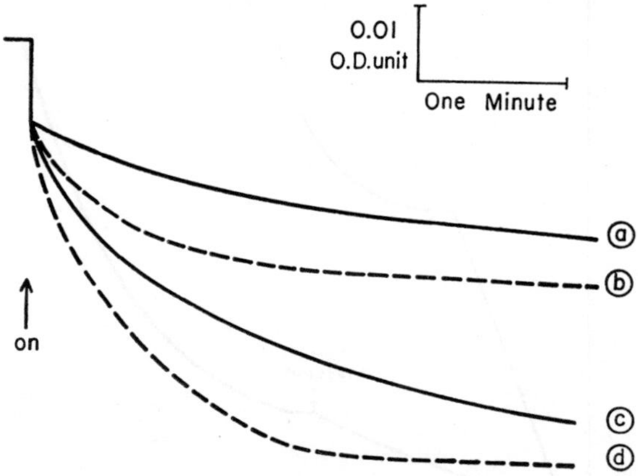

Fig. 2. The effects of aging on the rates and on the extent of the light-induced absorbancy changes of BTB at 6,080 Å in the presence of well washed chromatophores. Incubations and measurements carried out at 22°C.

Final concentration of bacteriochlorophyll: $3 \times 10^{-5} M$. All other concentrations represent final values. Final volume of reaction mixtures: 5.0 ml.

ⓐ–ⓑ Addition: $3 \times 10^{-5} M$ BTB; (no buffer added); pH approximately 6.7.

ⓐ Kinetics of light induced absorbancy change at start of experiment.

ⓑ Same preparation as in ⓐ. Kinetics of light-induced absorbancy change after 9 days of dark incubation.

ⓒ–ⓓ Additions: $3 \times 10^{-5} M$ BTB; $2 \times 10^{-3} M$ Pi; final pH 7.7.

ⓒ Kinetics of light-induced absorbancy change at start of experiment.

ⓓ Same preparation as in ⓒ. Kinetics of light-induced absorbancy change after 9 days of dark incubation.

aged chloroplasts,[7] also increased the PSP activity of unbuffered chromatophores incubated for 76 hours at 22° C to nearly its initial value (final conc. of PMS 1.3×10^{-6} *M*). Chromatophores incubated in potassium phosphate buffer (pH 7.7) for 76 hours at 22° C did not show any loss of PSP activity over this period of time.

Reactivation of the light-induced BTB reaction

At a time when the unbuffered reaction mixture containing BTB would show no further light-induced changes (approximately 15 days at 22° C), addition of PMS caused a complete reactivation of the light-induced BTB reaction. Addition of either phosphate or glycylglycine buffers produced effects similar to that of PMS but of lesser magnitude.

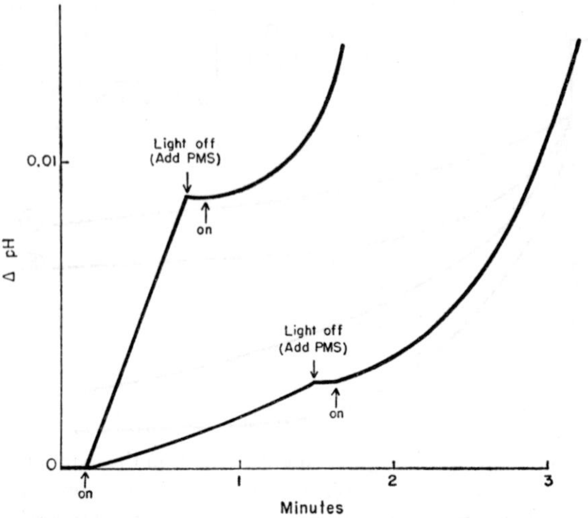

Fig. 3. Reactivation by PMS of photosynthetic phosphorylation activity in chromatophore preparations aged at 22°C.

Top curve: Fresh preparation of chromatophores. Bacteriochlorophyll concentration $3 \times 10^{-5} M$; additions: $2 \times 10^{-3} M$ orthophosphate; $2 \times 10^{-4} M$ ADP; (final concentrations). Final volume: 5.0 ml. pH 7.7. PMS added at arrow to give final concentration of $5 \times 10^{-6} M$.

Bottom curve: Aliquot of the same preparation as above but aged for 92 hours in the dark under Argon. PMS addition restores activity to nearly the same rates as in top curve.

Effect of Antimycin A

The light-induced BTB reaction in buffered or unbuffered reaction mixtures was always inhibited by Antimycin A ($4 \times 10^{-6} M$) even in reaction mixtures incubated as long as 9 days. This inhibition could be reversed by the addition of PMS, indicating that even in aged preparations the Antimycin A-sensitive portion of the electron transport system was still operative.

Light-induced pH changes of chromatophores treated with ethylene glycol

Chromatophores suspended in a mixture of ethylene glycol and 0.1 *M* glycylglycine buffer pH 7.75 (4/6; v/v) and stored at −10° C for 24 hours[8] did give pH changes (in the alkaline direction) upon illumination (Fig. 4), in contrast to chromatophores which had not been treated with ethylene glycol. Chromatophores from the same preparation suspended in 0.1 *M* glycylglycine buffer stored at −10° C for 24 hours also gave pH changes which were smaller than those of the preparations treated with ethylene glycol (Fig. 4).

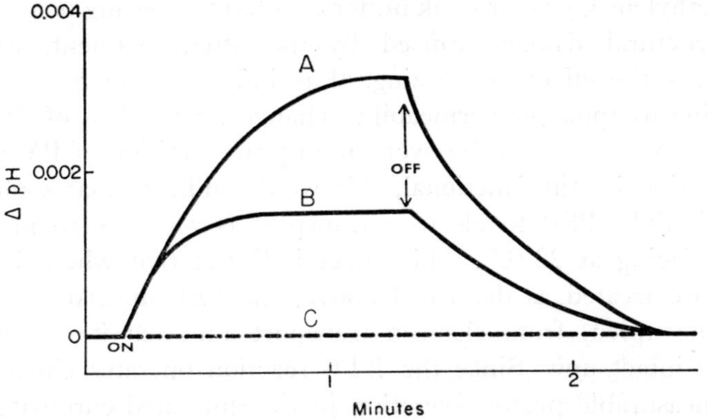

Fig. 4. Light-induced pH changes observed with ethylene-glycol treated chromatophores.

All measurements monitored with a glass electrode at 22°C. Bacteriochlorophyll, final concentration $= 3 \times 10^{-5} M$.

Curve A: Freshly prepared chromatophores pretreated in ethylene glycol: 0.1 *M* glycylglycine buffer pH 7.7 (40: 60 v/v) at −10°C for 24 hours. 0.2 ml of this preparation diluted with distilled water to 2.5 ml.

Curve B: 0.2 ml. of cold-treated preparation in 40% ethylene glycol diluted with 0.1 *M* glycylglycine buffer pH 7.7. Final volume 2.5 ml.

Curve C: Freshly prepared chromatophores suspended in distilled water, not subjected to cold pretreatment. Incubated at 22°C for 24 hours. Final volume 2.5 ml.

Crude unwashed preparations also exhibited light-induced pH changes which were not reversible and appeared to be due to the presence of residual components required for PSP.[9]

DISCUSSION

The above experiments, we feel, demonstrate that the light-induced BTB reaction can be more accurately described as a measure of cyclic electron flow rather than light-induced phosphorylation, or proton excretion into the surrounding medium. The light-induced BTB reaction is more stable, and functions even after the PSP activity in the absence of PMS has been lost. It is apparent that the light-induced absorbancy change of BTB does not require PSP activity for its existence, while all preparations which can carry out PSP also carry out the light-induced BTB reaction. The lack of any light-induced pH changes in untreated chromatophore preparations as opposed to the changes seen in chromatophores treated with ethylene glycol or weak buffer exposed to freezing temperatures points to structural damage caused by the latter treatment. Ethylene glycol enhances the effect of freezing, thus indicating further structural damage leading to apparent permeability changes. Since loss of PSP activity in these experiments can be overcome by the addition of PMS, while the BTB reaction is still functional, this would indicate that some component which links PSP to electron transport is sensitive to mild heat denaturation (aging at 22° C). This effect indicates that when the chromatophores are treated as described above, the PSP mechanism is only separated (uncoupled) from electron transport and that its components remain as an intact unit. Since the BTB reaction operates either in the absence of measurable proton excretion in the untreated chromatophores and also in the ethylene glycol-frozen chromatophores, the BTB reaction appears to be independent of proton changes be they "internal" or "external". Some reasonable explanations for this reaction are that it may be due to light-induced conformational changes in the chromatophore proteins; a light-stimulated unfolding of the protein could make weakly acid groups more accessible to the BTB. Furthermore, as a result of light-induced electron transport, weakly acid groups could be produced whose pK is appreciably lower than the pH of the ambient environment; such groups could then complex with BTB and result in a decreased absorbancy of the dye. Which one of these possibilities might be of importance here requires further elucidation. These possibilities seem plausible and it appears that investigations along this line may yield some useful information regarding the character and origin of the light-induced BTB reaction.

SUMMARY

The experiments described demonstrate that the light-induced absorbancy changes of bromthymol blue in the presence of chromatophores continues undiminished in preparations with or without added buffer when incubated at 22° C for nine days or longer, while the capacity of unbuffered preparations to carry out light-induced phosphorylation disappears in about four days.

Chromatophores capable of carrying out active photophosphorylation always have the capacity to bring about light-induced absorbancy changes of BTB; the latter reaction, however, in itself is not an adequate index of the capacity of the chromatophores to carry out photophosphorylation.

Phenazine methosulfate at low concentrations can fully restore the photophosphorylation activity of chromatophore preparations inactivated at 22° C.

REFERENCES

1. Cost, K., and Frenkel, A. W., *Biochemistry*, **6**, 663 (1967).
2. Chance, B., *J. Biol. Chem.*, **240**, 2729 (1965).
3. Antonini, E., Wyman, J., Moretti, R., and Rossi Fanelli, A., *Biochim. Biophys. Acta*, **71**, 124 (1963).
4. Nishimura, M., and Kadota, K., *Federation Proc.*, **26**, 610 (1967).
5. Frenkel, A. W., and Hickman, D. D., *J. Biophys. Biochem. Cytol.*, **6**, 285 (1959).
6. Nishimura, M., Ito, T., and Chance, B., *Biochim. Biophys. Acta*, **59**, 177 (1962).
7. Ash, O. K., *Ph. D. thesis*, p. 105, Brigham Young University (1961).
8. Parson, W. W., *Biochim. Biophys. Acta*, **131**, 154 (1967).
9. Frenkel, A. W., *J. Biol. Chem.*, **222**, 823 (1956).

Chairman's Remarks

One of the more important themes running through this symposium and one linking discussions on the structural, functional, and even the developmental and evolutionary aspects of the photosynthetic apparatus has revolved around the question of whether Photosystems I and II are localized within the lamellae in distinct and separable structural entities. Attempts were made in this Symposium to join, although as yet not too successfully, the two main experimental approaches used in this problem. These approaches were, on the one hand, electron microscopic examination of the chloroplast lamellae, especially by the freeze-etching technique (papers confined to Part I), and on the other hand, the extraction of the lamellae by detergents followed by fractionation into two lipoprotein particles, the one showing reactions indicative of Photosystem I and the other of Photosystem II. Several papers and much of the discussion in Part II centered around the use of detergents for separating the two photosystems and it appeared that improvements in this technique are reaching the point where a complete separation is possible of lipoprotein complexes containing the essential elements of Photosystem I and Photosystem II, respectively (H. Huzisige). D.I. Arnon also considered that the detergent method could be used to physically separate two photosystems, but he disagreed with generally accepted ideas on the identity of the electron transfer pathways associated with each photosystem. Instead, he proposed that Photosystem I should be identified with the process of ferredoxin-mediated cyclic photophosphorylation and Photosystem II with non-cyclic photophosphorylation.

The important question of whether treatment of chloroplast lamellae with detergents liberates two distinct particles, or whether it merely strips away a lipoprotein particle leaving behind a lamellar membranous fraction deficient in Photosystem I activity but still containing Photosystem II, remains to be resolved. In this connection, N.K. Boardman's paper which clearly described the existence of two distinct *b*-type cytochromes in chloroplasts is important (the existence of these cytochromes was originally recognised by Lundegåordh, but ignored in much of the subsequent work on photosynthetic electron transfer), since these cytochromes probably are the most convenient markers for distinguishing Photosystems I and II, the cytochrome-562 being associated with Photosystem I and cytochrome-559 with Photosystem II.

Related to considerations of the properties of detergent-solubilized chloroplast lamellae were studies on a soluble chlorophyll-protein complex which can be extracted from *Chenopodium* leaves. Further information on the photochemistry of this complex (A. Takamiya *et al.*) and on the splitting and reconstitution of the chlorophyll and protein moieties (E. Yakushiji *et al.*) was reported.

Understandably, most studies of the two light reactions of photosynthesis are confined to oxygen-evolving organisms, and it is often assumed that, since photosynthetic bacteria do not evolve oxygen and hence presumably lack Photosystem II, one light reaction only is involved in bacterial photosynthesis. The paper by S. Morita suggests this may be an over-simplification and that several light reactions may be involved.

The majority of the other investigations reported in Part II were also designed to further elucidate the pathway of photosynthetic electron transfer in bacteria and plants. Many interesting findings were described, especially on the role of cytochromes in photosynthesis (R.G. Bartsch, Y. Fujita, M. Nishimura), and on the discovery of secondary sites of inhibition by DCMU (S. Katoh and A. San Pietro, S. Izawa). A paper by E.C. Weaver elaborated on the already significant contributions made by studies of E.P.R. to our understanding of light-induced electron transfer reactions.

Although all of the papers given in this Session have provided new and valuable data on the mechanism of electron transfer in photosynthesis, very little new information that is fundamental to our understanding of the process was presented, and one is left with the general impression that new experimental approaches will be needed in this particular field of investigation before rapid advances can be expected. Studies on the physical separation of the two photosystems of photosynthesis, already mentioned above, may be considered as one such new approach, and the potential importance of another fairly recent experimental approach, namely, the study of light-induced proton movements, will become obvious from a perusal of the papers by N. Shavit *et al.*, M. Avron, and K. Cost and A.W. Frenkel.

Finally, mention should be made of the isolation from the blue-green alga *Anabaena cylindrica* of nitrate and nitrite reductases, both of which can couple directly to the photosynthetic electron transfer pathway via ferredoxin (A. Hattori). This observation demonstrates an interesting example of a change in enzymic specificity dictated by different degrees of structural complexity, since in higher plants the nitrite reductase, which is localized inside the chloroplast, can couple directly with ferredoxin, but in the case of the nitrate reductase, which is located outside the chloroplast, NADPH rather than reduced ferredoxin is the preferred substrate. The lamellae of blue-green algae are not separated from the cytoplasm by a limiting membrane and apparently reduced ferredoxin is the reductant for both reductases.

ROBERT M. SMILLIE

PART III

DEVELOPMENTAL ASPECTS

Differential Degradation of Photosynthetic Activities during the Process of Glucose-induced Degeneration of Chloroplasts in *Chlorella Protothecoides*

TAMIKO OH-HAMA, MITSUO MATSUKA AND EIJI HASE

Institute of Applied Microbiology, University of Tokyo, and Tokugawa Institute for Biological Research, Tokyo

INTRODUCTION

Recent studies in this laboratory[1] have demonstrated that reversible degeneration and regeneration of chloroplasts are induced in the cells of *Chlorella protothecoides* by controlling the concentration balance between carbon and nitrogen sources in culture media under certain light conditions. When the algal cells are grown in a medium rich in glucose and poor in nitrogen source (urea), apparently chlorophyll-less cells containing profoundly degenerated plastids—called "glucose-bleached" cells—are produced either in the light or in darkness. When the "glucose-bleached" cells are incubated in a medium rich in nitrogen source without added glucose, they turn normal green in the light and pale green in darkness. In the previous work,[2] the development of photosynthetic and related activities in the "glucose-bleached" cells was followed in relation to the formation of chlorophyll and chloroplast structures. In the present work, investigations were made on degradation of photosynthetic activities during the process of bleaching of green algal cells induced by glucose.

MATERIALS AND METHODS

Preculture

The "glucose-bleached" cells, which had been obtained by growing the alga in the light for 5 days in medium containing 1% glucose and 0.1% urea in addition to the basal mineral nutrients,* were transferred asepti-

Abbreviations: FMN, flavin mononucleotide; PMS, phenazine methosulfate; DPIP, 2,6-dichlorophenol-indophenol; CMU, 3-(*p*-chlorophenyl)-1,1-dimethylurea.

* Per liter: KH_2PO_4, 0.7 g; K_2HPO_4, 0.3 g; $MgSO_4 \cdot 7H_2O$, 0.3 g; $FeSO_4 \cdot 7H_2O$, 3 mg; Arnon's A_5 solution,[3] 1 ml; thiamine hydrochloride, 10 μg.

cally to medium containing 0.5% urea and the basal mineral nutrients but without added glucose, and incubated for 3 days in the light (2,000 lux). The chlorophyll-rich cells thus obtained were used as the starting green cells.

Bleaching experiments

The green cells were separated from the preculture medium by centrifugation, and transferred to the bleaching medium containing 1% glucose and the basal mineral nutrients but no nitrogen source. These procedures were performed under aseptic conditions. The incubation was made in Monod-type culture vessels. At intervals, portions of the cell suspension were taken out and centrifuged. The sedimented cells were washed twice in $M/500$ K_2SO_4 and used for further experiments. In some experiments, a portion of the green cells was incubated in the medium containing only the basal mineral nutrients, as a control. All bleaching experiments were done in total darkness.

Determination of chlorophyll

Chlorophyll was determined using an intact cell suspension (opal glass method),[4] as well as methanol extract.[5]

Photosynthetic reactions

The procedures for measurements of the Hill reaction (quinone), cyclic photophosphorylation and photosynthetic CO_2-fixation were the same as described before,[2] except that the Hill reaction was measured with a Clark oxygen analyzer in the presence of KCN (1×10^{-3} M) to exclude the photosynthetic CO_2-fixation. In the present study the Hill reaction using DPIP as oxidant (5×10^{-3} M) was also measured. In the measurement the cell suspension in the medium containing 1.0 M sucrose, 0.025 M phosphate buffer at pH 6.5 and 0.0125 M NaCl was subjected to sonic oscillation (10 kc) for 2 minutes under cooling at a temperature below 1C°, and then diluted three-fold with the medium containing NaCl and phosphate buffer at the same concentrations as above.

Source of monochromatic lights

To study the action spectra of CO_2-fixation, use was made of monochromatic lights with a half-band width of 10 mμ or less, which were obtained by the combined use of interference filters and appropriate sets of color filters. A projector lamp was used as light source. Thermal radiation was eliminated by insertion of a flat bottle filled with running water between the filters and the light source. The intensity of the incident lights was measured with a Kipp thermopile and adjusted to desired levels by regulating the voltage.

Fluorescence measurements

The fluorescence was determined in a Shimadzu recording spectrophotometer MPS model 50 with an attachment for fluorescence measurements using a filter (Corning No. 2030), which cuts off the light of wavelengths shorter than 660 mμ. All the fluorescence yields reported in this paper are relative values of the steady state emission, transmitting through the filter.

RESULTS AND DISCUSSION

Bleaching process

Previous experiments[6–8] have shown that when green cells are incubated, with shaking, in a medium containing a high concentration of glucose and mineral nutrients but no nitrogen source, they are profoundly bleached with degeneration of chloroplasts, accompanied by a considerable increase in cell mass. In these experiments, the opal glass method[4] was used to determine the amounts of chlorophyll, using intact cell suspensions. In Fig. 1 the amounts of chlorophyll measured by the opal glass method are compared with the total amounts of chlorophyll determined using methanol extracts. The latter decreased linearly, after a few hours' lag period, during the bleaching process. The quantity of chlorophyll (OD_{675}) determined by the opal glass method, on the other hand, changed in a markedly different manner, as seen in the figure. It showed an increase followed by a slow decrease during the initial 24-hr period, and subsequently it decreased linearly at a faster rate. Thus the OD_{675} at the 24th hr was at the same level as the starting one, whereas the total amount of chlorophyll (methanol extract) was about 60% of the initial level. It seems very probable that the slower decrease in the OD_{675} is due to an increase in the efficiency of light absorption by chlorophyll in the bleaching cells resulting from expansion of chloroplasts and loosening of lamellae, which have been found to occur at an early phase of the bleaching of algal cells.[9]

Degradation of photosynthetic abilities

Processes of degradation of photosynthetic abilities of the algal cells were followed in relation to degeneration of chloroplast structure. The results obtained are shown in Figs. 1–3 and 7. The electron micrographs will be presented elsewhere.[9] In Fig. 2 are reproduced the processes of

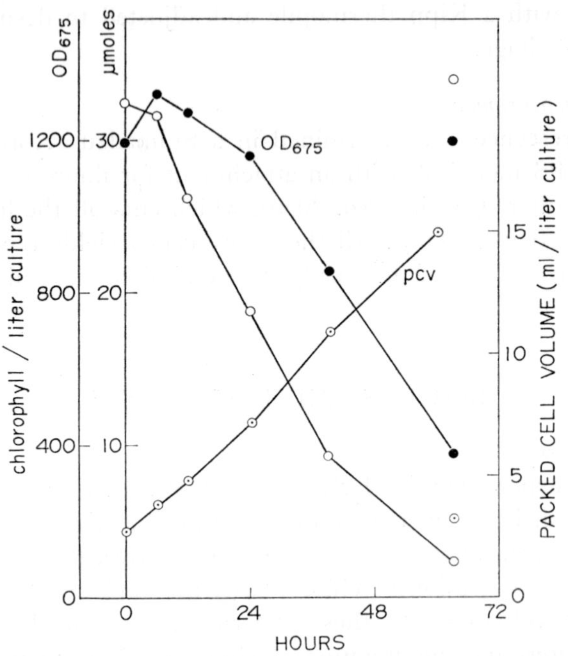

Fig. 1. Process of bleaching of algal cells induced by glucose. Open circles represent the amounts of chlorophyll determined using methanol extract, solid circles the amounts determined using an intact cell suspension (opal glass method, OD_{675}). Circles with a spot in the center stand for the packed cell volumes. Circles given at the 63rd hour not connected with curves indicate the controls (without glucose).

degradation of the activities of Hill reaction (quinone), FMN- and PMS-catalyzed photosynthetic phosphorylations and photosynthetic CO_2-fixation, measured at a high and a low light intensity, during bleaching of the green cells. It is seen that the photosynthetic activities, except for the activity for PMS-catalyzed photophosphorylation, decreased fairly rapidly and were almost completely lost while chlorophyll still remained at an appreciable level. The PMS-photophosphorylation activity measured at the high light intensity decreased in parallel with the disappearance of total chlorophyll (methanol extract), while the process of degradation of the same activity determined at the low light intensity was rather similar to the course of decrease in OD_{675} obtained by the opal glass method (see Fig. 1). The latter result can probably be explained by the increase in efficiency of light absorption by chlorophyll in the bleaching cells assumed earlier. It was confirmed in a previous study[2] that PMS-photophosphorylation in algal cells is strongly dependent on light. Apparently the light per chlorophyll in the bleaching cells available for PMS-photophosphorylation at the low light intensity increased and made up for the lowered chlorophyll content during the early 24-hr period of the bleaching process. The results in Fig. 3 show that chlorophyll *a*, which is largely associated with System I,[10] decreased more sharply than chlorophyll *b*, existing in a larger proportion in System II, during the initial 24-hr period.

Fig. 4 reproduces the processes of decay of the Hill reaction activities with quinone and with DPIP. The rate of decay of the activity was significantly slower with DPIP than with quinone. It is suggested that Systems I and II may be relatively rapidly cut off at a certain site(s) of the electron transport chain during the bleaching process, the disconnected systems, however, retaining activities for partial reactions of photosynthesis. These activities are maintained for some time, but destroyed sooner or later with the progress of bleaching. It has been demonstrated in our previous study[2] that in the greening process the PMS-catalyzed photophosphorylation activity was developed always in parallel with the chlorophyll formation under light-dark conditions as well as under continuous illumination,

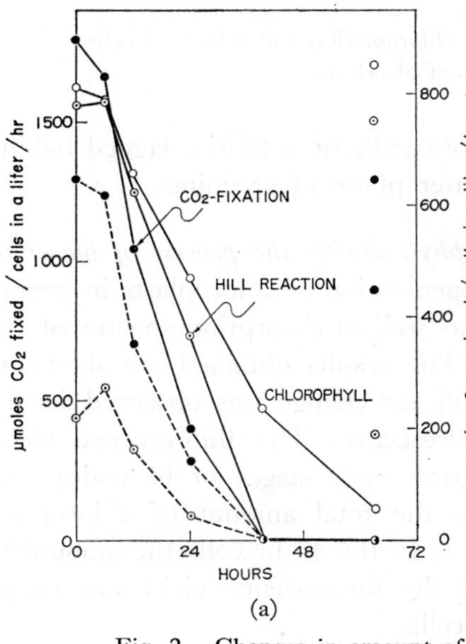

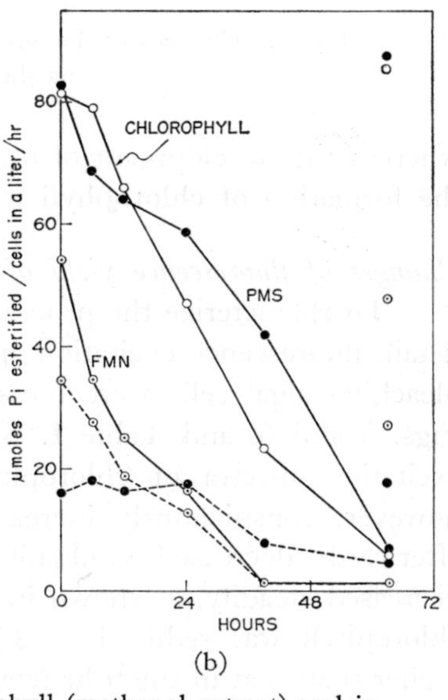

Fig. 2. Changes in amount of chlorophyll (methanol extract) and in activities of algal cells for the Hill reaction (quinone), PMS- and FMN-catalyzed photophosphorylations and photosynthetic CO_2-fixation during the process of bleaching. Solid lines, 20,000 lux; broken lines, 2,000 lux.

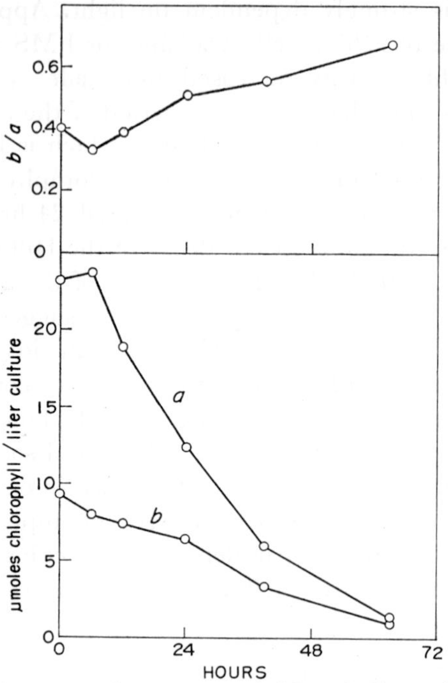

Fig. 3. Changes in the amounts of chlorophyll *a* and *b* in algal cells during the process of bleaching.

whereas the development of other photosynthetic activities lagged behind the formation of chlorophyll in the later phase of greening.

Changes of fluorescence yield of chlorophyll during the process of bleaching

To characterize the process of degeneration of chloroplasts in greater detail, fluorescence excitation spectra as well as absorption spectra of the bleaching algal cells were measured. The results obtained are shown in Figs. 5 and 6, and Table I. No significant change was observed in the excitation spectra of chlorophyll fluorescence. The fluorescence yield, however, conspicuously increased during early stages of bleaching and afterwards decreased gradually, while the total amount of chlorophyll decreased steadily, as shown in Table I. In the 48-hr cells the amount of chlorophyll was reduced to 8%, but the fluorescence yield was rather higher than that in the 0-hr (starting) cells.

During the bleaching process, the cells synthesize carotenoids, largely lutein, and their carotenoid content is not much lowered.[12] In considerably bleached cells, the absorbancy in the region of 400–500 mμ may be due,

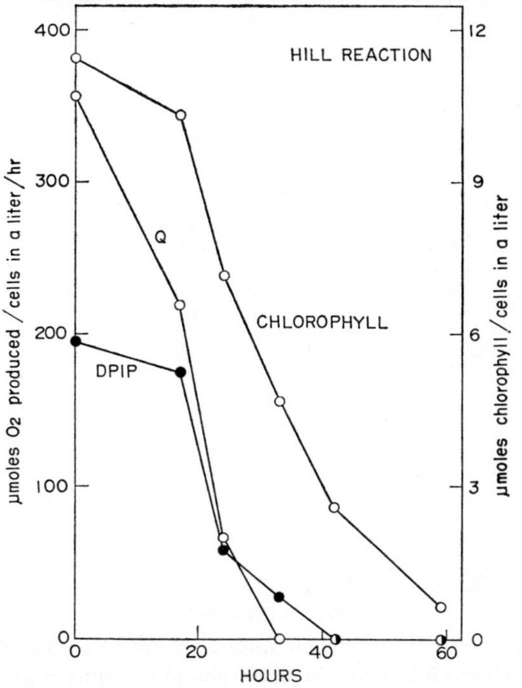

Fig. 4. Changes in activities of algal cells for the Hill reaction with DPIP and with quinone (Q) during the process of bleaching. Light intensity: 20,000 lux.

for the most part, to carotenoids (see Fig. 6). It is, however, not clear to what extent the carotenoids contribute to the fluorescence of chlorophyll in the bleaching algal cells. At any event, it is considered that the quantum yield of chlorophyll fluorescence is remarkably increased with the progress of bleaching. It has been reported that fluorescence yield of chloroplasts was increased by treatment with detergent.[11]

CO_2-fixing reaction under monochromatic lights

It was observed that the capacity of the algal cells for CO_2-fixation in darkness markedly increases during the bleaching process, as shown in Fig. 7. It has been shown previously that respiratory metabolism is also activated in this process.[7] Unexpectedly, it was found that at certain stages of the bleaching, the amount of dark CO_2-fixation by the algal cells was significantly decreased when they were illuminated with adequately weak light of a certain wavelength. This prompted us to study the effects of

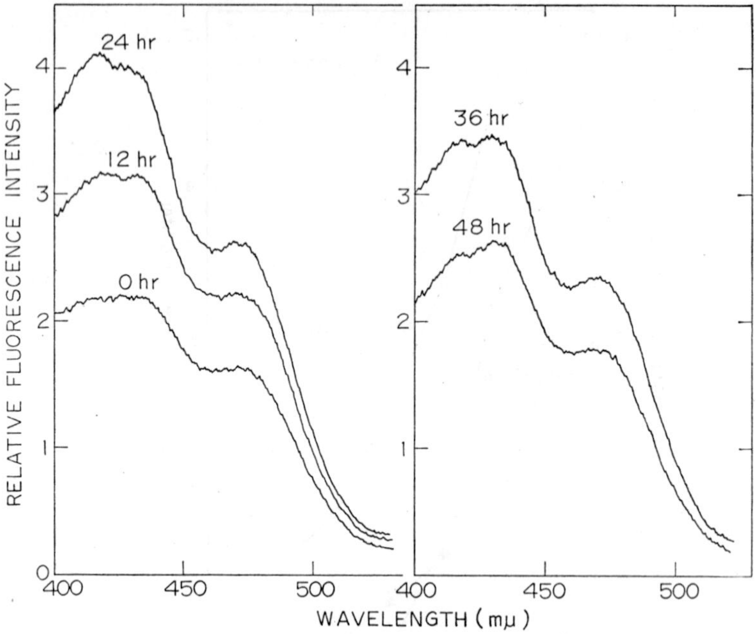

Fig. 5. Fluorescence excitation spectra of the bleaching algal cells. Cell density: 0.1 ml/liter in $M/100$ phosphate buffer at pH 7.0. Temperature: 22°C.

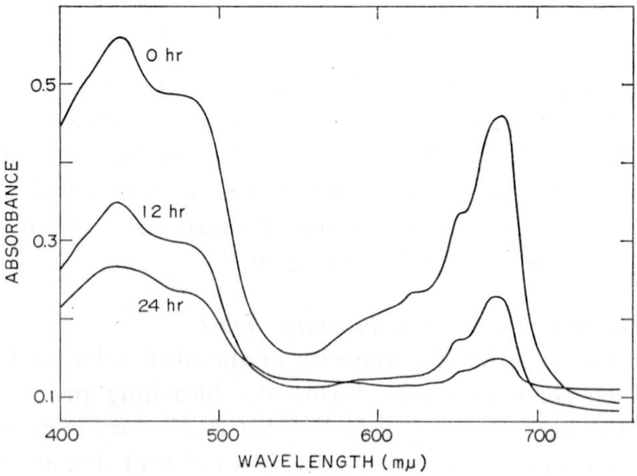

Fig. 6. Absorption spectra of the bleaching algal cells, measured by the opal glass method using an intact cell suspension. Cell density: 1.6 ml/liter.

TABLE I. Changes of fluorescence yield of chlorophyll during the process of bleaching.

(hr)	Chlorophyll (μmoles/ml cells)	chl *b*/*a*	Fluorescence intensity[a] excited at 415–30 mμ	470 mμ	F_{470}/F_{415-30}
0	6.44 (100)	0.36	2.2	1.6	0.73
12	4.05 (63)	0.43	3.2	2.2	0.69
24	1.91 (30)	0.49	4.1	2.6	0.63
36	0.83 (13)	0.50	3.4	2.4	0.71
48	0.50 (8)	0.54	2.6	1.8	0.69

a. Data taken from Fig. 5.

different monochromatic lights of low intensity on CO_2-fixation by bleaching algal cells. The results obtained so far are summarized in Figs. 8 and 9.

In the 0-hr cells CO_2 was fixed in large quantities, above the level of dark CO_2-fixation, when illuminated with red and blue lights, indicating that those cells performed normal photosynthetic CO_2-fixation. In the 12-hr cells, the action spectrum of CO_2-fixation was entirely different from that of the 0-hr cells: The 679 mμ light, under which photosynthetic CO_2-fixation was highest in the 0-hr cells, did not increase CO_2-fixation in the 12-hr cells; instead a sharp peak newly appeared at 523 mμ, which light produced about twice as much CO_2-fixation as the highest photosynthetic CO_2-fixation by the 0-hr cells at 679 mμ. Another peak was obtained at 600 mμ. The responses of the 18-hr cells to the 523 mμ and 600 mμ light were similar to those of the 12-hr cells, but these algal cells were considerably different in responses to other lights. In contrast to the above observations, no light stimulation of CO_2-fixation was observed in the 24-hr cells at the wavelengths examined; instead, the amounts of CO_2-fixation in the lights were significantly smaller than that in darkness. In the 53-hr cells, in which the bleaching had proceeded to a considerable extent, the increase of CO_2-fixation by light was recognized again, i.e. far-red, green (523 mμ) and blue lights caused considerable increases. When the response of the 0-hr (green) cells to light was measured in the presence of CMU, they gave an action spectrum strikingly similar to that of the 53-hr cells (Fig. 9). This suggested that those non-photosynthetic responses occur also in the normal green cells.

The results obtained above suggest that non-photosynthetic CO_2-fixation may be induced by light through several photoreceptors. As stated in a recent review by Nickerson and Bartniki-Garcia,[13] the role of carbon dioxide in morphogenesis is believed to be universal and of fundamental

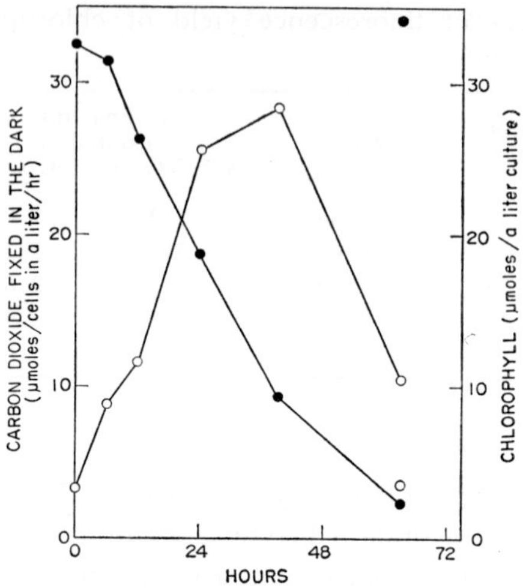

Fig. 7. Change in the capacity of algal cells for CO_2-fixation in darkness during the process of bleaching.

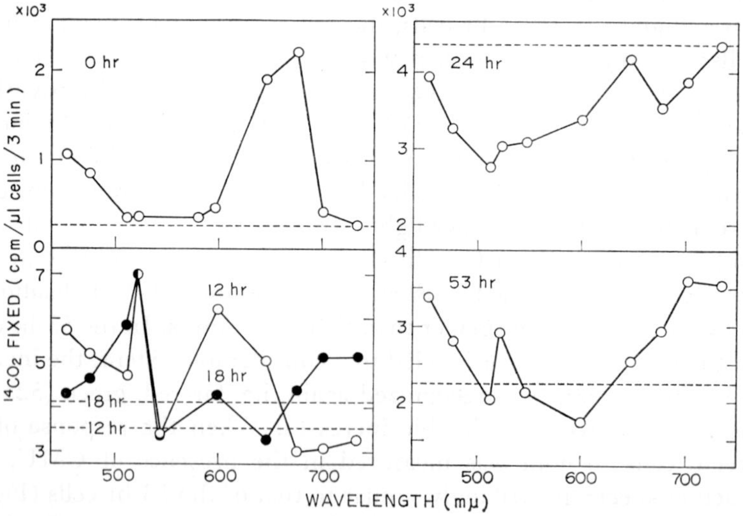

Fig. 8. Change in the action spectrum of CO_2-fixation during the process of bleaching of algal cells. Broken lines indicate the level of dark CO_2-fixation. Cells were preincubated for 3 min in the light or in darkness, and then $NaH^{14}CO_3$ ($2.5 \times 10^{-3} M$) was added to the suspension. After 3 min the cells were killed with acetic acid. Cell density, 1 µl/ml in M/100 phosphate buffer at pH 7.0; temperature, 22–23°C; light intensity, 300 ergs/cm²/sec.

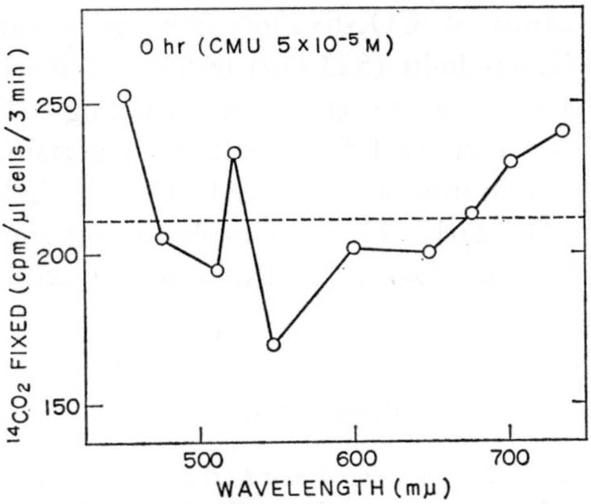

Fig. 9. Action spectrum of CO_2-fixation by the green cells in the presence of C MU.

importance, although little is known about its precise action. The light induced nonphotosynthetic CO_2-fixation found in this study may play a central role in the primary phase of photomorphogenesis.

This work was supported in part by grants from the Ministry of Education, Japan.

SUMMARY

The green cells of *Chlorella protothecoides* were induced to bleach with concurrent degradation of chloroplast structures by incubating (in the dark) the cells in a glucose-enriched medium without the addition of nitrogen source.

During the process of bleaching of algal cells, there occurred a marked decrease in the capacities for the PMS- and FMN-catalyzed photophosphorylations, the Hill reaction and photosynthetic CO_2-fixation. The decrease of the activity for PMS-photophosphorylation was slower than that of other photosynthetic activities. Chlorophyll *a* was broken down more rapidly than chlorophyll *b*. These results are discussed in relation to the change of chloroplast structures.

No significant change was observed in the excitation spectra of chlorophyll fluorescence in the region of 400–500 mμ during the bleaching process, while the fluorescence yield was strikingly increased, indicating a rise in the quantum yield of fluorescence.

The action spectrum of CO_2-fixation changed greatly during the course of bleaching. Green light (523 mμ) induced a marked increase in CO_2-fixation by the cells at several stages of bleaching. Blue and far-red light also brought about increased CO_2-fixation at certain stages. At an intermediary stage, the amounts of CO_2 fixed in all the lights tested were smaller than those in the dark. These non-photosynthetic effects of the chromatic lights on CO_2-fixation were also observed with normal green algal cells in the presence of CMU.

REFERENCES

1. Shihira-Ishikawa, I., and Hase, E., *Plant and Cell Physiol.*, **5**, 227 (1964).
2. Oh-hama, T., Shihira-Ishikawa, I., and Hase, E., *Ibid.*, **6**, 743 (1965).
3. Arnon, D. I., *Am. J. Bot.*, **25**, 322 (1938).
4. Shibata, K., *Methods of Biochemical Analysis*, **7**, 77 (1959).
5. Ogawa, T., and Shibata, K., *Photochem. Photobiol.*, **4**, 193 (1965).
6. Aoki, S., Matsuka, M., and Hase, E., *Plant and Cell Physiol.*, **6**, 487 (1965).
7. Matsuka, M., and Hase, E., *Ibid.*, **6**, 721 (1965).
8. Matsuka, M., Otsuka, H., and Hase, E., *Ibid.*, **7**, 651 (1966).
9. Yumoto, S., and Hase, E., in preparation.
10. Anderson, J. M., and Boardman, N. K., *Biochim. Biophys. Acta*, **112**, 403 (1966).
11. Sauer, K., and Park, R. B., *Biochim. Biophys. Acta*, **79**, 476 (1964).
12. Oshio, Y., unpublished data.
13. Nickerson, W. J., and Bartniki-Garcia, S., *Ann. Rev. Plant Physiol.*, **15**, 327 (1964).

Studies on Chlorophyllase of *Chlorella Protothecoides* in Relation to the Problem of Chlorophyll Formation

YASUTANE CHIBA, ICHIRO AIGA, KYOSUKE TAKAGI†
AND TSUTOMU SASA

Department of Biology, Faculty of Science, Kyushu University, Fukuoka, Japan

INTRODUCTION

Since Willstätter and Stoll[1] described how chlorophyll is formed from chlorophyllide and phytol on incubation with the air-dried meal of *Heracleum* leaves, the enzyme chlorophyllase had been assumed to catalyze both hydrolysis and esterification of the process in question. Holden[2] and Klein and Vishniac,[3] however, failed to repeat the phytylation experiment. On the other hand, Shimizu and Tamaki[4] reported that an enzyme obtained from tobacco leaves could esterify both chlorophyllide and pheophorbide with phytol *in vitro*, but we were unable to achieve this with an enzyme obtained from *Chlorella*.

In a previous paper,[5] we reported that, when methyl chlorophyllide *a* and phytol were incubated with chlorophyllase obtained from *Chlorella protothecoides*, both chlorophyll *a* and chlorophyllide *a* were formed. It was revealed that chlorophyllide *a* was derived from methyl chlorophyllide *a* and not from chlorophyll *a*. It was therefore concluded that this enzyme is capable of esterifying methyl chlorophyllide *a* with phytol but is unable to catalyze the phytylation of chlorophyllide *a*. In this experiment, paper chromatography was mainly employed for the identification of the pigments formed by the enzyme, since the spectrum in the visible region of chlorophyll *a* is identical with that of methyl chlorophyllide *a*.

The purpose of the present study is to obtain more precise information about the mode of exchange between the phytyl and the methyl groups on the C_7 propionic acid residue. Some investigations of the infrared spectra and labeling experiments with ^{14}C-methanol provided supplementary information.

* Contribution No. 160 from the Department of Biology, Faculty of Science, Kyushu University. Supported in part by a grant-in-aid for Fundamental Scientific Research from the Ministry of Education, Japan.

† On leave from the Faculty of Education, Yamaguchi University, Yamaguchi, Japan.

MATERIALS AND METHODS

The strain of *Chlorella protothecoides* and the method of preparation of the green cells were the same as those described previously[5]: the green cells were obtained by growing the precultured alga for 5 days in the light. The medium contained 0.5% glycine and 0.1% potassium acetate in addition to the fundamental inorganic medium which was reported by Shihira-Ishikawa and Hase.[6] The cells were harvested and washed by centrifugation, and suspended in distilled water. The suspension was stored in a deep-frozen condition until it was used for the extraction of the enzyme.

Chlorophyllase was extracted from the green cells and partially purified using *n*-butanol and cold acetone as reported in the previous paper.[5] After having been thawed, the cell suspension (50 ml) was sonicated in an Umeda 150 W 20 kc sonic oscillator. To the sonicated suspension were added 500 mg of sodium chloride and 70 ml of *n*-butanol, and the mixture was further sonicated for 10 min and centrifuged at $20{,}000 \times g$ for 30 min. The mixture was separated into three layers. The middle aqueous layer was sucked out and the enzyme was further purified with cold acetone. The 40–80% acetone fraction contained the enzyme and a large amount of a 260 mμ-absorbing substance. This substance was removed by gel-filtration with Sephadex G-50.

Chlorophyll *a* was prepared from spinach chloroplasts according to the method of Perkins and Roberts[7] and crystallized from either aqueous acetone or petroleum ether solution. The crystallization was repeated several times, and the crystals were dried over P_2O_5 *in vacuo*. Methyl chlorophyllide *a* was prepared enzymatically in aqueous methanol solution as will be described later.

The visible and infrared spectra were recorded with a Hitachi recording spectrophotometer model EPU-2 and a Hitachi infrared recording spectrophotometer model EPI-2, respectively.

RESULTS AND DISCUSSION

Enzymatic preparation of methyl chlorophyllide a

Methods of preparation of methyl chlorophyllide by enzymatic methanolysis in leaves[8] or in intact algal cells[5] have been reported. It is, however, difficult to obtain pigments free from colorless impurities by such methods. We therefore tried to obtain methyl chlorophyllide *a* from pure chlorophyll *a* in aqueous methanol solution by using chlorophyllase.

The reaction mixture contained 20 ml of chlorophyll *a* in acetone solution (about 20 mg chlorophyll), 20 ml of methanol and 60 ml of enzyme preparation. After having been incubated for 30 min at 30°, the reaction was stopped by adding 100 ml of acetone. The solution was decanted into a separating funnel containing 200 ml of ethyl ether. The ether solution, into which all the pigments were transferred, was washed with distilled water several times to remove the enzyme protein. The solution was allowed to stand for 1 hr and then passed through an anhydrous Na_2SO_4 column and dried *in vacuo*.

The dried pigments dissolved in 200 ml of 3% pyridine-petroleum ether were chromatographed on a powdered sugar column (5×30 cm) containing 3% corn starch. When the pigments were developed with the same solvent, chlorophyll *a* and methyl chlorophyllide *a* gradually moved downward, being separated from the tightly adsorbed band of chlorophyllide *a*. After chlorophyll *a* had been passed through the column, the band of

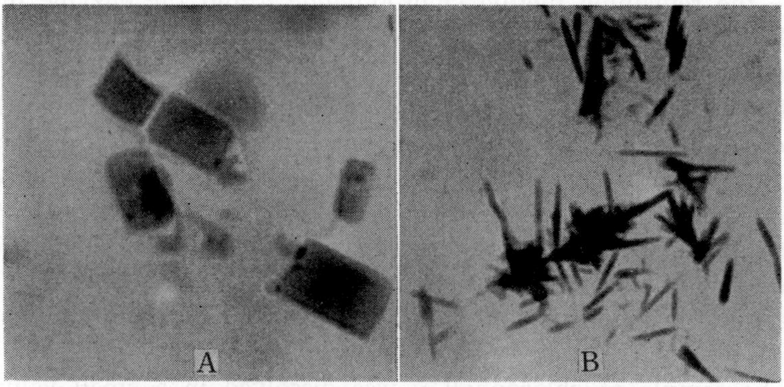

Fig. 1. Crystals of methyl chlorophyllide *a* used as substrate and chlorophyll *a* formed by chlorophyllase:
A, light-green plate-like crystals of methyl chlorophyllide *a*, $\times 1{,}000$;
B, fine needle-like crystals of chlorophyll *a*, $\times 430$.

chlorophyllide *a* was cut off, and methyl chlorophyllide *a* was eluted with acetone and dried *in vacuo*. The pigment was dissolved in a small volume of acetone and further precipitated by adding slowly an equal volume of water. The mixture was allowed to stand overnight in a cold room. The sedimented pigment was collected by centrifugation and dissolved in acetone. Distilled water was added to the acetone solution until a slight turbidity appeared. On leaving the suspension standing in a cold room for several hours, light-green plate-like crystals were found (Fig. 1A). The crystals were dried over P_2O_5, *in vacuo*, for 3 days before being subjected to infrared spectroscopic measurements.

Enzymatic preparation of chlorophyll a from methyl chlorophyllide a and phytol

In the previous paper,[5] chlorophyll *a* formed was identified by the Rf value and the visible spectrum. In order to obtain a crystalline sample from the chlorophyll *a* formed, a method of using an enzyme in solution was developed.

The reaction mixture consisted of 10 ml of methyl chlorophyllide *a* in acetone solution (about 20 mg methyl chlorophyllide *a*), 10 ml of 3% phytol in acetone solution, 20 ml of acetone and 60 ml of enzyme solution. After incubation lasting for 40 min at 30°C, 100 ml of acetone was added to the mixture. The pigments in aqueous acetone were then transferred into 200 ml of petroleum ether in a separating funnel. The ethereal solution was washed with 80% aqueous methanol and with distilled water to remove the remaining methyl chlorophyllide *a* and the enzyme protein. After having been passed through a Na_2SO_4 column, the solution was dried *in vacuo*.

The dried pigments were dissolved in 10 ml of 10% ethyl ether in petroleum ether. The solution was applied onto a sugar column and the adsorbed pigments were washed with a large volume of the same solvent for the complete removal of phytol. After development with 0.5% *iso*-propanol in petroleum ether, the bands of a small amount of methyl chlorophyllide *a* and chlorophyllide *a* were removed, and then chlorophyll *a* was eluted from the column with acetone. The chlorophyll *a* solution was dried *in vacuo*.

The sample of chlorophyll *a* was dissolved in a small volume of petroleum benzine and crystallized at −20°C. Amorphous crystals were collected by centrifugation and crystallization was repeated several times. The blue-black crystals thus obtained were dried over P_2O_5 *in vacuo*. They were recrystallized from aqueous acetone to compare them with the crystals of pure reference chlorophyll *a*. Fig. 1B shows fine needle-like crystals of chlorophyll *a* thus obtained, which were similar to those of pure chlorophyll *a*.

Infrared absorption spectra

To see whether phytol is attached to the C_7 propionic acid residue of the sample of chlorophyll *a* formed, the infrared absorption spectrum in the C-H stretching region, 3,000—2,700 cm^{-1}, was measured and the result was compared with the spectrum of methyl chlorophyllide *a*. Fig. 2 shows representative results. The absorption peak of methyl chlorophyllide *a* in this region was quite low (Fig. 2A), whereas the absorption peak of the sample of chlorophyll *a* formed as well as of pure reference chlorophyll *a* was high (Figs. 2B and 2C). Since it is known that the phytyl group contributes mainly to the absorption in this region of the spectra of chlorophylls,[9,10] the increase, at around 2940 cm^{-1}, in the spectrum of the sample of chlorophyll *a* may be attributed to the presence of phytol.

Time course of the enzymatic formation of chlorophyll a

Seventy ml of the reaction mixture comprised: 7 ml of methyl chlorophyllide *a* in acetone (O.D. = 50), 14 ml of acetone, 7 ml of 3% phytol in

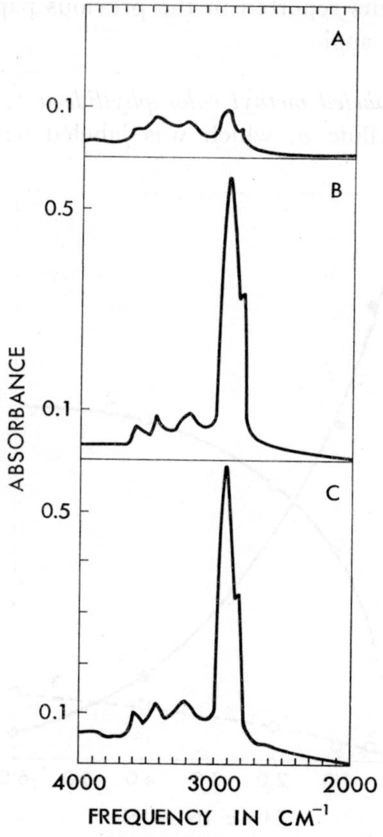

Fig. 2. Infrared absorption spectra in the C-H stretching region. A, methyl chlorophyllide *a*; B, chlorophyll *a* formed by chlorophyllase; C, pure reference chlorophyll *a*; KBr disc.

acetone, 42 ml of enzyme preparation (0.2 mg protein/ml) in 0.01% Triton X-100 and 0.005 *M* borate phosphate buffer (pH 7.0). The reaction was run at 30°C, and an aliquot (10 ml) was taken at appropriate intervals. To measure the concentrations of pigments in each aliquot, the pigments were separated on a sugar column, using 3% pyridine in *n*-hexane as described above. The pigment concentrations were determined by measuring the optical density at 663 mμ and the relative values per ml of the solvent were estimated. The concentration of chlorophyllide *a* was measured with a solvent of 10% methanol-3% pyridine in *n*-hexane. Fig. 3 shows the time course of the enzymatic formation of chlorophyll *a* from methyl chlorophyllide *a* and phytol in 40% acetone solution. After a 60 min incubation, a large amount of chlorophyll *a* and a small amount of chlorophyllide *a* were formed. The results reported in the previous paper[5] were thus confirmed from a different angle.

Experiments with ¹⁴C labeled methyl chlorophyllide a

Methyl chlorophyllide *a*, which was labeled with ^{14}C-methanol at

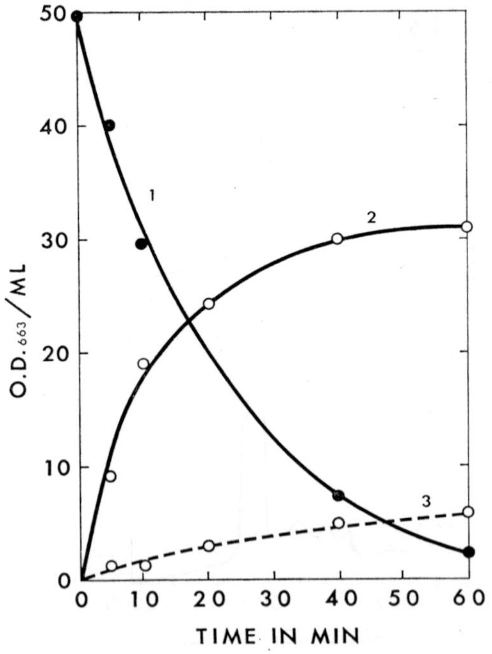

Fig. 3. Enzymatic formation of chlorophyll *a* (curve 2) and chlorophyllide *a* (curve 3) from methyl chlorophyllide *a* (curve 1) and phytol in 40% acetone solution.

the position of the propionic acid residue, was used as a substrate and the location of the label was examined after completion of the synthetic reaction.

The labeled methyl chlorophyllide *a* was prepared as follows. The reaction mixture consisted of 2 ml of methanol containing 0.1 mc of labeled carbon, 2 ml of chlorophyll *a* in acetone (about 5 mg) and 6 ml of the enzyme solution. The mixture was incubated for 30 min at 30°C. The labeled methyl chlorophyllide *a* was purified by the method described above. The pure methyl chlorophyllide *a* in ethyl ether solution (300 cpm/ml) was examined both by paper chromatography (Fig. 4A) and antoradiography (Fig. 4B). Then, the labeled methyl chlorophyllide *a* was incubated with phytol under the same conditions as described above, and

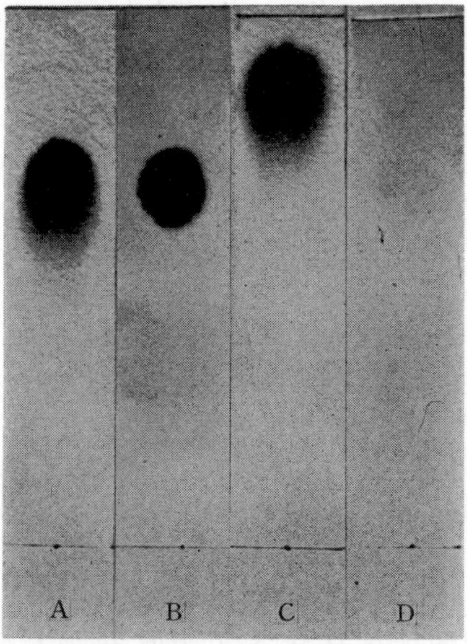

Fig. 4. Paper chromatogram (A) and autoradiogram (B) of ^{14}C-labeled methyl chlorophyllide *a*. Paper chromatogram (C) and autoradiogram (D) of chlorophyll *a* synthesized: solvent, *n*-hexane : benzene : acetone : methanol : water=130 : 70 : 50: 20 : 2.

chlorophyll *a* was isolated from the reaction mixture. The sample of purified chlorophyll *a* was also paper-chromatographed (Fig. 4C) and autoradiographed (Fig. 4D). As shown in Fig. 4D no radioactivity was found in the sample of chlorophyll *a* formed. This excludes the possibility that oxidative coupling had taken place between phytol and the carbomethoxy group of the C_7 propionic acid residue of methyl chlorophyllide *a*. It seems reasonable, therefore, to conclude that enzymatic transesterification takes place between phytyl and methyl groups on the propionic acid side chain of chlorophyll.

SUMMARY

1. Methyl chlorophyllide *a* was prepared by enzymatic methanolysis of pure chlorophyll *a* using a partially purified preparation of chlorophyllase obtained from *Chlorella protothecoides*.

2. The time courses of phytylation could be studied in detail by measuring the changes in concentration of chlorophyll *a* and methyl chlorophyllide *a*. These pigments were separated on a sugar column using 3% pyridine in *n*-hexane to determine the concentrations.

3. It was ascertained by infrared spectroscopy that phytol is attached to the C_7 propionic acid residue of the chlorophyll *a* molecule which was formed enzymatically.

4. Experiments with ^{14}C labeled methyl chlorophyllide *a* also confirmed that the enzyme catalyzes transesterification between methyl- and phytyl-groups on the propionic acid residue of chlorophyll *a*.

REFERENCES

1. Willstätter, R., and Stoll, A., in *Untersuchungen über Chlorophyll, Methoden und Ergebnisse*, pp. 172–193, Springer Verlag, Berlin, 1913.
2. Holden, M., *Biochem. J.*, **78**, 359 (1962).
3. Klein, A. O., and Vishniac, W., *J. Biol. Chem.*, **236**, 2544 (1961).
4. Shimizu, S., and Tamaki, E., *Bot. Mag. Tokyo*, **75**, 462 (1962).
5. Chiba, Y., Aiga, I., Idemori, M., Satoh, Y., Matsushita, K., and Sasa, T., *Plant and Cell Physiol.*, in press.
6. Shihira-Ishikawa, I., and Hase, E., *Plant and Cell Physiol.*, **5**, 227 (1964).
7. Perkins, H. J., and Roberts, D. W. A., *Biochim. Biophys. Acta*, **58**, 486 (1962).
8. Pennington, F. C., Strain, H. H., Svec, W. A., and Katz, J. J., *J. Amer. Chem. Soc.*, **86**, 1418 (1964).
9. Holt, A. S., and Jacobs, E. E., *Plant Physiol.*, **30**, 553 (1955).
10. Katz, J. J., Dougherty, R. C., and Boucher, L. J., in *The Chlorophylls* (L. P. Vernon and G. R. Seely, eds.), pp. 185–251, Academic Press, New York, 1966.

Aspects of Chloroplast Development: Transitory Pigment-Protein Complexes and Protochlorophyllide Regeneration*

LAWRENCE BOGORAD†, LARRY LABER‡, AND MERRILL GASSMAN§

Department of Botany, University of Chicago,
Chicago, Illinois

INTRODUCTION

At their most advanced state of development, plastids in etiolated leaves consist of a limiting membrane within which the most conspicuous structure is the para-crystalline prolamellar body. When etiolated leaves are illuminated, the prolamellar body dissociates into a group of loosely packed vesicles. This change occurs in a matter of seconds. Subsequently, as illumination is continued, the vesicles disperse through the plastid, become aligned in rows, and the members of each row fuse. Thus are formed several large flattened vesicles (primary lamellae) each of which traverses the entire plastid. At a later time, after several hours of illumination, additional vesicles form at various places along the primary lamellae and eventually a system of grana is built up.[1]

Another characteristic of proplastids is the presence of protochlorophyllide. This pigment is associated with a large protein and the entire complex has been termed the protochlorophyllide-holochrome by Smith (see ref. 2). *In vivo*, this complex absorbs in the blue and red regions of the spectrum; a strong maximum occurs at about 650 nm. When etiolated leaves are illuminated, the absorption at 650 nm decreases and a new absorption band with a maximum at about 684 nm appears. This change in absorption spectrum reflects the reduction of the protochlorophyllide *a*

* Supported in part by research grants from the National Institutes of Health and the National Science Foundation.

† Research Career Awardee of the National Institute of General Medical Sciences, N.I.H. Present address: Harvard University, The Biological Laboratories, Cambridge, Mass. 02138.

‡ Hutchinson Fellow of the Department of Botany, University of Chicago. Present address: Department of Biochemistry, University of Georgia, Athens, Ga.

§ National Aeronautics and Space Administration Fellow. Present address: The Rockefeller University, New York, N.Y. 10021.

to chlorophyllide *a*. Shibata[3] studied these changes in the absorption spectra of etiolated leaves during illumination and also observed that after the absorption at 684 nm appears, the maximum shifts back to 672 nm in a period of about 10 minutes and that subsequently a second shift occurs until the pigment protein comple *in vivo* has a maximum at about 677 nm. In red kidney beans (*Phaseolus vulgaris* var. Red Kidney) the chlorophyll *a* is esterified with phytol within 30 minutes after it is produced to form chlorophyll *a*.

Another consequence of brief illumination of etiolated leaves, even if followed by darkness, is the development of the capacity to form additional protochlorophyllide. Seven- to nine-day-old etiolated red kidney beans grown under our conditions form no additional protochlorophyllide for at least one hour even under continuous illumination.

As was shown by Krasnovsky and Kosobutskaya[4] and by Smith and his collaborators (e.g. ref. 5), alkaline buffered aqueous extracts of etiolated bean leaves contain protochlorophyllide-holochrome which can be transformed upon illumination into chlorophyllide (presumably still attached to a protein). The absorption maximum of protochlorophyllide-holochrome in aqueous solution is at 637 nm; upon illumination the maximum shifts to 675 nm. In addition, Boardman[2] has reported a transitory absorption at 680 nm following illumination of protochlorophyllide-holochrome in solution. The transformable pigment-protein complex can be partially purified by ammonium sulfate fractionation and some other procedures.[2] Capon and Bogorad[6] extended some of the purification procedures outlined by Smith[7]: Clarified aqueous extracts of etiolated bean leaves were fractionated to obtain protein which precipitated between 0 and 25% and between 26 and 45% saturation with ammonium sulfate. These fractions were designated A and B, respectively.

The A fraction, after being dissolved and dialyzed, was again treated with ammonium sulfate to yield a 0–25% of saturation fraction (Aa), and the remainder of the material was discarded. The B fraction was refractionated to obtain the protein which this time precipitated between 0 and 25% of saturation (fraction Ba) and another which precipitated between 26 and 45% of saturation (Bb). In these original investigations, transformable protochlorophyllide-holochrome was found in all three of the final fractions prepared from extracts of etiolated bean leaves. However, when similar extracts were made of bean leaves illuminated for varying periods of time, it was found that Bb fractions from leaves illuminated for 10 seconds no longer contained pigment-protein complex. After 60 minutes of illumination, pigment-protein complexes were absent from the Ba fractions as well. Finally, after 3 hours of illumination, no pigment could be found in any of the three ammonium sulfate fractions—i.e. all the pigment was associated with large membranes or membrane fragments. This, along with certain spectral differences, suggested either that there might be some progressive (i.e. aggregational or conformational) changes in the original pigment protein complex or that the pigment might be transferred from the holochrome protein and become associated with one or more different proteins in the course of membrane development.

In the present work, procedures for the preparation of ammonium sulfate fractions were somewhat modified from those of Capon and Bogorad and protochlorophyllide-holochrome from etiolated leaves was found only in the Ba and Bb fractions, but essentially three different ammonium sulfate fractions corresponding to those described above were studied. (A barium chloride treatment employed by Capon and Bogorad was omitted here.) The largest amount of transformable protochlorophyllide-holochrome from extracts of etiolated leaves was found in fraction Bb; from leaves illuminated for 5 minutes before extraction, most of the pigment complex was present in fraction Ba; and, after illumination for 60 minutes, extracts yielded pigment-protein complex which was found entirely in fraction Aa.

It is apparent that the pigment initially associated with soluble protein, as protochlorophyllide is, is included after conversion to chlorophyllide and certainly after esterification with phytol to form chlorophyll, in the insoluble lamellar system of the plastids. How does this come about? What does this process have to do with the buildup of photosynthetically active plastid lamellae? A series of possibilities can be outlined (Fig. 1).

One is that after conversion of the protochlorophyllide to chlorophyllide entire pigment-protein complexes aggregate to make an incomplete membrane fragment, etc. (Fig. 1A).

A second possibility (Fig. 1B) is that, upon reduction of the protochlorophyllide to chlorophyllide, the pigment molecule moves to another place on the protein (assuming that there is one active position per protein molecule at which the pigment can be photoreduced), and the active site becomes available for a newly formed protochlorophyllide. This, in turn, could be reduced, moved to a new position, etc. Finally, in this possibility, after a protein molecule had accumulated its maximum number of chlorophyll molecules, it could aggregate with other similar protein pigment complexes to form part of a membrane. This would be a snowballing method of buildup.

A third possible sequence of events (Fig. 1C) could be the separation

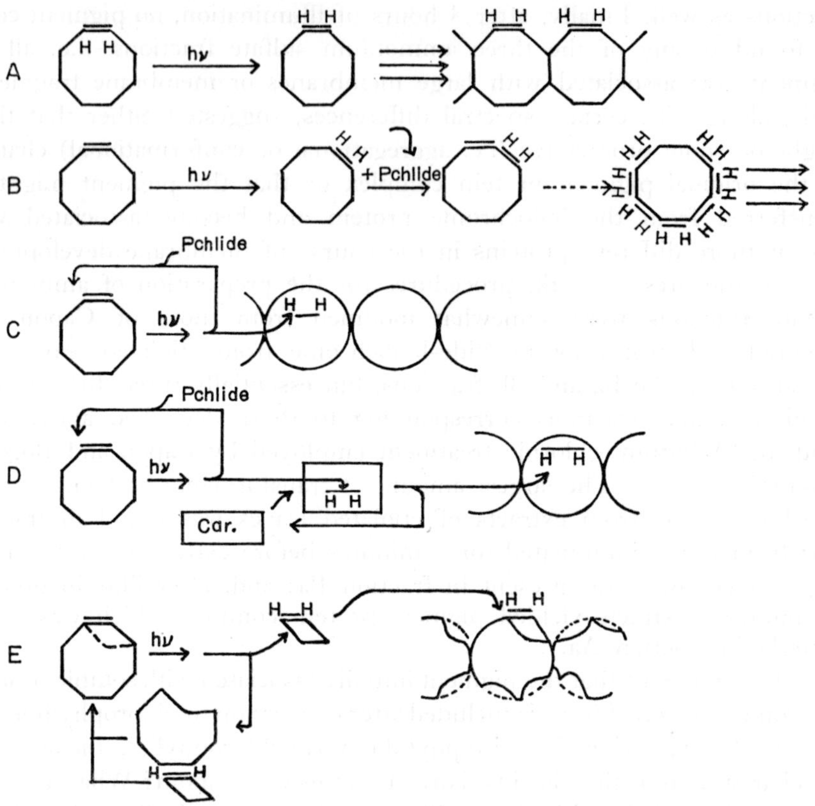

Fig. 1. Some possible relationships between protochlorophyllideo-holchrome and the formation of photosynthetic membranes in chloroplasts. Protochlorophyllide = ——. Chlorophyllide = $\overline{\text{H} \quad \text{H}}$.

of the pigment and protein immediately after photoreduction. According to this model, the chlorophyllide or chlorophyll, as soon as it is formed, is transferred directly to a preformed site in the lamella to make a chlorophyll-membrane complex and the protein portion of the protochlorophyllide-holochrome is re-used. The soluble protein picks up another protochlorophyllide molecule which is photoreduced, etc.

A fourth possibility (Fig. 1D) is very similar to the one just outlined except that a carrier for the chlorophyllide might exist which accepts the reduced pigment molecule from the holochrome protein and delivers it to some preformed site on the membrane. It is possible that, while attached to this carrier protein, phytolization might occur.

Still another possibility (Fig. 1E) is that the holochrome protein as identified to date really consists of two segments or two associated proteins. According to this scheme, after reduction of the protochlorophyllide, a segment of the "protochlorophyllide-holochrome," consisting of chlorophyllide and a bit of protein, separates from the larger protein of the complex. The pigmentless portion of the complex might then pick up another protochlorophyllide-small protein to reconstitute the large complex on which the photoconversion of another protochlorophyllide molecule could occur, etc. In the meantime, the smaller pigment's protein complex might be incorporated into the lamellar system, or a number of such small units might aggregate to form the lamellae. (The number of easily concocted hybrid schemes no more or less unlikely than A-E, e.g. that the "small protein" of E serves simply as a carrier or that the carrier of D becomes incorporated into the membrane with the pigment, is too great to enumerate here.)

RESULTS

Soluble Pigment-Protein Complexes

Experiments performed to investigate some of these possibilities[8] were extensions of those of Capon and Bogorad.[6] They consisted of sucrose density gradient analyses of extracts of etiolated and illuminated leaves and also of protochlorophyllide-holochrome. Solutions containing pigment protein complexes were centrifuged in gradients of 100–700 grams of sucrose per liter for 40 hours at 35,000 rpm in the SW 39 rotor of the Spinco preparative centrifuge. Solutions used both for extraction and centrifugation contained 0.05 *M* pH 9.5 glycine buffer. After centrifugation, the gradient was separated into 10-drop fractions, and each fraction was analyzed spectrophotometrically either for convertibility of protochlorophyllide to chlorophyllide or, in preparations containing pigment which had been converted prior to extraction or prior or subsequent to centrifugation, simply for its absorption at about 670–675 nm.

As is shown in Fig. 2, the photoconvertible protochlorophyllide-holochrome centrifuged as a homogeneous band in about the middle of the gradient. On the other hand, the pigment-protein complex obtained from leaves illuminated for 60 minutes was found in the denser part of the gradient (Fig. 3). However, sucrose density gradient analysis of Ba fractions from leaves illuminated for 5, 15 and 60 minutes (Fig. 4) indicated the presence of a transitory pigment-protein complex which centrifuges to a position higher in the gradient than that to which unconverted protochlorophyllide-holochrome moved. A similar pigment-protein complex,

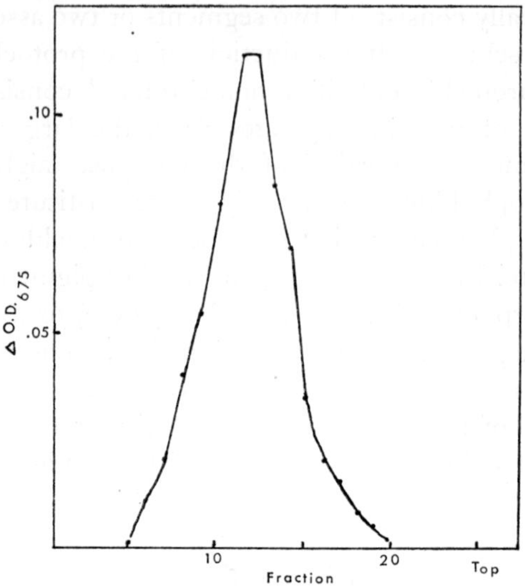

Fig. 2. The distribution in a sucrose density gradient of photoconvertible protochlorophyllide-holochrome from fraction Bb of etiolated bean leaves. The tube was punctured after centrifugation and 10-drop fractions were collected. The absorption spectrum of each fraction was determined using a microcuvette and a Cary spectrophotometer. All operations to this point were conducted under a green safelight. Each fraction was then illuminated with incandescent light (200–400 ft.c) for 2 minutes and the absorption spectrum of the solution was determined again. The increase in optical density at 675 nm after illumination was calculated and is plotted here against fraction number.

"lighter" than protochlorophyllide-holochrome, can be generated by illuminating the latter *in vitro* (Fig. 5). (The molecular weight of the protochlorophyllide-holochrome has been estimated to be about 500,000–700,000.[2] Molecular weight data for the "lighter" complex described here has not yet been obtained. It is entirely possible that some or all of the change in sedimentation properties might result from a conformational change following protochlorophyllide reduction, but in the discussion which follows it will be assumed that this is not the case.)

The last observation to be mentioned which is pertinent to an evaluation of the five proposals made earlier (Fig. 1) is that if leaves are irradiated for 30 minutes and then held in darkness for 24 hours, during which time they form considerable additional protochlorophyllide, the regenerated protochlorophyllide-holochrome comes to the same place in the gradient as material obtained from unilluminated etiolated leaves. Most (about 95%

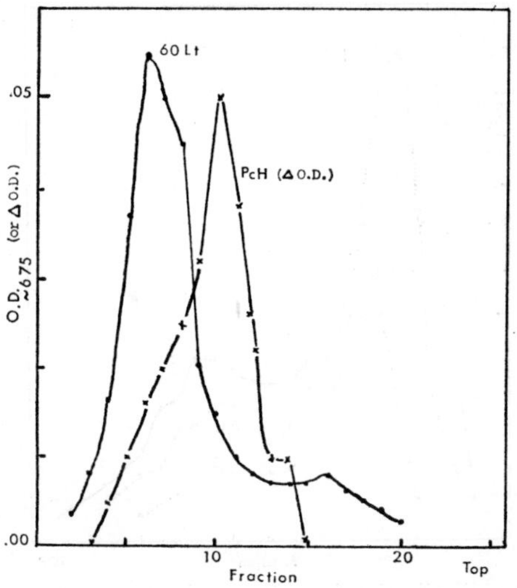

Fig. 3. The distribution in a sucrose density gradient of the photo-convertible protochlorophyllide-holochrome (PcH) from etiolated bean leaves and (60 Lt) the soluble chlorophyll-protein complex (Fraction Ba) prepared from leaves of dark-grown beans illuminated for one hour. (Leaves were harvested under a safelight and spread on moist filter paper. They were illuminated with light from a mixture of incandescent and cool-white fluorescent lamps at an intensity of 500–800 ft.c.). Fractions from the gradient containing protochlorophyllide were collected and assayed for convertibility (see Fig. 2 legend). Fractions of the 60 Lt (Ba) preparation were collected and absorption at 670–675 nm was determined.

or more) of the chlorophyllide formed during the initial illumination is, on the other hand, in a heavier fraction.

If we return now to the possibilities outlined in Fig. 1, it is apparent that the first of these, namely that immature membranes are formed by the aggregation of chlorophyllide-holochrome molecules (with one pigment molecule per protein), now seems untenable. If it were correct one would not expect to find any pigment-protein complex lighter than the protochlorophyllide-holochrome nor would it be expected that a complex smaller than a whole lamella (insoluble) would appear in leaf extracts. There are other complications about such a model (e.g. with regard to how new chlorophyllide molecules might be formed) and it seems just as well to be able to discard it.

The second possibility posed (Fig. 1B), that a protein molecule about the size of the holochrome might accumulate a large number of pigment molecules and then aggregate, also seems unlikely. For one thing,

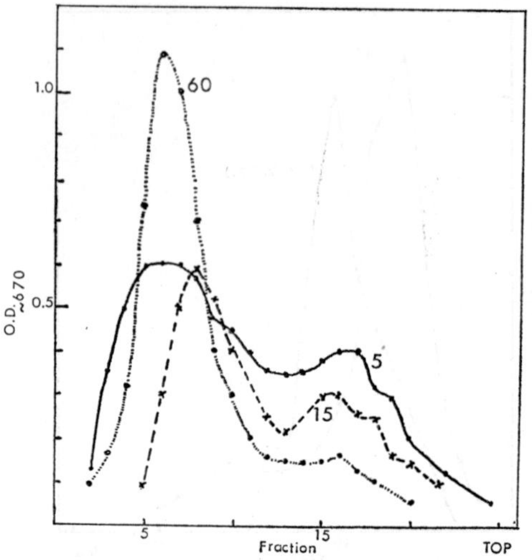

Fig. 4. The distribution in a sucrose density gradient of pigment-protein complexes in Ba fractions of leaves of dark-grown bean seedlings illuminated for 5, 15, or 60 minutes. See text and legends of preceeding figures for procedures used.

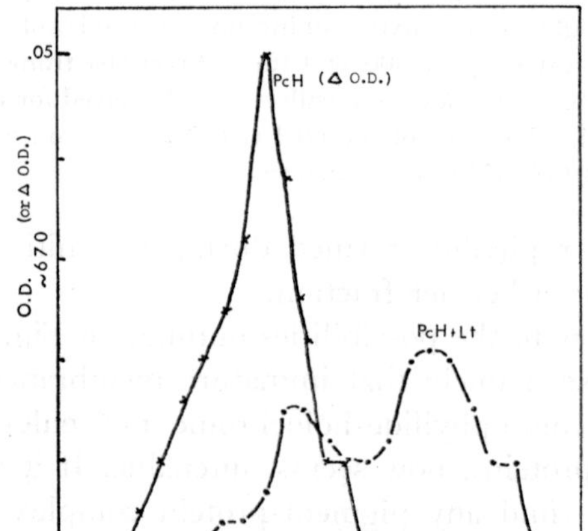

Fig. 5. Comparison of the behavior on centrifugation in a sucrose density gradient of unilluminated photoconvertible protochlorophyllide-holochrome (PcH) and of another fraction of the same preparation illuminated (*in vitro*) before centrifugation (PcH+Lt). Fractions from the centrifuge tube containing PcH were assayed as described in Fig. 2. The optical density of each 10-drop fraction from the "PcH+Lt" gradient was determined at 670 nm.

if this were the case one would expect to find protochlorophyllide and chlorophyll on the same protein molecules. The experiment mentioned, involving illumination of etiolated leaves for 30 minutes and return to darkness for 24 hours, argues against this possibility.

The third possibility (Fig. 1C), that the chlorophyllide or the chlorophyll formed from it might move directly from the original protein with which it was associated onto the membrane, leaves no room for the possibility of any soluble pigment-protein complex except one with the same characteristics as the protochlorophyllide-holochrome.

It is more difficult, and in fact impossible, to choose between possibilities D and E (Fig. 1) on the basis of the data presented. If the postulated carrier protein were smaller than protochlorophyllide-holochrome, this would fit with the facts observed. Similarly, a fraction of the original complex, separating as in proposal E, would also fit with the observations. Thus, at this early point in these investigations we have no evidence to contradict proposals D or E, nor do we have any information upon which to choose between them. In common with others described in the literature,[2, 9] the protochlorophyllide-holochrome preparations used here undoubtedly contained a number of proteins. It should be possible, with pure protochlorophyllide-holochrome, to distinguish experimentally between these two schemes by observing the consequence of illumination on subsequent centrifugal behavior and, if indicated, the effect of reconstituted protein mixtures on such behavior.

The soluble pigment-protein complex obtained after illuminating leaves for 60 minutes had no Hill reaction activity nor could any cytochromes be detected in this complex. Anderson and Boardman[10] found no photosynthetic activity in etiolated bean leaves illuminated for as long as 6 hours, and Butler[11] could find no change in the fluorescence yield of chlorophyll before etiolated bean leaves had been illuminated for 1.5 to 2 hours. Assuming that our conditions of growth of plants, illumination, etc., are at all similar to those of Butler, the data are in rough agreement—that is, that the chlorophyll formed initially is not anywhere near its functional environment after illumination for 1 hour—a time at which soluble pigment-protein complexes can still be obtained.

Protochlorophyllide Regeneration in Intact Leaves

All but one of the experiments described thus far have dealt with the disposition of pigment formed from protochlorophyllide present in etiolated leaves before illumination for, as already pointed out, the etiolated bean leaves used in these experiments appear not to form additional pig-

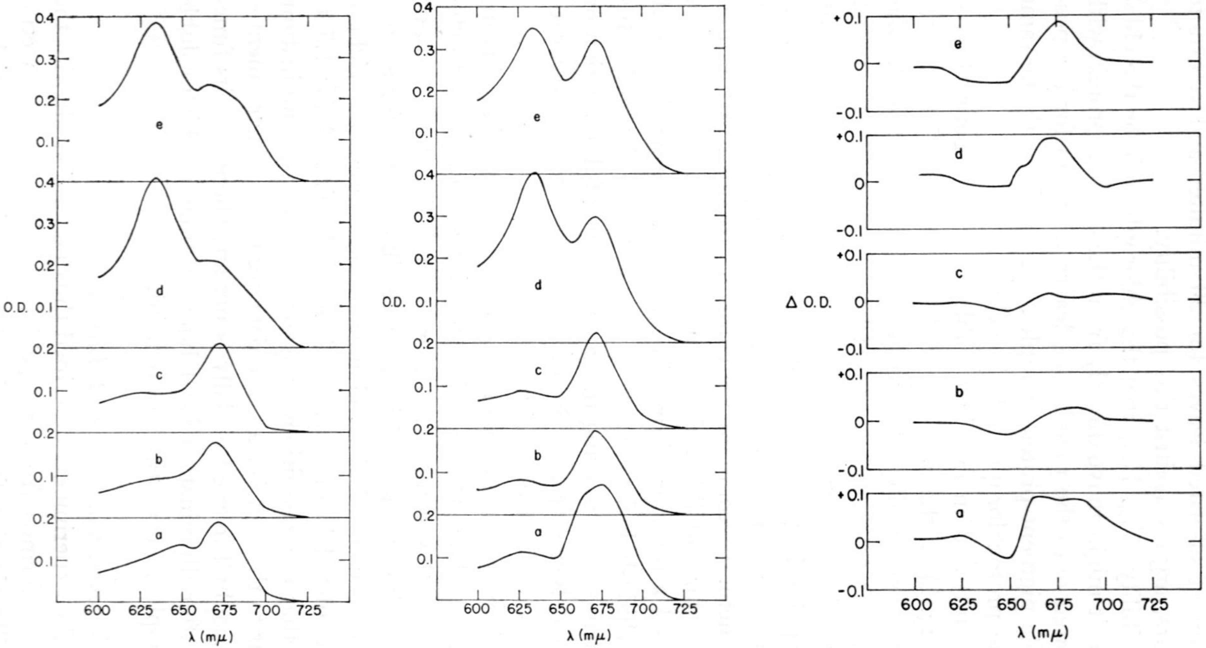

Fig. 6. Absorption spectra of 10 red kidney bean leaf halves measured in a Cary Recording Spectrophotometer Model 11 using the opal glass method of Shibata.[15] Left-hand column of figures: spectra of etiolated leaves illuminated for 1 minute and then maintained in darkness for 4 hours. Center column: the same sets of leaves as used to obtain the data in the left- hand column but absorption spectra shown here were made after leaves had been reilluminated for 1 minute. Right-hand column: difference spectra; i.e. left-hand column spectra minus right-hand column spectra. Frames "a": control; "b": incubated with 5 m*M* chloramphenicol 4 hours prior to first irradiation; "c": incubated with 1 m*M* puromycin prior to first irradiation; "d" and "e": as "a" and "b" but incubated with 10 m*M* δ-aminolevulinic acid during the 4-hour period in darkness following the initial 1 minute illumination.

ment within the first hour after illumination. The set of experiments to be described now[12] deal with the control of protochlorophyllide regeneration after brief illumination of etiolated leaves and touch on the problems of (a) the possible re-use of the protein portion of the holochrome in the photoreduction of protochlorophyllide and (b) the existence of preformed membrane sites for chlorophyll molecules.

In the experiments about to be described, etiolated red kidney bean leaves were illuminated with 1,000 ft.c. of incandescent light for 1 minute and then returned to darkness for 4 hours. The active protochlorophyllide present initially was reduced to chlorophyllide during the 1 minute illumination; additional active (i.e. photoconvertible) protochlorophyllide formed during the subsequent period in darkness. The newly formed pigment can be shown to be active by determining the absorption spectrum of leaves after the 4-hour dark period, then re-illuminating them briefly and determining their absorption spectrum again. The decrease in absorption at about 650 nm reveals that some protochlorophyllide-holochrome has been converted as a consequence of illumination; this conversion is also seen as increased absorption at about 684 nm. A set of absorption spectra and a computed difference spectrum are shown in the frames marked "a" of Fig. 6.[12] Frames "b" show that leaves treated with 5 m*M* chloramphenicol (CAM) for 4 hours prior to the initial irradiation regenerated only a very small amount of active pigment. The same sort of interference with regeneration of pigment occurs when leaves are treated with 1 m*M* puromycin for 4 hours before irradiation (Fig. 6, frames "c") or 90 μg per ml of actinomycin D for 16 hours prior to the initial irradiation (not shown here).

Etiolated leaves form much larger than normal amounts of protochlorophyllide when δ-aminolevulinic acid (ALA), a precursor of chlorophyll, is administered.[13] This shows that all of the enzymes of the biosynthetic chain of protochlorophyllide except one or more required for the synthesis of ALA are present and potentially active in etiolated leaves; the productivity of the entire chain is limited by the availability of ALA. As can be seen in Fig. 6 (frames "d", and compare with "a"), the bulk of this extra pigment absorbs at about 632 nm. The spectra in the "e" series of frames of Fig. 6 show clearly that CAM-treated leaves supplied with ALA form active as well as inactive protochlorophyllide in contrast to tissues treated with the protein synthesis inhibitor but not supplied with the porphyrin precursor.

(Cycloheximide inhibits the greening of Euglena[14] and of a number of etiolated higher plants but unlike CAM prolonged preincubation, even with very high levels of cycloheximide (50 μg/ml) it does not affect the initial regeneration of protochlorophyllide in experiments of the sort shown in Fig. 6. This suggests that protein synthesis inhibition by cycloheximide occurs outside of the plastid and that interference with massive greening is a consequence of plastid starvation.)

DISCUSSION

Altogether, these data, plus the fact that the regeneration of protochlorophyllide in briefly illuminated etiolated bean leaves occurs only after a delay of about 60 minutes, support the contention that new enzyme for the synthesis of ALA is formed (e.g. rather than existing molecules being activated) as a consequence of illumination and that some RNA metabolism is also needed for this to occur.

If we can assume—and there is nothing to assure us that this assumption is correct—that CAM interferes with the synthesis of all plastid proteins, there are a few additional conclusions to which we can come. First, it should be noted that this inhibitor of protein synthesis does not affect the initial conversion of protochlorophyllide to chlorophyllide and the spectral shifts which normally follow the reduction (compare frames "a" and "b" in the left-hand column of Fig. 6), so perhaps the sites to which the pigment molecules ultimately go in the membranes are present before illumination in sufficient number to accommodate the chlorophyll formed from protochlorophyllide present in etiolated leaves. Making the same assumption about the effect of CAM, the data presented in frame sets "b" and "e" of Fig. 6 suggest that the holochrome can function repeatedly (or at least twice) in the conversion of protochlorophyllide molecules.

Thus, with some assumptions, we can deduce from both the sucrose density gradient analyses and the data from the experiments just described, that the holochrome protein, or at least part of it, can act as an enzyme for picking up protochlorophyllide molecules and cooperating in their photoconversion to chlorophyllide molecules.

If we reexamine proposals D and E (Fig. 1) and retain the (possibly precarious) assumption about the uniformity and completeness of the effect of CAM on plastid protein synthesis, proposal E seems less attractive than before if the production of "small protein" is integrated with protochlorophyllide synthesis. It is obvious that with a few more assumptions almost any conclusion could be reached.

None of the chlorophyll deposition sequences proposed (Fig. 1), except "A," accounts for a protein-pigment complex which sediments ahead of protochlorophyllide holochrome. Among the obvious possibilities is that such a complex might include a large reusable protein (i.e. another carrier) or be an aggregate of, e.g. "small protein"-pigment complexes which later becomes part of the membrane. The possibility that these rapidly sedimenting complexes are fragments of forming or finished membranes should not be neglected.

One of the problems in the reduction of protochlorophyllide is the identification of the source of hydrogen atoms (2 are required for the formation of each chlorophyllide). It seems likely (see ref. 2 for discussion) that these atoms come from the protein with which the protochlorophyllide is associated. If the holochrome protein participates in the reduction of more than one protochlorophyllide molecule, some mechanism must exist to reduce the protein. In scheme D this could be done by the carrier protein; in E the hydrogen atoms to be transferred to protochlorophyllide could come from the "small protein."

SUMMARY

Five possible developmental relationships between the pigment and protein moieties of protochlorophyllide-holochrome and photosynthetically competent plastid membranes are outlined. They are evaluated on the basis of the data presented.

Protochlorophyllide-holochrome and pigment-protein complexes formed by illuminating it *in vitro* and for periods of time up to 60 minutes *in vivo* (etiolated red kidney bean leaves) are compared by sucrose density gradient centrifugation. One transitory complex sediments less rapidly than the holochrome; another, which appears only after longer illumination, sediments more rapidly.

Chloramphenicol, puromycin, and actinomycin D inhibit the regeneration of active protochlorophyllide by illuminated leaves of dark-grown beans. Inhibition by chloramphenicol is overcome by administration of δ-aminolevulinic acid.

REFERENCES

1. von Wettstein, D., *Brookhaven Symp. Biol.*, No. 11, p. 138 (1958).
2. Boardman, N. K., in *The Chlorophylls* (L. P. Vernon and G. R. Seely, eds.), pp. 437–479, Academic Press, New York, 1966.
3. Shibata, K., *J. Biochem.* (Tokyo), **44**, 147 (1957).
4. Krasnovsky, A. A., and Kosobutskaya, L. M., *Dokl. Akad. Nauk SSSR*, **85**, 177 (1952).

5. Smith, J. H. C., and Benitez, A., *Carnegie Inst. Wash. Year Book,* **52**, 151 (1953).
6. Capon, B., and Bogorad, L., *Bot. Gaz.*, **123**, 285 (1962).
7. Smith, J. H. C., *Carnegie Inst. Wash. Year Book*, **57**, 287 (1958).
8. Laber, L., *Ph. D. Thesis*, Univ. of Chicago (1967).
9. Capon, B., and Bogorad, L., *Bot. Gaz.*, **124**, 128 (1962).
10. Anderson, J. M., and Boardman, N. K., *Australian J. Biol. Sci.*, **17**, 93 (1964).
11. Butler, W. L., *Arch. Biochem. Biophys.*, **93**, 413 (1961).
12. Gassman, M., and Bogorad, L., *Plant Physiol.*, **42**, 781 (1967).
13. Granick, S., in *Proc. 5th Intern. Cong. Biochem. Moscow*, Vol. V, p. 176, Pergamon Press Ltd., Oxford, 1961.
14. Kirk, J. T. O., and Allen, R. L., *Biochem. Biophys. Res. Comm.*, **21**, 523 (1965).
15. Shibata, K., in *Methods of Biochemical Analysis* (D. Glick, ed.), p. 79, Interscience Publishers, Inc., New York, 1959.

Studies on the Induced Syntheses of Bacteriochlorophyll and δ-Aminolevulinate Synthetase in *Rhodopseudomonas Spheroides* with Special Reference to Chromatophore Formation*

GORO KIKUCHI

Department of Biochemistry, Tohoku University School of Medicine, Sendai, Japan

INTRODUCTION

Non-sulfur purple bacteria can grow either aerobically in the dark or anaerobically in the light, and aerobically grown cells contain no photosynthetic pigment and hence no chromatophore. However, chromatophore formation can be induced in dark-grown cells by simply transferring the cells to semianaerobic conditions in the dark.[1-6] Molecular oxygen but not light seems to play an essential role in the regulation of the chromatophore formation.

On the other hand, the first intermediate of the chlorophyll biosynthesis is δ-ALA, and the level of δ-ALA synthetase in dark-grown cells also increases several fold on transferring the cells to semianaerobic conditions.[2,5,6] Therefore, as an approach to clarify the mechanisms underlying the induced formation of chromatophore, we examined the induced syntheses of bacteriochlorophyll and δ-ALA synthetase in *R. spheroides* under dark-semianaerobic incubation conditions.

EXPERIMENTAL

R. spheroides was grown aerobically in the dark or anaerobically in the light, using malate and glutamate as the major nutrients, as described previously.[5] The incubation medium used for the induction of syntheses

Abbreviations: R., Rhodopseudomonas; δ-ALA, δ-aminolevulinic acid; Bchl, bacteriochlorophyll; FUDR, 5-fluorouridin deoxyribose; H. U., hydroxyurea; mit., mitomycin.

* This investigation was supported in part by research grants from the Ministry of Education of Japan, from the National Institutes of Health, U.S.A. (AM 08016), and from the Waksman Foundation of Japan.

of bacteriochlorophyll and δ-ALA synthetase contained potassium phosphate, ammonium phosphate, magnesium sulfate, manganese sulfate, calcium chloride, ferric citrate, glycine and fumarate, and the pH of the medium was adjusted to 6.8 (*cf*. ref. 5). Cells were suspended in the medium so as to give a cell concentration of about 6 mg dry weight per ml, and 20 ml portions of the suspension were placed in Erlenmeyer flasks of 100 ml capacity and the final volume was made up to 40 ml with other additions. The flasks were shaken gently (about 80 rev./min) in a water bath at 30°C in the dark in air.

For the estimation of the amount of bacteriochlorophyll, the pigment was extracted with a mixture of acetone and methanol (7 : 2, v/v) from the cells collected by centrifugation, and absorbancies of the extracts were measured at 775 mμ. The specific extinction ($E_{1cm}^{1\%}$) of bacteriochorophyll at 775 mμ was assumed to be 0.457×10^3 (ref. 1).

For the assay of δ-ALA synthetase activity, sonic extracts of the cells were centrifuged usually for 90 minutes at $105{,}000 \times g$, and the resulting supernatants were employed. Incubation mixtures for δ-ALA synthetase assay contained in a final volume of 2.0 ml: 50 μmoles of succinate, 50 μmoles of glycine, 5 μmoles of ATP, 5 μmoles of $MgCl_2$, 0.25 μmole of pyridoxal phosphate, 0.02 μmole of CoA, 2 μmoles of mercaptoethanol, 100 μmoles of potassium phosphate buffer (pH 6.8), and cell-free extracts (*cf*. ref. 7). Incubation was carried out for 30 minutes at 30°C, and δ-ALA formed was determined according to Urata and Granick.[8] Methods of other experiments will be described for respective tables and figures.

RESULTS AND DISCUSSION

Correlation between bacteriochlorophyll synthesis and chromatophore formation

Under our standard incubation condition, the dark-aerobically grown cells commenced to synthesize bacteriochlorophyll in the manner shown in Fig. 1 (*cf*. ref. 6). The level of δ-ALA synthetase also increased rapidly. The induced formation of δ-ALA synthetase, however, immediately stops when the 0_2 tension of the cell suspension is increased by vigorous aeration, and the level of δ-ALA synthetase falls rapidly (Fig. 1). Subsequently the synthesis of bacteriochlorophyll also ceases. Cell growth during 6 hours of semianaerobic incubation was less than 20% as judged by the viable count. Electron-microscopic examination of the semianaerobically incubated cells has demonstrated the appearance of chromatophores in the cells (Fig. 2), and the number of chromatophores in the ultra-thin

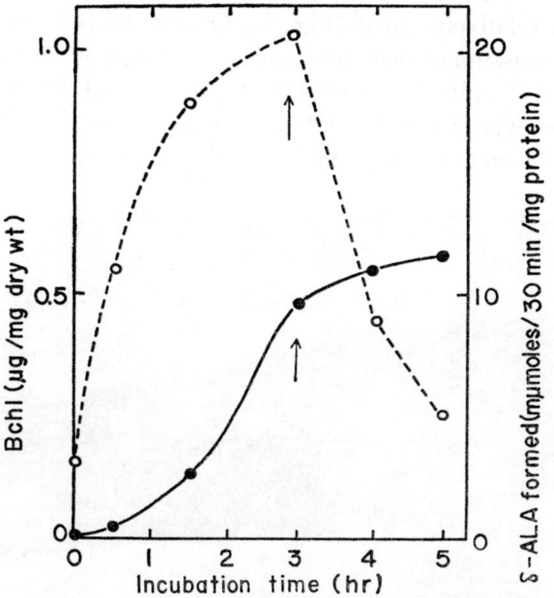

Fig. 1. Effect of oxygen tension on the induced syntheses of bacteriochlorophyll and δ-ALA synthetase.[6]

Cells were incubated with low aeration for an initial 3 hrs, then transferred to Sakaguchi flasks and further incubated with vigorous shaking in air. Solid line, bacteriochlorophyll; broken line, δ-ALA synthetase activity. (Cell-free extracts used were those prepared by 60 min centrifugation at 10,000×*g*).

sections of these cells were roughly proportional to the amounts of pigment synthesized. Further, the particle preparation prepared from these cells after induction in the dark showed active photophosphorylation, although the specific activity of these particle preparations was far lower than those of light-anaerobically grown cells when compared on the bases of bacteriochlorophyll contents and of protein amounts employed (Table I). These observations would justify the assumption that an increase in amount of bacteriochlorophyll may be taken as an indication of chromatophore formation.

Effect of inhibitors of DNA synthesis

The induced syntheses of δ-ALA synthetase and bacteriochlorophyll are inhibited by various inhibitors of protein and nucleic acid synthesis. Previously we reported[5] that the induced synthesis of bacteriochlorophyll was also strongly inhibited by mitomycin C, as well as phleomycin, whereas the inhibition by mitomycin of the induced formation of δ-ALA synthetase was relatively small (Fig. 3). On the basis of these and other observations, we assumed that the induced synthesis of bacteriochlorophyll might require not only the synthesis of protein and RNA, but also the synthesis or turnover of DNA, while this might not necessarily be required for the induced formation of δ-ALA synthetase. Therefore, we attempted

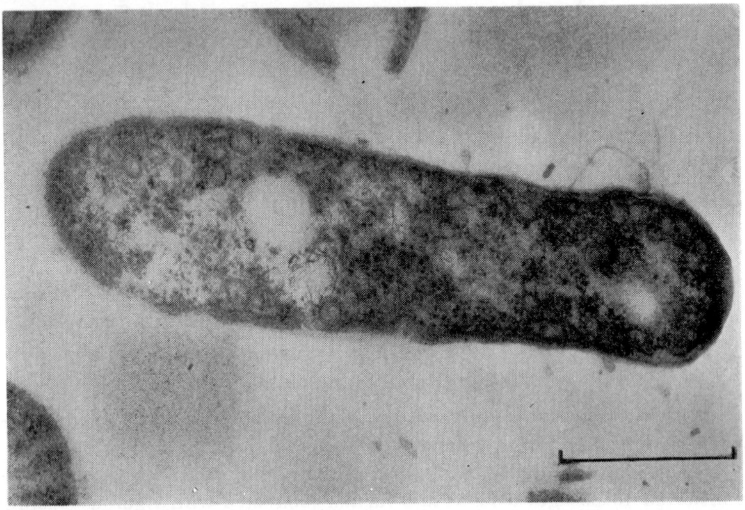

Fig. 2. Electron micrograph of *R. spheroides* cell (wild type strain, grown dark-aerobically), after 6-hr incubation under a dark-semianaerobic condition. Fixative, 1% osmium tetraoxide. The scale in the figure represents 0.5 μ.

TABLE I. Photophosphorylation by particles prepared from *R. spheroides* cells grown under different conditions.

Source of Particles	Bacteriochlorophyll μg/mg protein	Photophosphorylation	
		μmoles Pi esterified/mg Bchl/hr	μmoles Pi esterified/mg protein/hr
(1) Light-anaerobically grown cells	151	97.4	14.4
(2) Dark-aerobically grown cells which were subsequently incubated semianaerobically in the dark for 6 hr	100	32.0	3.2

Particles used were those sedimented between 10,000 and 105,000×g after 90 min centrifugation of cell-free extracts. Reaction mixtures contained, besides particles, succinate, Pi, ADP, glucose, hexokinase, $MgCl_2$ and Tris-HCl buffer (pH 7.4), and were incubated for 30 min at 30°C. For details see ref. 9.

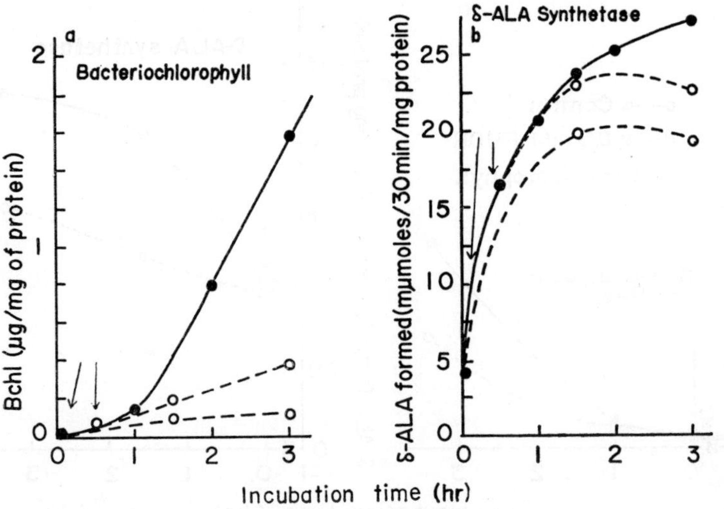

Fig. 3. Effects of addition of mitomycin C, at different times during the incubation, upon the induced formation of bacteriochlorophyll (a) and δ-ALA synthetase (b). Solid line, without mitomycin C; broken line, with 2 μg/ml mitomycin C. (Cell-free extracts used were those prepared by 60 min centrifugation at 10,000× *g*).[5]

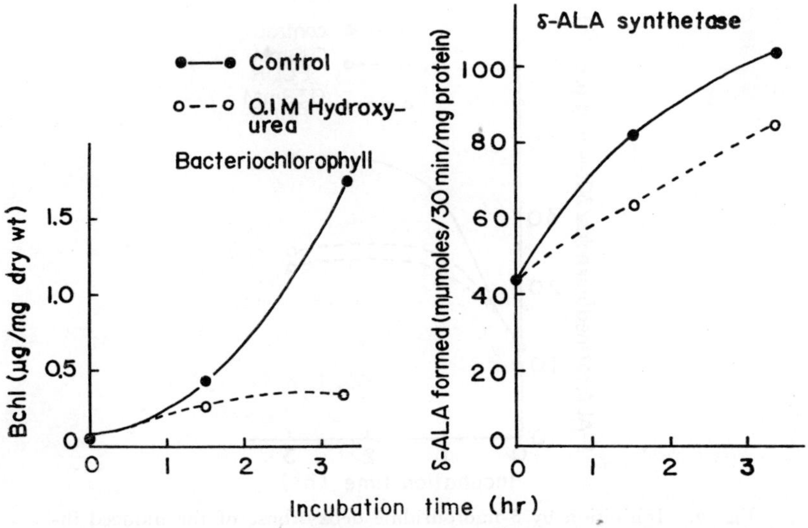

Fig. 4. Effect of hydroxyurea on induced syntheses of bacteriochlorophyll and δ-ALA synthetase. Incubation conditions are described in *Experimental*.

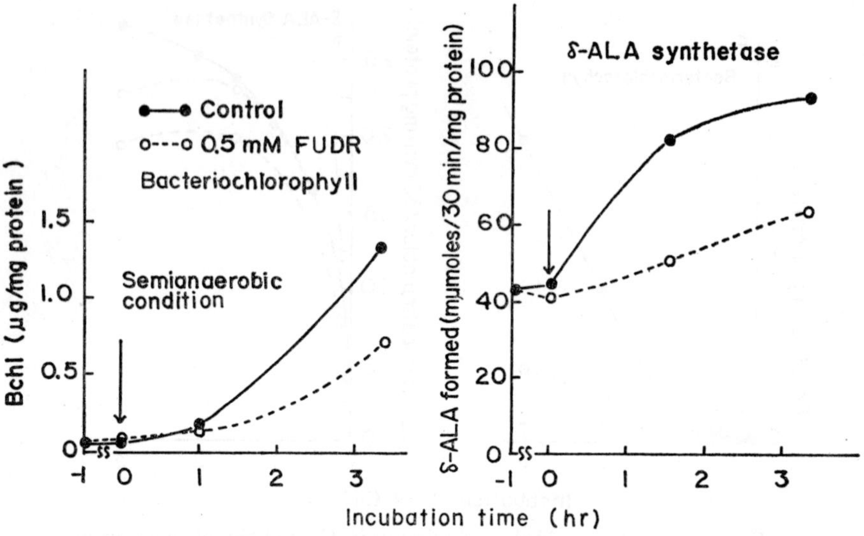

Fig. 5. Effect of 5-fluorouridine deoxyribose on induced syntheses of bacteriochlorophyll and δ-ALA synthetase.

Cell suspensions, with or without added FUDR, were preincubated for 1 hr under aeration, then transferred to semianaerobic conditions at the time indicated by arrows. Other conditions were similar to those described for Figs. 3 and 4.

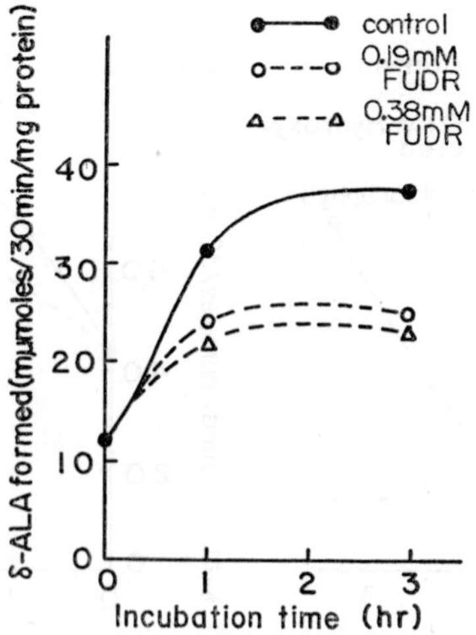

Fig. 6. Inhibition by 5-fluorouridine deoxyribose of the induced increase of δ-ALA synthetase in L-57 strain (colorless mutant) of *R. spheroides*. Cells were incubated in the standard incubation mixture semianaerobically in the dark at 30°C.

similar experiments with hydroxyurea and 5-fluorouridine deoxyribose (FUDR), which are known to be specific inhibitors of DNA synthesis. As shown in Fig. 4, however, hydroxyurea significantly inhibited the induction of the synthesis of both bacteriochlorophyll and δ-ALA synthetase. Similar results were obtained with FUDR. The inhibition by FUDR of the induction of δ-ALA synthetase was more apparent when FUDR was added to the cell suspension 30 to 60 minutes prior to the start of semianaerobic incubation (Fig. 5).

Further, as shown in Fig. 6, FUDR inhibited the induced increase of δ-ALA synthetase even in a colorless mutant, L-57 strain of *R. spheroides*, which lacks the ability to synthesize bacteriochlorophyll. The mutant strain was kindly supplied by Dr. J. Lascelles.

Various possibilities could be considered as to the mechanism of the

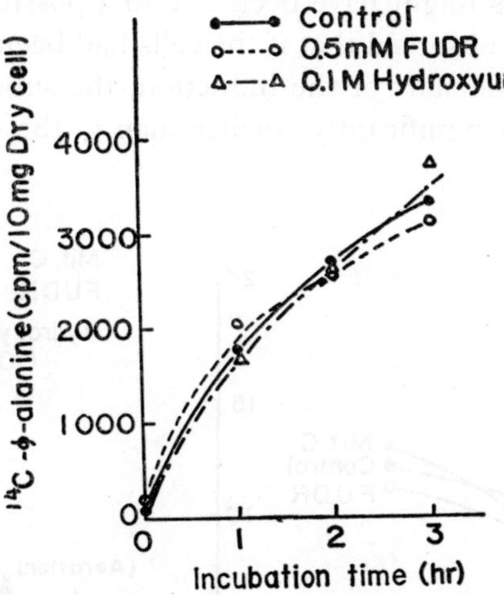

Fig. 7. Effects of hydroxyurea and 5-fluorouridine deoxyribose on the incorporation of ^{14}C-phenylalanine into cell proteins.

Dark-grown *R. spheroides* cells were incubated semianaerobically in the dark in the standard incubation medium supplemented with 1 mg (0.5 μc)/ml ^{14}C-phenylalanine. Inhibitors, when employed, were added right before the addition of ^{14}C-phenylalanine. Two ml each of the incubation mixture (3 mg dry weight of cells per ml) were taken at the times indicated, and mixed with 2 ml of 10% trichloroacetic acid, then heated for 30 min at 100°C. After cooling, the suspension was passed through a glass-fiber filter, and the residues on the filter were washed 5 times with 3 ml each of 5% trichloroacetic acid. The residues and filter were dried under an infra-red lamp, transferred onto a planchet, and the radioactivity was counted with a thin-window gas-flow counter.

inhibition by these inhibitors of DNA synthesis of the induced formation of δ-ALA synthetase. However, the observed inhibition does not appear to be due to a non-specific inhibition of protein synthesis or RNA synthesis in general, since the incorporation of ^{14}C-phenylalanine into protein components under comparable incubation conditions was not appreciably affected by FUDR or hydroxyurea for as long as 3 hours (Fig. 7). Also we have observed that the induced increase of catalase, which could be brought about by transferring the light-grown cells to aerobic incubation conditions, was not significantly affected by inhibitors of DNA synthesis even when the inhibitors were added to the cell suspensions one hour prior to the start of vigorous aeration (Fig. 8). These results point to the possibility that the induced formation of δ-ALA synthetase, too, might require the synthesis or turnover of DNA. Although the induced formation of δ-ALA synthetase appeared to be practically insensitive to mitomycin in our previous experiments, this might have been due to a possibly slower action of mitomycin. In fact we observed that if the cells had been preincubated with mitomycin C before the start of the induction, the subsequent increase of δ-ALA synthetase was significantly smaller than in the control experiment without mitomycin C.

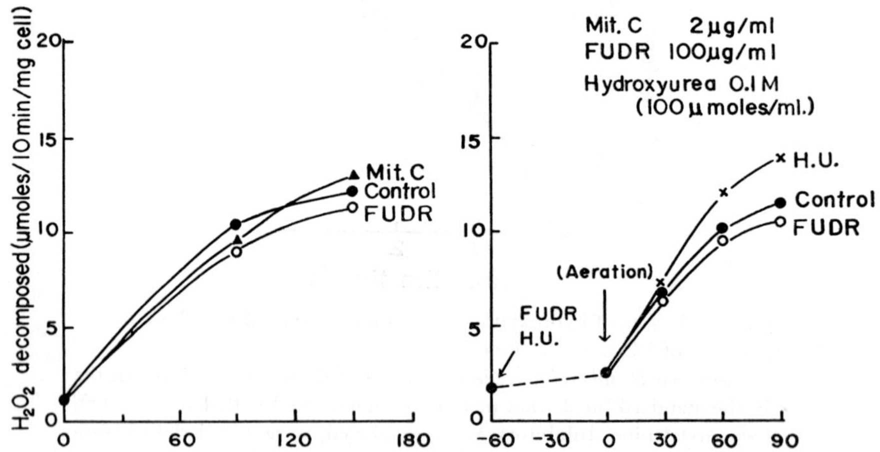

Fig. 8. Effects of inhibitors of DNA synthesis on induced formation of catalase in *R. spheroides*.

Light-anaerobically grown cells were incubated in standard incubation medium under vigorous aeration. Catalase activity was assayed according to Clayton's method,[10] except that permanganate titration was employed instead of iodometry.

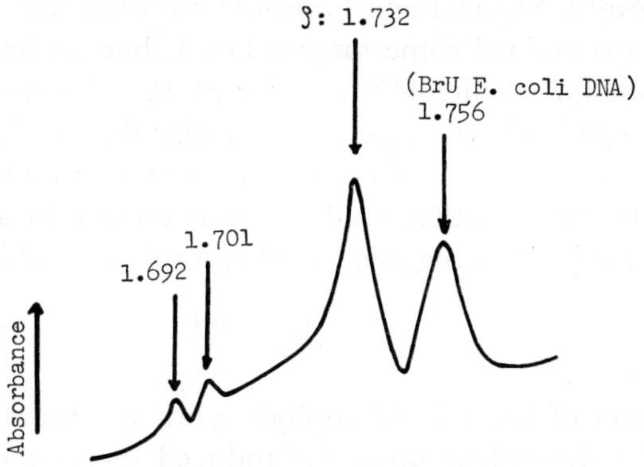

Fig. 9. Cesium chloride density gradient centrifugation (in analytical ultracentrifuge) of DNA prepared from light-anaerobically grown *R. spheroides*.

DNA was extracted by the method of Murmer[11] and purified according to Saito and Miura.[12] Absorbancies at 258 mμ were recorded directly. BrU stands for "bromouracil labelled".

Examination of DNA preparations

It then appeared worth examining whether *R. spheroides* cells might contain some minor DNA components which would turn over rapidly under the particular conditions for the induction. First, we prepared DNA from light-grown and dark-grown cells, and compared the physical properties of the respective DNA preparations by applying methylated albumin column chromatography, CsCl density equilibrium centrifugation, and sucrose gradient centrifugation. Sometimes we observed differences between the two DNA preparations in the patterns of their chromatographic or sedimentation profiles. For instance, as shown in Fig. 9, there were small peaks besides the major DNA peak showing the ρ value of 1.732 in the sedimentation pattern of DNA from light-grown *R. spheroides* cells, but these peaks were later shown to be attributable to polysaccharides contaminating these DNA preparations; the light-grown cells usually contain larger amounts of polysaccharides than the dark-grown cells. Secondly, we incubated the dark-grown cells seminanaerobically in the dark for 1 hour in a medium containing ^{32}P, then isolated the DNA from these cells and subjected it to CsCl density equilibrium centrifugation. However, the radioactivity was evenly distributed in the whole DNA range as judged by the ratio of $^{32}P/A$ 260 mμ in each fraction. Apparently these results are unfavorable to the suspected occurrence of a particular minor DNA component in *R. spheroides*, but the final conclusion must await more detailed examination. Recently Suyama and Gibson[13] reported that various photosynthetic bacteria contained some satelite DNA, but we have been unable to obtain evidence for satelite DNA as far as *R. spheroides* is concerned. As an alternate hypothesis, we may also consider the possibility that some loci of the chromosomal DNA might rapidly turn over under the inducing conditions. At any rate, further studies are necessary to account for the observed high sensitivity of pigment synthesis towards inhibitors of DNA synthesis.

Effect of hemin

The induction of bacteriochlorophyll synthesis, however, appears to be not necessarily dependent upon the induced increase of δ-ALA synthetase. We have observed that the induced increase of δ-ALA synthetase was strongly inhibited by the addition of hemin to the incubation medium, probably due to a feedback repression mechanism, but the induced synthesis of bacteriochlorophyll was not appreciably inhibited by hemin, regardless

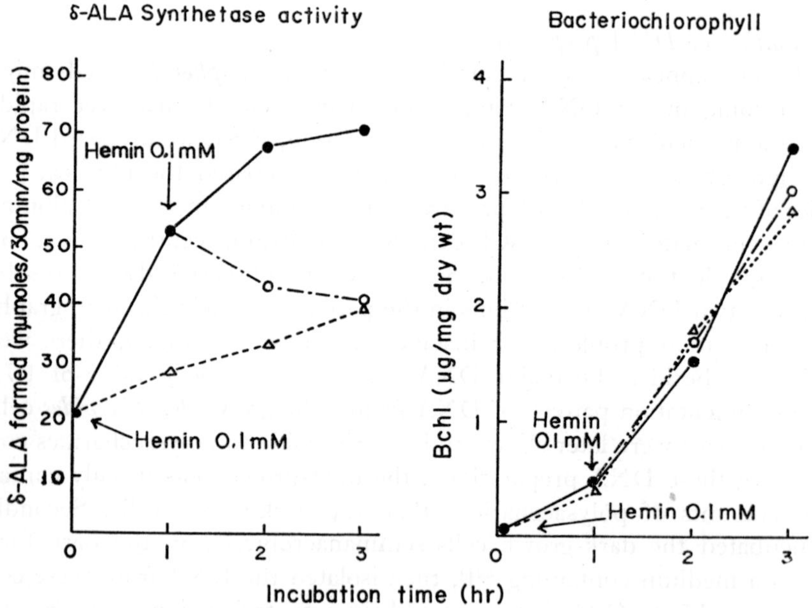

Fig. 10. Effect of hemin on the induced syntheses of δ-ALA synthetase and bacteriochlorophyll.

Hemin was added at the times indicated by arrows so as to give a final concentration of 0.1 mM in the incubation mixtures. Solid lines, control without hemin.

of the time of addition of hemin (Fig. 10). Apparently similar results were obtained when δ-ALA was added in the incubation mixture.[6] These results agree with the observations made with cell cultures growing photosynthetically,[14] and suggest that although there may be some correlation between the mechanisms involved in the induction of synthesis of both δ-ALA synthetase and bacteriochlorophyll, such correlation if it exists may be rather low.

Acknowledgement

I would like to acknowledge the participation of Drs. M. Higuchi, K. Goto, H. Sakai and T. Oba and Mr. H. Kumagai in this investigation.

SUMMARY

The induced syntheses of bacteriochlorophyll and δ-ALA synthetase in dark-aerobically grown *R. spheroides* were studied with the aim of clarifying the mechanisms underlying the formation of chromatophore. The data obtained suggested that the induction of synthesis of bacteriochlorophyll, and possibly of δ-ALA synthetase as well, would require the synthesis or turnover of DNA. However, so far no experimental evidence could be obtained for the occurrence of stelite DNA in *R. spheroides*. On the other hand, the correlation, if it exists, between the mechanism involved in the induction of synthesis of both δ-ALA synthetase and bacteriochlorophyll appeared to be rather low.

REFERENCES

1. Cohen-Bazire, G., Sistrom, W. R., Stanier, R. Y., *J. Cell. Compt. Physiol.*, **49**, 25 (1957).
2. Lascelles, J., *Biochem. J.*, **72**, 508 (1959).
3. Bull, M. J., and Lascelles, J., *Biochem. J.*, **87**, 15 (1963).
4. Higuchi, M., and Kikuchi, G., *Nature*, **200**, 1191 (1963).
5. Higuchi, M., Goto, K., Fujimoto, M., Namiki, O., and Kikuchi, G., *Biochim. Biophys. Acta*, **95**, 94 (1965).
6. Goto, K., Higuchi, M., Sakai, H., and Kikuchi, G., *J. Biochem.*, **61**, 186 (1967).
7. Kikuchi, G., Kumar, A., Talmage, P., and Shemin, D., *J. Biol. Chem.*, **233**, 1214 (1958).
8. Urata, G., and Granick, S., *J. Biol. Chem.*, **238**, 811 (1963).
9. Sato, H., Takahashi, K., and Kikuchi, G., *Biochim. Biophys. Acta*, **112**, 8 (1966).
10. Clayton, R. K., *Biochim. Biophys. Acta*, **36**, 35 (1959).
11. Marmur, J., *J. Mol. Biol.*, **3**, 208 (1961).
12. Saito, H., and Miura, K., *Biochim. Biophys. Acta*, **72**, 619 (1963).
13. Suyama, Y., and Gibson, J., *Biochem. Biophys. Res. Comm.*, **24**, 549 (1966).
14. Lascelles, J., *J. Gen. Microbiol.*, **23**, 487 (1960).

Control of the Photosynthetic Apparatus in Bacteria

JANE GIBSON

Microbiology Section, Division of Biological Sciences, Cornell University, Ithaca, New York

INTRODUCTION

The photosynthetic bacteria are capable of very extensive cellular changes in response to alterations in the environment. Many of the Athiorhodaceae are able to grow as rapidly aerobically in the dark as they are able to grow photosynthetically, and during dark growth the characteristic photosynthetic apparatus can be completely repressed. Even among those which are not capable of aerobic growth, and among the sulfur bacteria, there are major changes in the pigment content of the cells in response to such other factors as light intensity and presence of oxygen in the environment. Consequently, these organisms have been extensively studied in attempts to discover the processes involved in the development of the photosynthetic mechanism, and the means by which they are controlled. The picture which emerges is at present still very incomplete. No genetic analysis is yet possible, since neither conjugation, transformation nor transduction has been demonstrated. Indeed, no bacteriophage active against photosynthetic bacteria has been reported. The comparative ease with which the concentration of photosynthetic pigments can be measured has, however, led to a partial understanding of the operative control mechanisms, and some work on DNA metabolism suggests analogies with control systems in higher organisms.

RESULTS AND DISCUSSION

Nature of the photosynthetic apparatus

It is now clear that the photosynthetic apparatus of purple photosynthetic bacteria, whether vesicular or lamellar in arrangement, is derived from and continuous with the peripheral membrane of the cell.[1–5] Whether there is any difference in composition between the peripheraland invaginated membrane is still under discussion. The dimensions of the membranes are the same, and indeed also the same as those of organisms which are non-photosynthetic.[6] A substantially different arrangement is found in green bacteria.[7–9] The photosynthetic pigments appear to be borne in vesicles which are bounded by a membrane thinner (about 30 Å wide) than the peripheral one, and although contiguous with the latter, they do not appear to communicate with the space between the cell wall and the peripheral membrane.

Description of the chemical composition of photosynthetic membranes is far from complete.[10] About 70% of the isolated membranous fraction from either green or purple bacteria is protein in nature, but probably no more than 20% can be accounted for in terms of known functional constituents. By analogy with higher plants, a structural protein like that isolated from chloroplasts[11] and responsible for binding and orienting the functional components of the photosynthetic apparatus has been suggested, but no specific studies have been reported. A substantial difference between green and purple bacteria in the content of galactolipid has been found;[12] *Chloropseudomonas ethylicum* contains much greater quantities than a number of Athiorhodaceae examined, in some of which galactolipids were barely detectable.

There are therefore marked variations between the structure and composition of the phtosynthetic apparatus in the two main divisions of the photosynthetic bacteria. Fractionation with the detergent Triton X-100[13,14] (Vernon, this volume) has also shown that the detailed structure of the membranes of different purple bacteria must vary, since different fractions contain specific components in preparations from *Chromatium* and *Rhodospirillum rubrum*. Such studies emphasize that the photosynthetic apparatus may have substantially different properties in the various photosynthetic bacteria, and that it is not possible to extrapolate from the behavior of one type to another.

Studies involving the control of the development of the photosynthetic bacteria have been rather restricted in the kinds of organisms used, and because of the relative ease with which they can be followed, changes in pigment content of bacterial cells have been the index used in many studies. However, derepression of the photosynthetic apparatus also must involve an increase in rate of lipid synthesis and production of new protein components of the membrane, since aerobically-grown cells do not contain membranous invaginations, as well as new enzyme activities. Induction of photosynthesis is thus an enormously complex event, of which only a small number of contributing factors have been studied to date.

Control of pigment synthesis

The pigment content of photosynthetic bacteria is governed both by light intensity and by the oxygen tension in the growth medium. Pigment synthesis is increased both on going from high light intensity to a lower one in photosynthetically grown cells or, in aerobically growing cells, by lowering the oxygen tension, without the need for any stimulus from illumination. It was therefore suggested[15] that a common triggering event might be involved, and that the control of development of the photosynthetic apparatus was mediated by the redox state of some cellular component, which could be oxidized either by a light-induced reaction, or by molecular oxygen. The nature of the triggering substance has not been completely clarified, but its behavior would suggest that it is closely connected with the electron transport components such as cytochromes, which behave in this way.

The kinetics of induction of pigment synthesis in aerobically-grown cells of either *Rsp. rubrum*[1,15] or of *Rps. spheroides*[16,17] indicate the operation of the two types of control familiar from studies of the metabolism of *E. coli*. The admission of oxygen causes an immediate halt in pigment synthesis, suggesting some means of feed-back regulation which would bring about rapid adjustments to environmental changes. The effect is too rapid to be accounted for by dilution of enzyme activities. Induction of pigment synthesis in aerobically grown cells on transfer to lower oxygen tension, on the other hand, requires a period of about one hour before the full rate of synthesis is established. During this period, transcription of the previously repressed genetic elements occurs, and pigment formation can be prevented by inhibitors such as Actinomycin D.[16,18] The enzyme responsible for the formation of δ-aminolaevulinic acid (ALA synthetase) appears to be intimately connected with both mechanisms of control, since Burnham and Lascelles[19] have shown it to be strongly inhibited by hemin, which is an end product of the iron porphyrin branch of the pigment synthesising system. ALA synthetase also increases greatly in activity before the onset of bacteriochlorophyll in aerobic cultures adapting to pigment synthesis.[16,17,20] It has been clearly established that pigment synthesis does not proceed in the absence of protein synthesis, since pigment formation stops abruptly on the exhaustion of the requirements of amino-acid auxotrophs, and can also be stopped by the addition of such inhibitors of protein synthesis as chloramphenicol and puromycin.[16,18,21,22] This effect is again more rapid and complete than could be accounted for by inhibition of enzyme synthesis, and has led to the suggestion that the synthesis of the whole photosynthetic apparatus is regulated by the rate of formation of some structural protein component of the membrane, responsible for the ordering of the pigment molecules into functional photosynthetic units.

A number of mutants have been obtained having nutritional and photosynthetic defects, especially from *Rps. spheroides*,[21,23] an organism in which even the spontaneous mutation rate is quite high. The detection of photosynthetic mutants is, of course, facilitated by the capacity of the organism to grow well aerobically; mutants of the strictly anaerobic forms have not been described. Several of the mutant strains of *Rps. spheroides* have had multiple nutritional requirements, and have been produced with relatively heavy doses of mutagenic agents, so that there is a strong possibility of multiple lesions. Nevertheless, the occurrence of as drastic a change as that shown by Lascelles' albino mutant L-57 suggests an analogy with the permanently bleached mutants of *Euglena*. Evidence has accumulated to demonstrate the semiautonomous replication of plastids in this organism, and suggestions have been made that something similar could occur in bacteria. Convincing evidence for this is still lacking, but it is perhaps worth considering such evidence and analogies as can be put forward.

Structural considerations indicate that a constant number of proplastids, as is found in *Euglena*, is unlikely to be present in bacteria. In Athiorhodaceae capable of aerobic growth, such conditions result in the restriction of the membrane to the periphery of the cell, where it would be equally partitioned between the daughter organisms at division. In those types in which aerobic growth is impossible, septa form across the lamellae or partition the vesicles.[10] Continuity of repressed or expressed photosynthetic material from generation to generation does therefore exist in purple bacteria; whether the vesicles of green bacteria actually divide has not been clarified, but at least the possibility of self-perpetuating bodies must be considered.

More direct evidence for analogy with higher forms comes from the observation that interference not only with protein synthesis, but also with nucleic acid formation, will prevent the adaptive formation of bacteriochlorophyll in *Rps. spheroides*. This has been shown both by the effect of limitation of adenine in adenine-less mutants[21] and by the experiments of Higuchi *et al.*[16] in which a number of inhibitors of replication of DNA, of transcription and of translation were investigated. Mitomycin, which inhibits DNA polymerase, as well as phleomycin, which has less tendency to bring about DNA degradation, have characteristic effects on adaptive pigment formation in non-proliferating suspensions of *Rps. spheroides*. Addition of the inhibitors at any stage after inducing pigment formation by lowering the oxygen tension allows pigment synthesis to proceed for about one hour before it is halted. Acriflavin, on the other hand, can prevent pigment synthesis entirely, provided that it is added at the start of the induction period, although at a later stage synthesis continues for a period, as with mitomycin. These experiments imply that new synthesis of DNA is required for full development of the photosynthetic apparatus.

The presence of autonomously replicating elements in bacteria was originally inferred from the rapid spread of genetic traits from certain donor strains of *E. coli* through a population of recipient cells. The properties of non-chromosomal elements have subsequently been defined through the study of the sex factor in *E. coli*, resistance transfer factors in *Salmonella*, and the behavior of lysogenic phages like λ of *E. coli*. Apart from the ability to replicate more rapidly than the main genome, these factors also promote conjugation in the host, so that their spread through the population becomes possible. There is also evidence for sufficient homology between the main genome and the episome for incorporation to occur, and in several cases fragments of the main chromosome can remain attached to the episome on its detachment. As is the case with plastids, however, the replication of the episome is normally held in check, and cells contain only one or at most a small number of episomal elements.

The guanosine-cytosine content of the F factor in *E. coli* is the same as that of the main genome, and therefore DNA preparations from episome-bearing strains do not show any evidence of this in the ultracentrifuge. However, when the *E. coli* F factor is transferred by conjugation into *Proteus*, which has an appreciably higher % G-C content, the presence of the episome can be demonstrated by buoyant density measurements as well as by genetic means.[24]

DNA has been prepared from a number of photosynthetic bacteria, and the composition found is summarized in Tables I and II. Reexamination of preparations from a number of species using higher than usual concentrations in the ultracentrifuge cell[25] showed that two species, *Chromatium* and *Rps. spheroides*, contained a small amount of material with a buoyant density appreciably different from that of the main band. The cultures from which the preparations were made were carefully checked for purity, and at least with *Chromatium*, with its characteristic morphology, it is hardly possible for the satellite band to be the result of contamination. It was possible to show that both UV absorbing bands were sensitive to DNAase. Heating followed by rapid cooling led to the splitting of both main and satellite bands, indicating that both components in the preparation were initially native. DNA from more than a thousand species of bacteria has now been examined,[26] and all preparations have been unimodal except for those from these two species of photosynthetic bacteria, some

TABLE I.

Organism	DNA base composition % G-C
Chlorobium	51–58 (from ref. 34)
Chromatium	64–66 (from ref. 35)
Rsp. rubrum	60–70 (from ref. 36)
Rps. spheroides	71
Rps. palustris	66

TABLE II.

	Chromatium	*Rps. spheroides*
Main component		
ρ g/cm^3	1.725	1.730 (from ref. 25)
% G-C	66	71
Satellite		
ρ g/cm^3	1.713	1.724 (from ref. 25)
% G-C	54	66

Halobacteria[27] and *Pseudomonas stutzeri*.[28] In certain, but not all, photosynthetic bacteria there is therefore evidence for the existence of a second type of DNA, in about 10% of the amount found in the main band; the possibility exists, of course, that a second species of DNA exists in the others, but with a buoyant density too close to that of the main band to allow detection.

There is at present no direct evidence to connect the presence of this second type of DNA with the material suggested by the experiments of Higuchi *et al.*[16] with inhibitors of DNA synthesis. Indeed, all analyses of purified "chromatophores" indicate that the DNA content is very low.[16, 29–31] However, the isolation procedures generally involve the use of buffers at physiological pH, and since they are based on differential centrifugation procedures, take quite a long time. Therefore, any DNA associated with the pigmented fraction might well have been broken down by cellular DNAases during preparation, although Tewari, Vötsch, Mahler and Mackler[32] have reported that the DNA of yeast mitochondria can be isolated with electron transport particles, and is resistant to DNAase. Other methods of "chromatophore" preparation need to be developed to settle this point.

Episomal DNA has in general a different sensitivity from the main genome material to some inactivating agents. Heat and UV are both more effective against the plastids of *Euglena* than in killing, and *E. coli* can be "cured" of F-factor by growth in the presence of some acridine derivatives, notably acridine orange.[33] The conditions under which this "curing" can affect high percentages of the population are, however, quite specific in terms of growth conditions and cell concentration. We have examined a large number of cultures of *Rps. spheroides* treated with acridines or subjected to varying heat treatments, but have not yet found any set of conditions which will induce a defined and stable phenotypic change, such as would be expected from elimination of an episome, in a substantial number of colonies obtained by subsequent plating. Clearly, however, "curing" efficiencies of at least 10% would be required if they were to be recognized phenotypically against the spontaneous mutation rate, and the possibility also exists that any phenotypic changes might be very hard to detect. Mandel[28] has reported that *Pseudomonas stutzeri* can be "cured" of its satellite DNA by growth in acriflavine, but Palleroni was unable to detect any phenotypic changes in the "cured" clones (R. Y. Stanier, personal communication).

Control of the development of the photosynthetic apparatus of photosynthetic bacteria by an episomal element, together with the inhibitor experiments, might suggest that the contribution of satellite DNA to the total in the cell might vary with growth conditions, and be greater when the pigment concentration of the cells was high. Very careful experiments will be needed to check this possibility. Cohen-Bazire and Sistrom[10] have shown the importance of using cells growing at constant rates in determing the relationship between light intensity and pigment content, and the DNA experiments also will need to be done with cells in balanced growth if changes caused by irrelevant factors are to be eliminated.

The present position, therefore, is little more than suggestive of an analogy between the control mechanisms of photosynthetic bacteria and algae, attractive as such an analogy would be. The nucleic acid metabolism of *Rps. spheroides* differs from that of *E. coli* in that labeled uracil enters into DNA to a much greater extent,[16] and in that label from thymine and thymidine is incorporated into protein and RNA as well as DNA (Gibson, unpublished experiments), so that the interpretation of the inhibitor studies may be complex. Should it transpire that the satellite DNA is not associated with the control of photosynthetic apparatus, there is still some hope that the finding may open up a system whereby the genetics of these organisms may be investigated.

Acknowledgement

Work reported in this article was supported in part by Grant GB 5793 from the National Science Foundation.

REFERENCES

1. Cohen-Bazire, G., and Kunisawa, R., *J. Cell. Biol.*, **16**, 401 (1960).
2. Drews, G., and Giesbrecht, P., *Zbl. Bakt. Abt. I Orig.*, **190**, 508 (1963).
3. Giesbrecht, P., and Drews, G., *Arch. Mikrobiol.*, **54**, 297 (1966).
4. Holt, S. C., and Marr, A. G., *J. Bacteriol.*, **89**, 1413 (1965).
5. Holt, S. C., and Marr, A. G., *Ibid.*, **89**, 1421 (1965).
6. Murray, R. G. E., and Watson, S. W., *J. Bacteriol.*, **89**, 1594 (1966).
7. Cohen-Bazire, G., in *Bacterial Photosynthesis* (H. Gest, A. San Pietro and L. P. Vernon, eds.), pp. 89–110, Antioch Press, Yellow Springs, 1963.
8. Cohen-Bazire, G., Pfennig, N., and Kunisawa, R., *J. Cell. Biol.*, **22**, 207 (1964).
9. Holt, S. C., Conti, S. F., and Fuller, R. C., *J. Bacteriol.*, **91**, 311 (1966).
10. Cohen-Bazire, G., and Sistrom, W. R., in *The Chlorophylls* (L. P. Vernon and G. R. Seely, eds.), pp. 313–341, Academic Press, New York, 1966.
11. Criddle, R. S., in *Biochemistry of Chloroplasts* (T. W. Goodwin, ed.), pp. 203–232, Academic Press, London and New York, 1966.
12. Constantopoulos, G., and Bloch, K., *J. Bacteriol.*, **93**, 1788 (1967).
13. Garcia, A., Vernon, L. P., and Mollenhauer, H., *Biochemistry*, **5**, 2399 (1966).
14. Garcia, A., Vernon, L. P., and Mollenhauer, H., *Ibid.*, **5**, 2408 (1966).
15. Cohen-Bazire, G., Sistrom, W. R., and Stanier, R. Y., *J. Cell. Comp. Physiol.*, **49**, 25 (1957).
16. Higuchi, M., Goto, K., Fujimoto, M., Namiko, O., and Kikuchi, G., *Biochim. Biophys. Acta*, **95**, 94 (1965).
17. Lascelles, J., *Biochem. J.*, **72**, 508 (1959).
18. Biedermann, M., Drews, G., Marx, R., and Schröder, J., *Arch. Mikrobiol.*, **56**, 133 (1967).
19. Burnham, B. F., and Lascelles, J., *Biochem. J.*, **87**, 462 (1963).
20. Lascelles, J., *J. Gen. Microbiol.*, **23**, 487 (1960).
21. Bull, M. J., and Lascelles, J., *Biochem. J.*, **87**, 15 (1963).
22. Sistrom, W. R., *J. Gen. Microbiol.*, **28**, 599 (1962).
23. Lascelles, J., *Biochem. J.*, **100**, 175 (1966).
24. Falkow, S., Wohlhieter, J. A., Citarella, R. V., and Baron, L. S., *J. Bacteriol.*, **87**, 209 (1964).
25. Suyama, Y., and Gibson, J., *Biochem. Biophys. Res. Comm.*, **24**, 549 (1966).
26. Hill, L. R., *J. Gen. Microbiol.*, **44**, 419 (1966).
27. Joshi, J. G., Guild, W. R., and Handler, P., *J. Mol. Biol.*, **6**, 34 (1963).
28. Mandel, M., *J. Gen. Microbiol.*, **43**, 273 (1966).
29. Gibson, K. D., *Biochemistry*, **4**, 2042 (1965).
30. Newton, J. W., and Newton, G. A., *Arch. Biochem. Biophys.*, **71**, 250 (1957).
31. Sykes, J., Gibbon, J. A., and Hoare, D. S., *Biochim. Biophys. Acta*, **109**, 409 (1966).
32. Tewari, K. K., Vötsch, W., Mahler, H. R., and Mackler, B., *J. Mol. Biol.*, **20**, 453 (1966).
33. Hirota, Y., *Proc. Nat. Acad. Sci.* (Wash.), **46**, 57 (1960).
34. Mandel, M., Bergendahl, J. C., and Pfennig, N., *J. Bacteriol.*, **89**, 917 (1965).
35. Marmur, J., Falkow, S., and Mandel, M., *Ann. Rev. Microbiol.*, **17**, 329 (1963).
36. Schildkraut, C. L., Marmur, J., and Doty, P., *J. Mol. Biol.*, **4**, 430 (1962).

Biogenesis of Chloroplasts: Roles of Chloroplast DNA and Chloroplast Ribosomes

ROBERT M. SMILLIE, N. STEELE SCOTT, AND D. GRAHAM

Plant Physiology Unit, C.S.I.R.O., Division of Food Preservation and School of Biological Sciences, The University of Sydney, Australia

INTRODUCTION

Chloroplasts contain both DNA and RNA, and recently, the synthesis of DNA, RNA, and protein has been demonstrated in isolated preparations of purified chloroplasts (see ref. 1 and reviews by Kirk[2] and Brawerman.[3]) Biochemists are now confronted with the problem of how these organelle-localized synthetic systems cooperate with each other, and with other cellular synthetic systems, to form a functional organelle of the complexity of the chloroplast. The results presented in this paper are directed towards two main questions. First, what cellular systems are directly involved in the synthesis of chloroplast proteins (which for convenience are divided into three groups: soluble proteins including enzymes of the photosynthetic CO_2 reductive cycle, electron transfer proteins, and membrane proteins)*? Second, what is the role of chloroplast DNA in the synthesis of chloroplast protein?

RESULTS AND DISCUSSION

Function of chloroplast ribosomes in differentiating chloroplasts

Electron micrographs of thin sections of chloroplasts fixed in formaldehyde-osmium tetroxide or glutaraldehyde-osmium tetroxide show numerous ribosomes which are prominent in the interlamellar spaces.[4–6] Like ribosomes from other sources, purified preparations of chloroplast ribosomes support the incorporation of amino acids into protein.[7,8] However, they possess several features which distinguish them from the neighboring cytoplasmic ribosomes. Perhaps the most obvious difference is their size; they are smaller and usually have a sedimentation constant of about 70S compared with a value of about 80S for cytoplasmic ribo-

* The term membrane protein as used here includes all the insoluble proteins of organelle membranes other than those directly catalyzing electron transfer reactions.

somes.[9] In some organisms, such as *Euglena gracilis*, differences between the ratios of the nitrogenous bases of the RNA of chloroplast and cytoplasmic ribosomes have been noted.[3]

Although the obvious biological role of these ribosomes would appear to be the synthesis of chloroplast proteins, direct evidence of their involvement in the synthesis of even one identifiable chloroplast protein is lacking. It is not known if protein synthesis in chloroplasts is restricted to only one ribosomal system. Nor is it known if some chloroplast proteins are synthesized by the cytoplasmic microsomal system and subsequently transferred to the chloroplasts. The latter possibility is quite feasible, especially since recent studies point to the microsomal system and not the mitochondrion as the site of synthesis of mitochondrial cytochrome *c* and some of the enzymes of the tricarboxylic acid cycle.[10–12]

We have investigated the intracellular sites of synthesis of individual chloroplast proteins by the use of two selective inhibitors of protein synthesis, chloramphenicol and cycloheximide. Before considering their effect on the synthesis of chloroplast protein *in vivo*, it is necessary to consider the specificity of their action on cells containing differentiating chloroplasts and on other systems involving the synthesis of new protein.

Action of chloramphenicol on chloroplast differentiation

Chloramphenicol (Fig. 1) is an antibiotic which inhibits protein synthesis in bacteria. Its site of action is the ribosome.[13] Although protein synthesis in higher organisms is relatively insensitive to chloramphenicol,[14] our earlier experiments[15] clearly demonstrated an inhibition of chlorophyll synthesis by chloramphenicol during the light-dependent differentiation of chloroplasts in *E. gracilis* (Fig. 2). This sensitivity of chloroplast development towards chloramphenicol has since been confirmed for *E. gracilis* and for several other photosynthetic organisms.[16–20] Although the effective concentration range for inhibition of chloroplast differentiation in *E. gracilis* (Fig. 2) is much higher than for inhibition of bacterial growth, the difference may be due partly to low permeability of the *E. gracilis* cells to chloramphenicol, since amino acid incorporation by isolated chloroplasts is sensitive to lower concentrations of the inhibitor.[8, 21]

One of the most interesting aspects of the action of chloramphenicol on the light-dependent differentiation of chloroplasts, and one that is of crucial importance to its use as an experimental tool in studies described below, is the apparent specificity of its action towards the synthesis of chloroplast proteins (Table I). Thus, while chloramphenicol at 1 mg/ml inhibits chlorophyll synthesis, the synthesis of cytochrome-552 (a *c* type

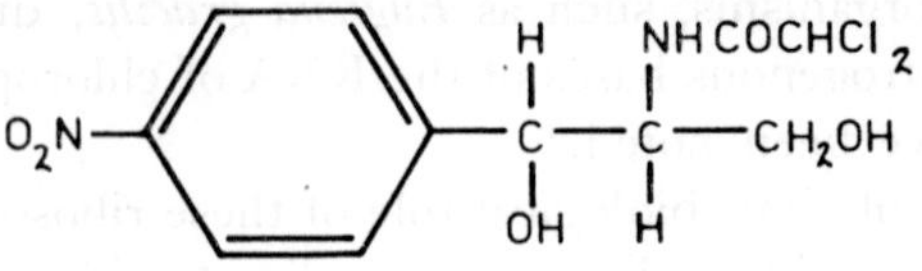

Fig. 1. Chloramphenicol.

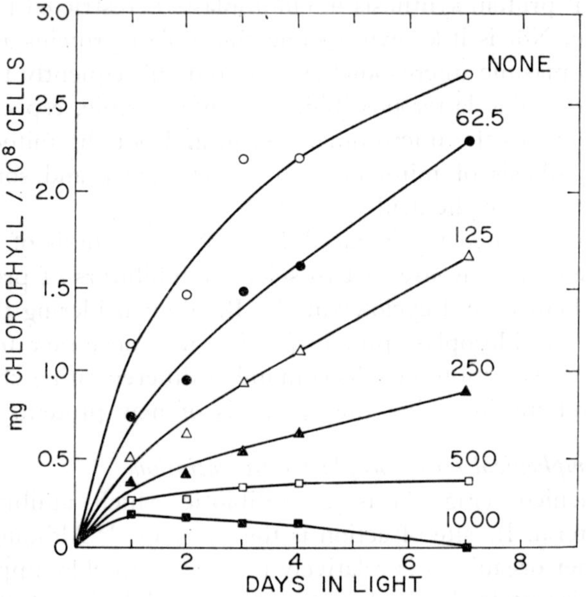

Fig. 2. Inhibition of chlorophyll synthesis in *E. gracilis* by chloramphenicol (concentrations in μg/ml are shown on graph). Dark-adapted cells were illuminated with white light (150 ft-candles) in a medium containing salts, but no carbon growth source. The chloramphenicol was added immediately prior to exposing the cells to light. Data from ref 15.

TABLE I. Specificity of inhibition by chloramphenicol in photosynthetic cells.

	Inhibited	Not inhibited
Pigment production	Chlorophyll synthesis	
Growth (*E. gracilis*)	Photoautotroph	Heterotroph (light or dark)
Amino acid incorporation	Chloroplast ribosomes	Cytoplasmic ribosomes

cytochrome), and the incorporation of manganese into chloroplast lamellae,[15] it does not inhibit essential cytoplasmic protein synthesis as shown by the capacity of chloramphenicol-treated cells to divide when supplied with glucose (see below). This selective action of chloramphenicol is dramatically shown by its inhibition of the autotrophic growth of cells of *E. gracilis* at concentrations which fail to inhibit heterotrophic growth, either in light or in the dark.[15] The inference from these studies was that protein synthesis occurring on chloroplast ribosomes was more sensitive to chloramphenicol than that taking place on cytoplasmic ribosomes. This conclusion is supported by studies with cell-free systems. Protein-synthesizing systems isolated from chloroplasts are sensitive to low concentrations of chloramphenicol,[7, 8, 21] whereas cytoplasmic systems from photosynthetic cells are comparatively insensitive.[7] Possible reasons for the sensitivity of chloroplast ribosomes to chloramphenicol are considered in a following section.

Inhibitory action of cycloheximide on protein synthesis

The other inhibitor of protein synthesis used in our studies was cycloheximide (Fig. 3). This compound was originally employed as an anti-fungal agent,[22] its primary action being on protein synthesis. Although cycloheximide, like chloramphenicol, inhibits peptide formation on the ribosome, a perusal of the phylogenetic range of its inhibitory effect on growth (Table II) shows up a curious relationship: organisms whose growth is inhibited by cycloheximide are relatively insensitive to chloramphenicol,

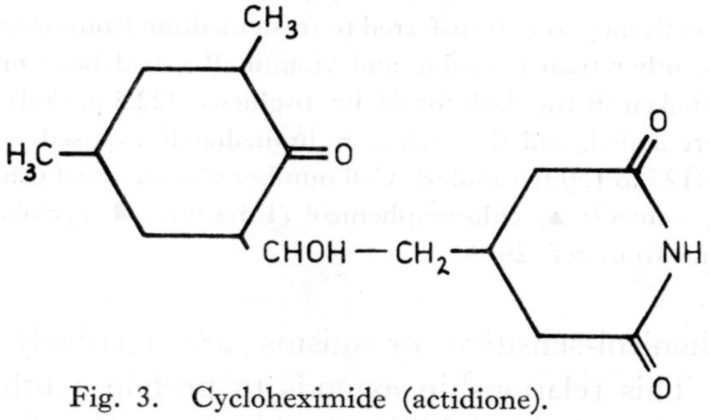

Fig. 3. Cycloheximide (actidione).

TABLE II. Action of chloramphenicol and cycloheximide on the growth of various organisms.

Organism	Chloramphenicol	Cycloheximide
Bacteria	+	−
Blue-green algae	+	−
Protozoa	−	?
Fungi	−	+
Plants and green algae	−	+
Mammals	−	+

+ indicates inhibition of growth.

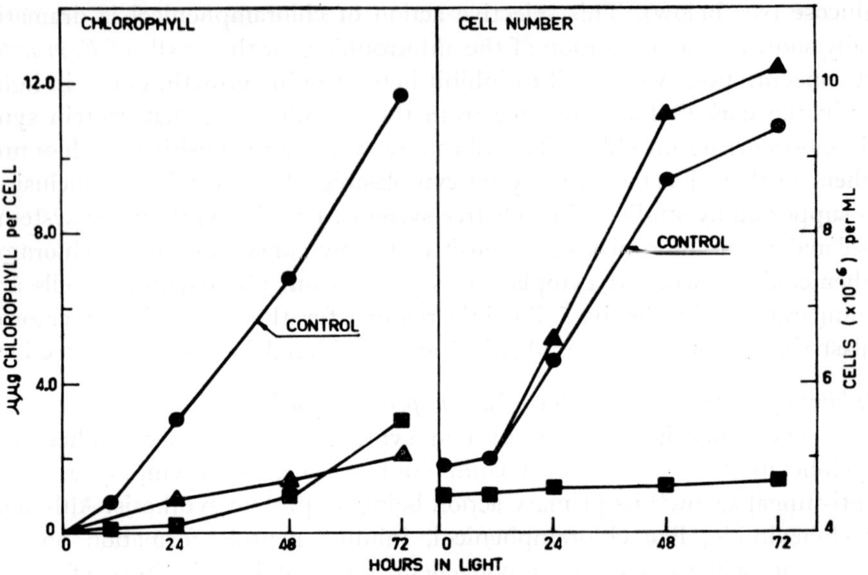

Fig. 4. The effect of chloramphenicol and cycloheximide on chlorophyll synthesis and cell division. Cells of *E. gracilis*, strain Z, were grown heterotrophically in continuous darkness, and while still in the exponential phase of growth they were transferred to fresh medium from which all carbon compounds, other than thiamine and vitamin B_{12}, had been omitted. The cells were shaken in the dark for 48 hr, D-glucose (22.5 mg/ml) and the inhibitors were added, and the cells were immediately exposed to continuous white light (125 to 150 ft-candles). Cell number was recorded using a Coulter counter. ●, control; ▲, chloramphenicol (1 mg/ml); ■, cycloheximide (15 μg/ml). Data from ref. 26.

and chloramphenicol-sensitive organisms are relatively unaffected by cycloheximide. This relationship extends to protein synthesis in cell-free systems since cycloheximide inhibits the incorporation of amino acids into protein by ribosomes from yeast[23] and mammalian cells,[24] but not by bacterial ribosomes.[24] Because of these striking differences in the effectiveness of chloramphenicol and cycloheximide as inhibitors of protein synthesis in various organisms, both were employed in our studies on the *in vivo* synthesis of chloroplast proteins.

Effect of chloramphenicol and cycloheximide on chlorophyll synthesis and cell division in E. gracilis

The growth of *E. gracilis* is very sensitive to cycloheximide. Cell division in the presence of glucose was almost completely inhibited by a concentration of 1 μg/ml. Chlorophyll synthesis is also inhibited by cycloheximide, as was previously noted by Kirk and Allen,[25] but higher concentrations of cycloheximide are necessary. Fig. 4 compares the effects of chloramphenicol and cycloheximide on chlorophyll synthesis and cell division. Both processes were studied simultaneously by adding glucose to starved, dark-adapted cells and immediately exposing the cells to continuous white light. The inhibitors were added with the glucose. In the control cultures, chlorophyll synthesis and cell division commenced after a lag of about 9 to 12 hours. Cycloheximide at 15 μg/ml completely stopped cell division and inhibited chlorophyll synthesis by 75% after 72 hours of illumination. The addition of chloramphenicol at 1 mg/ml had a similar effect on chlorophyll synthesis (82% inhibition after 72 hours), but resulted in stimulation of the rate of cell division rather than inhibition. As chloramphenicol can also inhibit the synthesis of certain mitochondrial proteins,[27] this result is important since it shows that the concentration of chloramphenicol used was not high enough to be detrimental to the respiratory system. At higher concentrations (3 to 5 mg/ml), growth was inhibited, possibly resulting from an inhibition in the synthesis of mitochondrial protein.

Effect of chloramphenicol and cycloheximide on the synthesis of chloroplast and cytoplasmic proteins in vivo

(i) *Proteins of the photosynthetic electron transfer pathway:* Table III shows the light-induced changes in the cellular levels of some of the proteins of the photosynthetic electron transfer pathway in cells incubated with

TABLE III. Action of chloramphenicol and cycloheximide on the synthesis *in vivo* of proteins of the photosynthetic electron transfer pathway.

Conditions	Increase during illumination for 72 hr: Cytochrome-552 (mμmoles/10^9 cells)	Ferredoxin-NADP-reductase (μmoles substrate/min/10^9 cells)	Cytochrome-561 (mμmoles/10^9 cells)
Control	46.4	8.55	44
Chloramphenicol	7.7	1.58	0.9
Cycloheximide	11.4	3.06	13.3
% inhibition with chloramphenicol	84	81	98
% inhibition with cycloheximide	76	64	70

Dark-adapted cells were illuminated as described in Fig. 4. Data from Smillie *et al.*[26]

TABLE IV. Action of chloramphenicol and cycloheximide on the synthesis *in vivo* of enzymes of the Calvin cycle.

Conditions	Increase during illumination for 72 hr: Ribulose-1,5-bisphosphate carboxylase (μmoles substrate/min/10^9 cells)	Increase during illumination for 72 hr: NADP$^+$-glyceraldehyde-3-phosphate dehydrogenase (μmoles substrate/min/10^9 cells)
Control	0.62	13.0
Chloramphenicol	0.07	−0.6
Cycloheximide	1.34	15.3
% inhibition with chloramphenicol	89	100
Cyclocheximide treatment as % of control	216	118

Dark-adapted cells were illuminated as described in Fig. 4. Data from Smillie *et al.*[26]

chloramphenicol or cycloheximide. The synthesis of cytochrome-552 and the lamellar-bound proteins, ferredoxin-NADP$^+$-reductase and cytochrome-561 (a *b* type cytochrome), was inhibited by both chloramphenicol and cycloheximide.

(ii) *Calvin cycle enzymes:* Table IV shows the effects of chloramphenicol and cycloheximide on the synthesis of two enzymes of the Calvin cycle. The response of these enzymes to chloramphenicol was similar to that obtained for proteins of the photosynthetic electron transfer pathway, the light-induced increases of both ribulose-1,5-bisphosphate carboxylase and NADP$^+$-glyceraldehyde-3-phosphate dehydrogenase being suppressed. In the presence of cycloheximide the results are especially revealing since the synthesis of both enzymes continued at the same or even higher rates when compared with uninhibited cells. The results may be interpreted to mean that the two Calvin cycle enzymes are synthesized within the chloroplasts on the chloramphenicol-sensitive, chloroplast ribosomes. Cycloheximide-sensitive, cytoplasmic ribosomes do not appear to be involved in the synthesis of these two enzymes. The similarities between chloroplast and bacterial ribosomes in their response towards chloramphenicol has already been noted (see Tables I and II) and it is interesting that this similarity can now be extended to include their lack of sensitivity to cycloheximide.

(iii) *Fraction I protein*: A possible criticism of our interpretation of the results with the two enzymes of the Calvin cycle is that the light-induced increases in the activities of these enzymes may be due to some activation process rather than to a net synthesis of protein. For this

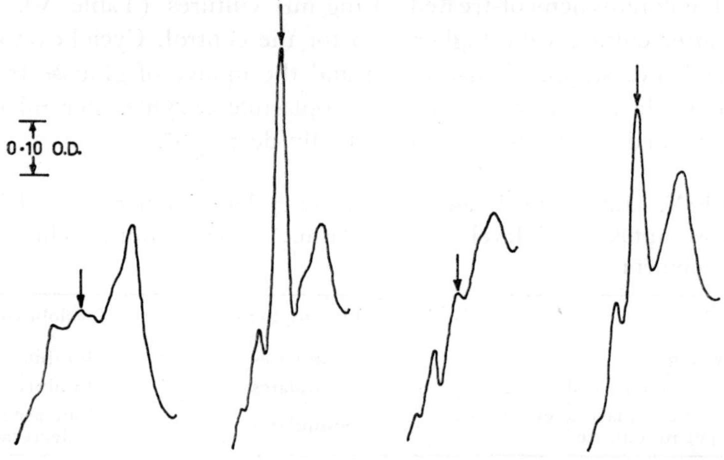

Fig. 5. Densitometer traces of a protein extract of *E. gracilis* fractionated electrophoretically on polyacrylamide gel. The separated proteins were stained with amido black. The traces show the region on the gel occupied by Fraction I protein (arrows point to the Fraction I peaks). Experimental conditions were as in Fig. 4. Data from ref. 26.

reason it was considered important to be able to measure the absolute amount of protein produced for at least one soluble chloroplast protein. Fraction I protein is a convenient soluble protein to measure since it is the site of activity of ribulose-1,5-bisphosphate carboxylase and makes up over 50% of the soluble protein of the chloroplast. A procedure for isolating and assaying Fraction I protein by electrophoresis of extracts of *E. gracilis* through polyacrylamide gels was developed by Mr. A.M. Grieve in our laboratory, and some results are shown in Fig. 5. Chloramphenicol strongly inhibited the synthesis of Fraction I protein, but cycloheximide produced only a slight inhibition.

(iv) *Cytoplasmic enzymes:* No indications have been obtained of inhibition of the synthesis of cytoplasmic enzymes by chloramphenicol. The activities per cell of several cytoplasmic enzymes have been followed during chloroplast development in *E. gracilis*. During the 72 hours of illumination following the addition of glucose to the medium (see Fig. 4), glucose was taken up and there were increases per ml of culture in cell number (see Fig. 4), enolase, glucose-6-phosphate dehydrogenase, $NADP^+$-isocitrate dehydrogenase and NAD^+-glyoxylate reductase in both the control and chloramphenicol-treated (1 mg/ml) cultures (Table V). Values for the latter culture were higher than for the control. Cycloheximide, on the other hand, stopped cell-division and the uptake of glucose from the medium, while the activities of the cytoplasmic enzymes per ml culture either remained about the same or actually decreased.

TABLE V. Effect of chloramphenicol and cycloheximide on cell division, glucose uptake, and levels of cytoplasmic enzymes during chloroplast development.

	Chloramphenicol	Cycloheximide
Cell division	Stimulates	Inhibits
Glucose uptake by cells	Stimulates	Inhibits
Increase in cytoplasmic enzymes (see text) per ml culture	Stimulates	Increase nil, or decrease

The experimental conditions are the same as described for Fig. 4.

(v) *Nitrate and nitrite reductases:* Recent studies on the enzymes nitrate and nitrite reductase, both of which are induced in corn seedlings by nitrate,[28] are especially relevant to the results given in Tables III-V. Nitrite reductase is probably localized in chloroplasts, where it can directly couple with the photosynthetic electron transfer pathway through ferredoxin.[29] Studies in our laboratory and also by Losada and Paneque[30] indicate that, in contrast to nitrite reductase, nitrate reductase cannot couple directly with a protein component of the electron transfer system, but can accomplish this indirectly via $NADP^+$. Schrader *et al.*[28] found that chloramphenicol inhibited the induction of the nitrite reductase but not the nitrate reductase. This suggests that the synthesis of nitrite reductase occurs on the chloroplast ribosomes, while the nitrate reductase, on the other hand, is synthesized in the cytoplasm.

Synthesis of chloroplast structural protein

Although inhibition of the synthesis of both the soluble and electron transfer proteins by chloramphenicol implicates chloroplast ribosomes in their synthesis, the question arises as to why the synthesis of the electron transfer proteins, in contrast to that of the soluble proteins, is at the same time inhibited by cycloheximide. One possible explanation is that one of the inhibitors acts indirectly by interfering with the formation of the lamellar membrane proteins to which electron transfer proteins are bound. Accordingly, an investigation of the action of the two inhibitors of protein synthesis on the formation of chloroplast structural protein was commenced and some results obtained with chloramphenicol are presented below.

(i) *Synthesis of chloroplast structural protein in the presence of chloramphenicol:* Criddle[31] has isolated a membrane protein, which he has called structural protein, from spinach chloroplasts. Structural protein accounts for about 40% of the membrane proteins of chloroplasts and appears to be a single protein, showing only one *N*-terminal amino acid. Mr. Barry Taylor in our laboratory has succeeded in isolating an analogous protein from the chloroplasts of *E. gracilis* by closely following Criddle's procedures. The two questions considered here are, first, does light induce a synthesis of structural protein of the same magnitude as that of Calvin cycle and electron transfer proteins and second, is the synthesis of structural protein inhibited by chloramphenicol?

Cells of *E. gracilis* were grown through several generations either in white light (700 ft. candle) or in darkness and in the presence or absence of chloramphenicol (1 mg/ml). Glucose (22.5 g/*l*) was included in the medium as a source of carbon. Cells were harvested while still in the logarithmic phase of growth, a plastid fraction was isolated, and the structural protein content of this fraction determined (Table VI). Although the amount of structural protein is highest in cells grown in light, dark-grown cells contain from one third to two thirds the amount found in the illuminated cells. This suggests that the synthesis of structural protein does not

TABLE VI. Formation of structural protein in the presence of chloramphenicol.

Growth conditions	mg structural protein in plastid fraction per 10^9 cells
Dark	1.6
Dark+Chloramphenicol	3.2
Light	4.6
Light+Chloramphenicol	4.4

Structural protein was extracted by Criddle's[31] procedure and the amount of protein estimated by the biuret reaction. The values are corrected to equal number of cells. Chloramphenicol, 1 mg/ml.

have the same dependence on light as does the synthesis of Calvin cycle and electron transfer proteins, which usually show increases of 10-fold or greater. However, it has not yet been possible to estimate accurately the extent of the possible contribution to the values shown in Table VI of structural protein originating from contaminating organelles such as mitochondria.

Another difference between the results for structural protein and electron transfer protein is the response to chloramphenicol. No reduction in the amount of structural protein occurred in the presence of chloramphenicol, either in the light or in the dark. Indeed, dark-grown cells when grown in the presence of the inhibitor contained significantly higher amounts of structural protein.

(ii) *Electron microscopy:* Electron micrographs of thin sections of cells grown in the presence or absence of chloramphenicol revealed structural differences which were consistent with data shown in Table VI (the electron microscopy was carried out by Dr. Joan Bain). Cells grown through several generations either in the light or dark and in the presence of chloramphenicol contained well-defined plastids, indicating that the synthesis

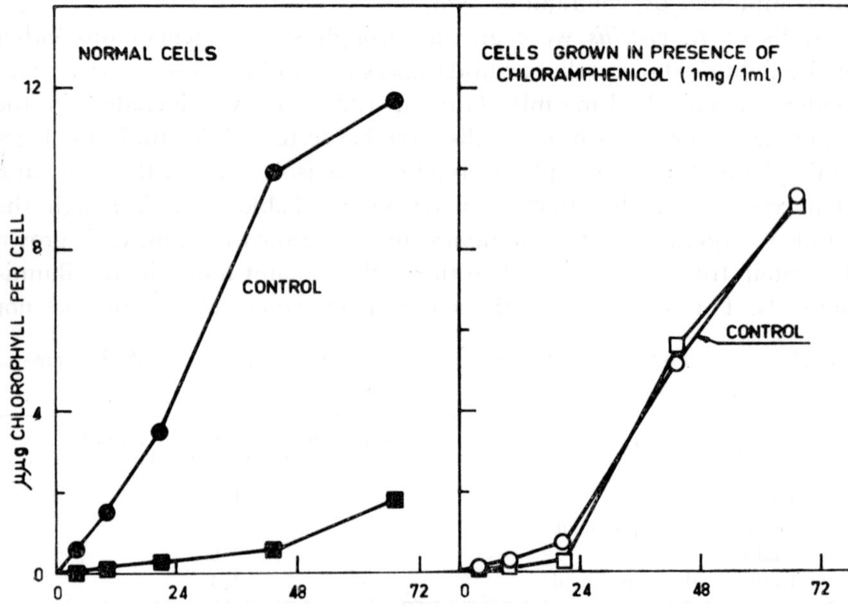

Fig. 6. Reversal of cycloheximide inhibition of chlorophyll synthesis by pre-treatment with chloramphenicol. Cells were grown in the dark as described in Fig. 4, with and without chloramphenicol (1 mg/ml). The control and chloramphenicol-treated cells were washed in a chloramphenicol-free medium and suspended in the same medium to which glucose (22.5 mg/ml) was added as a carbon growth substrate. The cells were then exposed to continuous white light in the presence or the absence of cycloheximide (15 μg/ml). Samples were taken at intervals for determinations of chlorophyll and cell number. Normal cells: ●, control; ■, cycloheximide. Chloramphenicol-pretreated cells: ○, control; □, cycloheximide.

of the protein of the outer membrane of the plastid was not inhibited by chloramphenicol. Further, internal membranes were observed in plastid in chloramphenicol-treated cells grown either in the light or in darkness. Hence, not only the synthesis of the plastid outer membrane, but also the synthesis of at least some of the plastid inner membranes, continued in the presence of chloramphenicol.

Reversal of cycloheximide inhibition of the synthesis of proteins of the photosynthetic electron transfer pathway by pre-treatment with chloramphenicol

Higher amounts of structural protein are apparently produced in cells grown in the dark in the presence of chloramphenicol (Table VI). If cycloheximide inhibits the synthesis of electron transfer protein indirectly by interference with the formation of other lamellar components, it is possible the chloramphenicol-treated cells accumulate sufficient structural and other membrane proteins and lipids to render the light-induced synthesis of their electron transfer proteins less susceptible to inhibition by cycloheximide. To test this hypothesis, cells were grown in the dark with chloramphenicol (1 mg/ml), harvested, washed several times with fresh medium to remove chloramphenicol, and then illuminated in the presence or absence of cycloheximide (15 μg/ml). The pretreatment with chloramphenicol had little effect on the subsequent rate of chlorophyll production, although there was an increase in the lag period before chlorophyll synthesis commenced. When these cells were illuminated in the presence of cycloheximide, not only was the inhibition of chlorophyll

TABLE VII. Summary of results with inhibitors.

Protein Group	Protein	Synthesis inhibited by Chloramphenicol	Synthesis inhibited by Cycloheximide
Cytoplasmic	—	—	+
Chloroplast: "Soluble"	RuDP carboxylase	+	—
	Ga-3-P dehydrogenase	+	—
	Fraction I	+	—
	Nitrite reductase (Data from Schrader *et al.*[28])	+	?
Electron transfer	NADP$^+$-ferredoxin-reductase	+	—
	Cytochrome-552	+	—
	Cytochrome-561	+	—
Membrane	Structural protein	—	?
	Outer membrane	—	?

+, inhibited; —, not inhibited.

synthesis reduced, it was completely eliminated (Fig. 6)! The same was true for the synthesis of electron transfer proteins. Hence, it may be concluded that the synthesis of electron transfer proteins was not directly inhibited by cycloheximide. These and other results with the inhibitors are summarized in Table VII.

Binding of chloramphenicol by chloroplast ribosomes

The interpretation of the experiments cited above hinges on the preferential inhibition by chloramphenicol of protein synthesis on chloroplast ribosomes. The question naturally arises as to why chloroplast ribosomes are more sensitive to chloramphenicol than are cytoplasmic ribosomes. Chloramphenicol is thought to inhibit protein synthesis in bacteria by attaching to the ribosome and interfering with the function of messenger RNA.[13] Vazquez[32] has compared the sensitivity of growth of various organisms towards chloramphenicol with the capacity of ribosomes isolated from the same organisms to bind chloramphenicol. His results, some of which are summarized in Table VIII, showed a decided preference of ribosomes from chloramphenicol-sensitive organisms to bind chloramphenicol. Thus 70S ribosomes isolated from the several bacteria tested all bound chloramphenicol (19–33 mμg chloramphenicol bound per mg ribosomes), whereas the 80S ribosomes isolated from higher organisms, which are relatively chloramphenicol-insensitive (see also Table II), showed a much reduced capacity to bind chloramphenicol(less than 1 mμg chloramphenicol per mg ribosomes). The inference from these studies was that the 70S ribosomes from chloroplasts might also exhibit a greater capacity

TABLES VIII. Binding of chloramphenicol by ribosomes.

Organisms and tissues	Ribosomal preparation	Binding of Chloramphenicol
Staphylococcus aureus, Duncan	70S	+
Bacillus megaterium, KM	70S	+
Escherichia coli, B	70S	+
Pseudomonas striata	70S	+
Saccharomyces fragillis, Jorgenson	80S	−
Crithidia oricopelti	80S	−
Tetrahymena pyriformis, W	80S	−
Rat liver	80S	−
Rat liver	Polysomes	−
Pea seedlings (etiolated)	80S	−

Results summarized from Vazquez.[32]
+ indicates strong binding.

TABLE IX. Binding of ^{14}C-chloramphenicol by ribosomes from various sources.

Source	Fraction	Chloramphenicol bound (mμg/mg RNA)
E. coli	total	88
Yeast	"	6
Euglena gracilis	"	10
	chloroplast	13
	cytoplasm	6
Pea leaf : etiolated	total	4
green	"	11
	chloroplast	22
	cytoplasm	7
Wheat leaf : etiolated	total	9
green	"	17
	chloroplast	28
	cytoplasm	11

Data from Anderson and Smillie[33]. The binding of chloramphenicol by ribosomes was estimated using Vazquez's[32] procedure.

to bind chloramphenicol than the 80S plant cytoplasmic ribosomes. Accordingly, the capacities of chloroplast and cytoplasmic ribosomes from various photosynthetic organisms to bind ^{14}C-chloramphenicol were compared (Table IX). The results confirmed Vazquez's[32] findings that ribosomes from *Escherichia coli* bind chloramphenicol more than do ribosomes from yeast and further, demonstrated that preparations enriched in chloroplast ribosomes have a greater capacity to bind chloramphenicol when compared with either whole cell ribosomal preparations, or those enriched for cytoplasmic ribosomes. As with bacterial ribosomes,[13, 32] the binding of chloramphenicol to chloroplast ribosomes was inhibited by erythromycin.[33]

These experiments suggest a possible explanation for the inhibition by chloramphenicol of chloroplast protein synthesis either *in vivo* or *in vitro*, namely, that chloroplast ribosomes more effectively bind chloramphenicol than do cytoplasmic ribosomes. The observation of Eisenstadt and Brawerman[7] that chloroplast ribosomes of *Euglena* bind messenger RNA less tightly than do cytoplasmic ribosomes may be related.

Inhibition of the synthesis of chloroplast proteins by 5-fluorouracil

The synthesis of ribosomal RNA in *Euglena* plastids is induced by light.[15, 34] This synthesis, however, can be inhibited by 5-fluorouracil and hence this compound effectively inhibits chloroplast development in *E.*

gracilis,[4] and also in *Chlorella*.[16] The action of 5-fluorouracil during chloroplast development in non-dividing cells is probably highly selective towards chloroplast ribosomes, since it is only inhibitory when added during the first 10 to 12 hours of illumination, a period preceding the synthesis of chloroplast protein and corresponding to the period of light-induced ribosomal RNA synthesis. If 5-fluorouracil selectively inhibits the synthesis of chloroplast ribosomes under our experimental conditions, its action on the synthesis of chloroplast proteins should parallcl that of chloramphenicol. Our previous experiments[15] have demonstrated an in-

TABLE X. Effect of 5-fluorouracil on the synthesis of some chloroplast proteins.

	Increase during 72 hrs of illumination	
	Control	5-Fluorouracil
Chlorophyll (mg)	11.6	5.4
Ferredoxin-NADP+-reductase (μmoles substrate/min)	8.5	2.9
Cytochrome-561 (mμmoles)	69.5	19.5
Ribulose-1,5-bisphosphate carboxylase (μmoles substrate/min)	1.00	0.06

The experiment was carried out as described in Fig. 4, except that the addition of glucose was omitted. 5-Fluorouracil (5×10^{-3} *M*) was added at the time the cells were placed in light. All values shown are per 10^9 cells.

hibition by 5-fluorouracil of the synthesis of one of the proteins of the photosynthetic electron transfer pathway of *Euglena*, cytochrome-552. Under the same conditions, a light induced increase in the activities of several cytoplasmic enzymes was not affected. In more recent experiments, inhibition of the production of ferredoxin-NADP$^+$-reductase, cytochrome-561, and ribulose-1,5-bisphosphate carboxylase by 5-fluorouracil has been demonstrated (Table X).

Origin of chloroplast ribosomal RNA and the role of chloroplast DNA in chloroplast development

The present results and those obtained in many other laboratories indicate that chloroplasts contain a distinctive ribosomal system which functions in the synthesis of chloroplast proteins. One of the most important questions concerning the mechanism of organelle biosynthesis centres on the relative roles of nuclear DNA and organelle DNA in the differentiation of complex organelles such as chloroplasts and mitochondria. In this context it can be asked whether the RNA of ribosomes in chloroplasts originates from chloroplast DNA rather than from DNA in the nucleus. This hypothesis was tested in the case of the RNA of chloroplast ribosomes by investigating the possibility of forming a specific DNA-RNA hybrid between a purified preparation of chloroplast DNA and chloroplast ribosomal RNA. The formation of such a hybrid could indicate coding of the RNA by DNA.[35]

The results in Table XI show that chloroplast DNA from *E. gracilis* can form a ribonuclease-resistant hybrid with ribosomal RNA isolated from chloroplasts labeled with ^{32}P. The amount of RNA used in this experiment saturated all binding sites on the chloroplast DNA available to the ribosomal RNA. Hybridization between chloroplast DNA and ribosomal RNA isolated from dark-grown heterotrophic cells was less than 1% of that found between chloroplast DNA and chloroplast ribosomal RNA. It has not been established whether all of the ribosomal RNA found in chloroplasts can form a hybrid with chloroplast DNA.

The possibility of interference from messenger RNA in these experiments is discussed by Scott and Smillie.[36]

The very low amount of hybrid formed between chloroplast DNA and ribosomal RNA from dark-adapted heterotrophic cells suggests that the dark-adapted cells contain quite small amounts of chloroplast ribosomal RNA and consequently the latter is mainly formed by transcription from chloroplast DNA when the cells are exposed to continuous light. These conclusions agree with the observed inhibition of chloroplast development by 5-fluorouracil mentioned above, and with the results of other experiments demonstrating a considerable net increase in ribosomal RNA upon exposure of dark-adapted cells of *E. gracilis* to light.[15,34]

If a specific hybrid is formed between chloroplast DNA and chloroplast ribosomal ^{32}P-RNA, then the addition of preparations containing

TABLE XI. Hybridization of chloroplast DNA with ribosomal RNA from chloroplasts.

Source of ribosomal RNA	Counts per 100 min in hybrid
Chloroplasts	9600
Dark-grown heterotrophs	130

3 μg of chloroplast DNA ($\rho = 1.686$) was annealed with 29 μg of ribosomal ^{32}P-RNA according to the method of Gillespie and Spiegelman[43] except that the incubation time was extended to 48 hrs. The chloroplast DNA was purified by repeated density gradient fractionations in $CsCl_2$. Details of the preparation of this DNA and labeled ribosomal RNA are given by Scott and Smillie.[36] The counts indicate the extent of formation of ribonuclease-resistant hybrids.

unlabeled chloroplast ribosomal RNA to the annealing mixtures should result in a decrease in the number of counts bound to DNA. On the other hand, dilution by unlabeled ribosomal RNA which had not been coded for by chloroplast DNA should not influence the number of counts bound. The addition of unlabelled ribosomal RNA from either autotrophic *E. gracilis* or from a preparation enriched for chloroplast ribosomal RNA resulted in decreases in the amount of ^{32}P-RNA bound to DNA (Table XII). Ribosomal RNA from dark-adapted cells, or from cells of a bleached mutant ZUV-1, which lacks chloroplast DNA, did not significantly change the number of counts bound.

The hybridization experiments indicate that chloroplast DNA functions in the formation of a protein-synthesizing system within the chloroplast by coding for RNA of chloroplast ribosomes. These results constitute the first direct evidence for a specific biological role for chloroplast DNA.

CONCLUSIONS

Cellular sites of synthesis of organelle protein

Similarities between the genetics, origin, structure, and function of chloroplasts and mitochondria have often been stressed, see e.g. ref. 37, and it is now possible to begin to compare cellular mechanisms for the synthesis of the proteins of these organelles. Amino acids are incorporated into only the membrane protein of isolated mitochondria,[10, 12, 38] a process which is inhibited by chloramphenicol.[38, 39] The synthesis of many soluble mitochondrial proteins, including cytochrome *c*, catalase, and enzymes catalyzing the cyclic oxidation of carbon compounds to CO_2 is extra-mitochondrial. These enzymes are synthesized by the cytoplasmic endoplasmic reticulum and transferred to the mitochondria by a process requir-

TABLE XII. Specificity of binding of chloroplast DNA to chloroplast ribosomal ^{32}P-RNA.

Source of unlabeled RNA	Counts per 100 min in hybrid	% inhibition
None	9,600	—
Autotrophic cells	7,200	25
Chloroplasts	5,900	38
Dark-adapted heterotrophs	9,100	5
Bleached mutant (ZUV-1)	10,500	0

Experimental conditions are given in Table XI except that 100 μg of unlabeled ribosomal RNA from various sources was included in the annealing mixtures. Data from Scott and Smillie.[36]

TABLE XIII. Comparison of mechanisms for the synthesis of *c* type cytochrome in chloroplasts and mitochondria.

	Mitochondrion	Chloroplast
Cytochrome	Cytochrome *c*	*c* type cytochrome (cytochrome *f*)
Function	Oxidative electron transfer	Photosynthetic electron transfer
Extractability	Soluble	Varies, easily extractable in *E. gracilis* and many algae
Site of synthesis	Cytoplasmic ribosomes[11,12]	Chloroplast ribosomes
Action of chloramphenicol on *in vivo* synthesis	Not inhibited[27]	Inhibited[15,26]
Action of cycloheximide on *in vivo* synthesis	?	Not inhibited*
Probable location of structural gene	Nucleus[40]	Nucleus[41]

* In chloramphenicol-pretreated cells (see Fig. 6)

ing adenosine 5′-triphosphate and probably guanosine 5′-triphosphate.[12] In contrast, the sites of synthesis of enzymes catalyzing the cyclic reduction of CO_2 in photosynthesis are located inside the chloroplast, and this is also true for proteins of the photosynthetic electron transfer pathway. The contrast between the mitochondrial and chloroplast systems is perhaps best illustrated by comparing the formation of the *c*-type cytochrome of chloroplasts with that of mitochondrial cytochrome *c* (Table XIII).

The site of synthesis of the "insoluble" electron transfer proteins of mitochondria, including cytochromes a, a_3, b, and c_1 is not so clear. Kadenbach[12] has obtained evidence from *in vitro* experiments suggesting that these proteins are also synthesized outside the mitochondrion. On the other hand their synthesis *in vivo* is inhibited by chloramphenicol.[27]

A second protein-synthesizing system in the chloroplast?

As was discussed above, the protein-synthesizing system of the mitochondrion appears to be directed primarily towards the synthesis of the structural protein and other membrane proteins. The basic similarities in composition and probable functions between the structural proteins of chloroplasts and mitochondria[31] suggest that structural protein is also synthesized within the chloroplast. Yet the synthesis of chloroplast structural protein shows several differences when compared with the synthesis of the other chloroplast proteins investigated. It did not appear to have the same degree of dependence upon light, nor was it inhibited by chloramphenicol at 1 mg/ml. These differences could be explained by the ex-

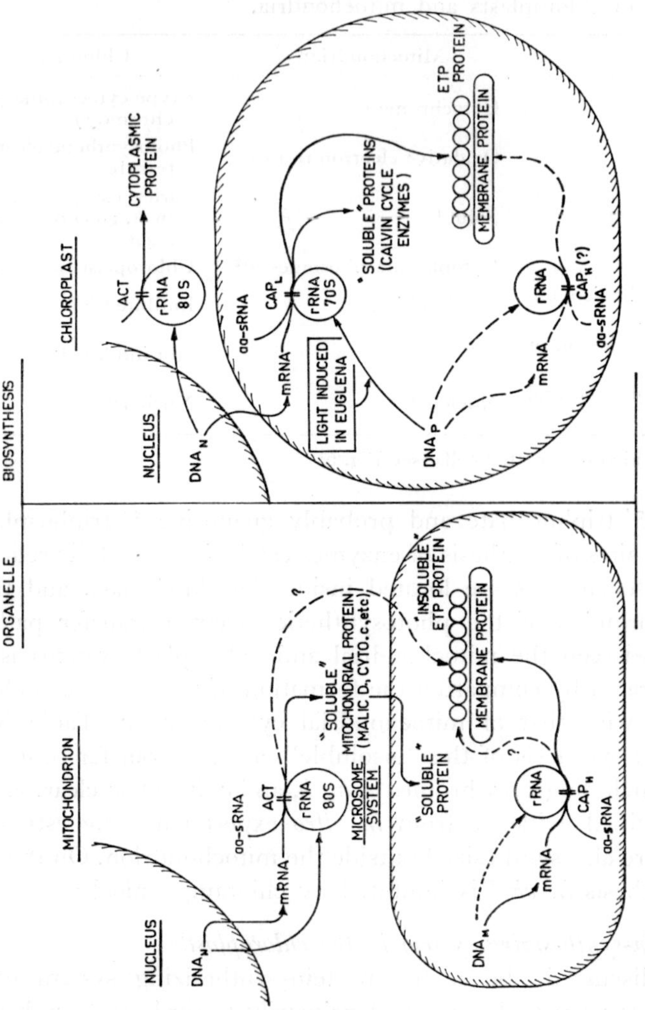

Fig. 7. Comparison of possible processes involved in the development of chloroplasts and mitochondria. CAP_L: low chloramphenicol, inhibits chloroplast synthesis only (1 mg/ml in *E. gracilis*). CAP_H: high chloramphenicol, inhibits chloroplast *and* mitochondrial synthesis (4 to 5 mg/ml). ACT: inhibited by actidione (cycloheximide).

istence of a second protein-synthesizing system in chloroplasts, similar to the one present in mitochondria, and mainly responsible for the synthesis of chloroplast structural and other membrane proteins. Although chloramphenicol inhibits the synthesis of mitochondrial membrane proteins, the concentrations used in our experiments were insufficient to inhibit cellular growth (Fig. 4), and therefore presumably, the synthesis of the *Euglena* mitochondrial structural protein. At a higher concentration

(4 mg/ml) growth is inhibited. Since this concentration corresponds with that necessary to inhibit the synthesis of mitochondrial proteins in the yeast *Saccharomyces cerevisiae*,[27] the mitochondrial system of *E. gracilis* (and possibly, the postulated system in chloroplasts responsible for the formation of membrane protein) may also be inhibited by chloramphenicol, but only at concentrations in excess of those required to inhibit the synthesis of proteins which function directly in photosynthesis. If a second protein-synthesizing system with characteristics similar to the mitochondrial system does exist in chloroplasts, its presence would probably not be obvious from the analytical and electron microscopic studies carried out so far. Electron micrographs of mitochondria have seldom revealed ribosomes, most observations of ribosomes in mitochondria being confined to abnormal cells such as tumour cells.

If a second "mitochondria-like" system is responsible for the formation of membrane protein in chloroplasts, then the main difference between the mechanism for protein synthesis associated with differentiating mitochondria on the one hand and differentiating chloroplasts on the other is that whereas many soluble mitochondrial proteins (e.g. malate dehydrogenase and cytochrome *c*) are synthesized outside mitochondria, soluble chloroplast proteins are synthesized within chloroplasts on an additional ribosomal system, which is either absent or is less well developed in the mitochondrion.

While our results are consistent with the existence of a second protein-synthesizing system in chloroplasts for the synthesis of membrane protein, they are also consistent with a cytoplasmic synthesis of membrane protein sub-units which are subsequently transferred to and assembled within the chloroplasts.

Role of chloroplast DNA

Nuclear mutations giving rise to cells with impaired photosynthesis are exceedingly common, and in some instances the defects have been traced either to single enzymic steps in the synthesis of chloroplast components or to the absence of a particular protein required for photosynthesis (for a discussion of these mutants, see ref. 2). The most obvious conclusion is that the structural genes for many, if not all, of the proteins and other components active in photosynthesis are located in the nucleus. What then is the function of chloroplast DNA?

Two possible functions are proposed. Firstly, our hybridization studies suggest chloroplast DNA carries information for the formation of chloroplast ribosomes. Secondly, Woodward and Munkres[42] have shown that mitochondrial DNA codes for structural protein in the mitochondrion, and chloroplast DNA possibly fulfills a similar function in the chloroplast.

Many of the concepts discussed above relating to the biosynthesis of proteins of the chloroplast and mitochondrion are summarized in Fig. 7.

Acknowledgements

The authors thank Mrs. J. Conroy and Miss Ruth Burley for their excellent technical assistance. Part of this study was made possible through the generosity of the Charles F. Kettering Research Foundation.

SUMMARY

Concepts are developed that (i) enzymes of the photosynthetic CO_2 reductive cycle and proteins of the photosynthetic electron transfer pathway are synthesized *in vivo* within chloroplasts on chloroplast ribosomes and (ii) chloroplast DNA is involved in the formation of this chloroplast ribosomal system by coding for chloroplast ribosomal RNA.

Evidence on the sites of synthesis of chloroplast proteins was obtained by studying the differential inhibition by two inhibitors of protein synthesis, chloramphenicol and cycloheximide, on the *in vivo* synthesis of Calvin cycle enzymes and photosynthetic electron transfer proteins. Chloramphenicol, but not cycloheximide, inhibited the formation of these proteins; in contrast, cell division was inhibited by cycloheximide, but not by chloramphenicol. The inhibitory effect of chloramphenicol on the synthesis of chloroplast protein was attributed to the greater capacity for chloroplast ribosomes to bind chloramphenicol compared with cytoplasmic ribosomes.

Evidence on the function of chloroplast DNA was obtained by demonstrating the formation of a specific hybrid between chloroplast DNA and chloroplast ribosomal RNA. Significant hybridization between chloroplast DNA and nonchloroplast ribosomal RNA could not be demonstrated.

The possibility is raised that chloroplast structural protein is synthesized by a system different from the one synthesizing Calvin cycle enzymes. Comparisons are made between cellular mechanisms for the synthesis of mitochondrial proteins and those for chloroplast proteins.

REFERENCES

1. Spencer, D., and Whitfeld, P. R., *Biochem. Biophys. Res. Comm.*, **28**, 538 (1967).
2. Kirk, J. T. O., in *The Biochemistry of Chloroplasts* (T. W. Goodwin, ed.), Vol. I, pp. 319–340, Academic Press, London, 1966.
3. Brawerman, G., *Ibid.*, pp. 301–317.
4. Smillie, R. M., *Can. J. Bot.*, **41**, 123 (1963).

5. Jacobson, A. B., Swift, H., and Bogorad, L., *J. Cell Biol.*, **17**, 557 (1963).
6. Gunning, B. E. S., *Ibid.*, **24**, 79 (1965).
7. Eisenstadt, J. M., and Brawerman, G., *J. Mol. Biol.*, **10**, 392 (1964).
8. Spencer, D., *Arch. Biochem. Biophys.*, **111**, 381 (1965).
9. Lyttleton, J. W., *Exptl. Cell Res.*, **26**, 312 (1962).
10. Roodyn, D. B., *Biochem. J.*, **85**, 177 (1962).
11. Beattie, D. S., Basford, R. E., and Koritz, S. B., *Biochemistry*, **5**, 926 (1966).
12. Kadenbach, B., *Biochim. Biophys. Acta*, **134**, 430 (1966).
13. Wolfe, A. D., and Hahn, F. E., *Biochim. Biophys. Acta*, **95**, 146 (1965).
14. Petermann, M. L., *The Physical and Chemical Properties of Ribosomes*, Elsevier Pub. Co., Amsterdam, 1964.
15. Smillie, R. M., Evans, W. R., and Lyman, H., *Brookhaven Symp. Biol.*, No. 16, p. 89 (1963).
16. Aoki, S., Matsubara, J. K., and Hase, E., *Plant and Cell Physiol.*, **6**, 475 (1965).
17. Parthier, B., *Z. Naturforshg.*, **206**, 1191 (1965).
18. Pogo, B. G. T., and Pogo, A. O., *J. Protozool.*, **12**, 96 (1965).
19. Margulies, M. M., *Plant Physiol.*, **41**, 992 (1966).
20. Aaronson, S., Ellenbogen, B. B., Yellen, L. K., and Hutner, S. H., *Biochem. Biophys. Res. Comm.*, **27**, 535 (1967).
21. Goffeau, A., and Brachet, J., *Biochim. Biophys. Acta*, **95**, 302 (1965).
22. Whiffen, A. J., Bohonos, N., and Emerson, R. L., *J. Bact.*, **52**, 610 (1946).
23. Siegel, M. R., and Sisler, H. D., *Biochim. Biophys. Acta*, **87**, 83 (1964).
24. Ennis, H. L., and Lubin, M., *Science*, **146**, 1474 (1964).
25. Kirk, J. T. O., and Allen, R. L., *Biochem. Biophys. Res. Comm.*, **21**, 523 (1965).
26. Smillie, R. M., Graham, D., Dwyer, M. D., Grieve, A., and Tobin, N. F., *Ibid.*, **28,** 604 (1967).
27. Huang, M., Biggs, D. R., Clark-Walker, C. D., and Linnane, A. W., *Biochim. Biophys. Acta*, **114**, 434 (1966).
28. Schrader, L. E., Beevers, L., and Hageman, R. G., *Biochem. Biophys. Res. Comm.*, **26,** 14 (1967).
29. Ramirez, J. M., Del Campo, F. F., Paneque, A., and Losada, M., *Biochim. Biophys. Acta*, **118**, 58 (1966).
30. Losada, M., and Paneque, A., *Ibid.*, **126**, 578 (1966).
31. Criddle, R. S., in *The Biochemistry of Chloroplasts* (T. W. Goodwin, ed.), Vol. I, pp. 203–231, Academic Press, London, 1966.
32. Vazquez, D., *Nature*, **203**, 257 (1964).
33. Anderson, L. A., and Smillie, R. M., *Biochem. Biophys. Res. Comm.*, **23**, 535 (1966).
34. Brawerman, G., Pogo, A. O., and Chargaff, E., *Biochim. Biophys. Acta*, **55**, 326 (1962).
35. Hall, B. D., and Spiegelman, S., *Proc. Natl. Acad. Sci., U.S.*, **47**, 137 (1961).
36. Scott, N. S., and Smillie, R. M., *Biochem. Biophys. Res. Comm.*, **28,** 598 (1967).
37. Gibor, A., and Granick, S., *Science*, **145**, 890 (1964).
38. Wheeldon, L. W., and Lehninger, A. L., *Biochemistry*, **5**, 3533 (1966).
39. Mager, J., *Biochim. Biophys. Acta*, **38**, 150 (1960).
40. Sherman, F., Stewart, J. W., Margoliash, E., Parker, J., and Cambell, W., *Proc. Natl. Acad. Sci., U.S.*, **55**, 1498 (1966).
41. Gorman, D. S., and Levine, R. P., *Plant Physiol.*, **41**, 1648 (1966).
42. Woodward, D. O., and Munkres, K. D., *Proc. Natl. Acad. Sci. U.S.*, **55**, 872 (1966).
43. Gillespie, D., and Spiegelman, S., *J. Mol. Biol.*, **12**, 829 (1965).

DNA Species in Chloroplasts of *Chlorella*

TATSUICHI IWAMURA AND SEIKO KUWASHIMA

The Tokugawa Institute for Biological Research, Tokyo and
The Institute of Applied Microbiology, Tokyo University, Tokyo

INTRODUCTION

In an earlier work of one of the present authors,[1] it was demonstrated that the chloroplasts of *Chlorella* contain DNA. Both the base composition and the metabolic behavior of the DNA in the chloroplast fraction were found to differ markedly from those of the supernatant DNA of the homogenates, which derived from nuclei. The characteristic metabolic behavior of the chloroplast DNA is an apparent rapid turnover of the nucleotides under photosynthesizing conditions.[2] Subsequently[3] we showed, by the technique of CsCl equilibrium density-gradient centrifugation, that the DNA species showing the characteristic metabolic behavior differs in its buoyant density from the major, nuclear DNA.

In this paper, we present some data suggesting that the chloroplasts of *Chlorella* contain not one but at least two kinds of DNA, a major and a minor species, and that the former may be the genetic DNA of the organelle, while the latter may be a metabolic DNA showing the characteristic metabolic behavior mentioned above.

EXPERIMENTAL

Chlorella ellipsoidea was grown synchronously by the method developed in our laboratory.[4,5] Algal cells were taken at those stages of the growing phase[4] where the most vigorous increase in cell mass takes place, and where the cells have been found to be most suitable for obtaining chloroplast preparations, i.e., the D$\sim$L-stage or L_1-stage.[5]

The algal cells harvested were washed with deionized water and resuspended in a sucrose-buffer-protein solution containing 0.5 *M* sucrose, 0.07 *M* KCl, 0.02 *M* Tris-HCl (pH 8.0 at 0°C), 0.002 *M* $CaCl_2$, and 1% bovine serum albumin. All the operations thereafter were carried out at temperatures near 0°C. The cell suspension was passed once through a French pressure cell at 80–100 kg/cm^2, and the homogenates obtained were centrifuged at 20,000×*g* for 10 min. The green pellet obtained was resuspended in the same sucrose-buffer-protein solution as above, except that the concentration of the protein was reduced to 0.01%. This sucrose-buffer-protein solution was used also for the density-gradient centrifugation by varying the concentration of sucrose alone.

Two kinds of discontinuous sucrose density-gradient centrifugation were employed: one with layers of 1.0 *M*, 1.5 *M*, 2.0 *M*, and 2.5 *M* sucrose (SDG-1), and the other with 1.0 *M*, 1.25 *M*, 1.5 *M*, and 1.75 *M* sucrose (SDG-2). About 6 ml of the green pellet suspension containing 1–2 ml wet cell volume equivalent was layered on top of the SDG-1 gradient in a centrifuge tube of 30 ml capacity and spun at 25,000 RPM for 60 min in a Hitachi RPS 25 swinging bucket rotor. After centrifugation, the green layer between 1.0 *M* and 1.5 *M* sucrose, hereafter referred to as L 2-1, was collected, washed with the sucrose(0.5 *M*)-buffer-protein(0.01%), and centrifuged in SDG-2. The two layers were separated, one between 1.0 *M* and 1.25 *M* sucrose, hereafter designated as L 2_1-1, and the other between 1.25 *M* and 1.5 *M* sucrose, L 2_2-1. These two fractions were further centrifuged in SDG-2, yielding the purified L 2_1 and L 2_2, which are designated as L 2_1-2 and L 2_2-2, respectively. Electron microscopical examinations showed that L 2-2 (L 2-1 spun again in SDG-1) does not contain any distinct particles other than the chloroplasts.

DNase digestion was carried out on aliquots of the fractions thus obtained. They were washed once with a medium containing 0.5 *M* sucrose, 0.04 *M* KCl, 0.02 *M* Tris-HCl (pH 7.4 at 0°C), 0.005 *M* $MgCl_2$, and 0.01% bovine serum albumin, and resuspended in the same medium. DNase I (Sigma, Stock No. DN-C) was added to the suspension at the concentration of 20 μg/ml and the incubation was done at 0°C for 30 min. The condition of the digestion was such that free DNA in an amount exceeding that of the DNA in those fractions was completely digested.

The nucleic acids were obtained from the fractions by successive extractions with 1% sodium dodecylsulphate in 0.1 *M* Tris-HCl buffer (pH 8.0 at 20°C) containing 0.05 *M* EDTA and with buffer-saturated phenol, and this extraction was repeated on the insoluble residue after it had been ground with quartz sand. The crude nucleic acids thus obtained were deproteinized with a chloroform-isoamyl alcohol mixture (96 : 4), digested with RNase I and RNase T_1, and finally subjected to the isopropanol treatment of Marmur to remove polysaccharides.[6] Upon this treatment, fibrous DNA was precipitated from the DNA preparations obtained from intact cells but, as will be mentioned later, no fibrous material or a very small amount of DNA, if any, was precipitated from those obtained from the chloroplast fractions. The purified DNA was finally dissolved in a saline-citrate solution (0.15 *M* NaCl and 0.015 *M* sodium citrate, pH 7.0).

Analysis of the CsCl equilibrium density-gradient ultracentrifugation was performed with a Spinco E analytical ultracentrifuge equipped with UV-optics. The marker DNA was that of *Micrococcus leisodeikticus*. 800 mg of recrystallized CsCl (Merck Co., Germany) was dissolved in 0.625 ml of the DNA solution. The centrifugation was run at 44,770 RPM for over 20 hr at 25°C. The photodensitometry was done with a Shimadzu microphotometer for the measurement of line spectra.

RESULTS AND DISCUSSION

In Fig. 1 are shown the analytical patterns of the DNAs obtained from DNase-digested and non-digested L 2_1-2 and L 2_2-2. In each of these four cases, there appeared at least two DNA species, one major and the

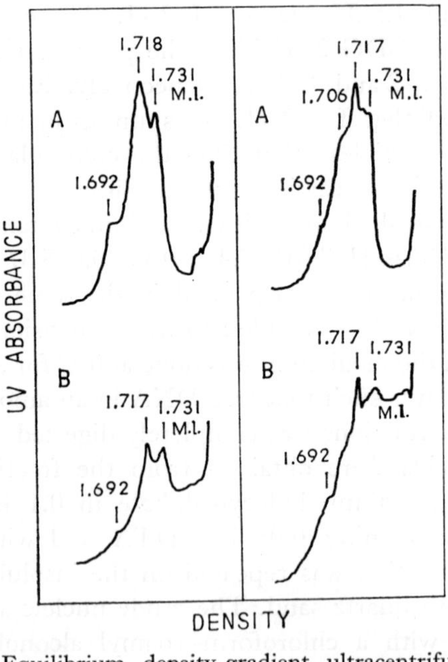

Fig. 1. Equilibrium density-gradient ultracentrifugations in CsCl of DNAs isolated from the chloroplasts of *Chlorella ellipsoidea*. *Left*: L 2_1–2 DNA; *right*: L 2_2-2 DNA. A: DNA isolated from the chloroplast fractions untreated with DNase I; B: DNA isolated from those treated with DNase I. M. l.: The marker DNA, the DNA of *Micrococcus leisodeikticus*, the buoyant density of which is 1.731 gm/ml.

other minor, having buoyant densities of 1.717 and 1.692 gm/ml, respectively. The 1.692 component is the DNA species showing the apparent rapid turnover of its nucleotides mentioned above. This component has been designated as μ_2 in a recent review of Iwamura.[7] The 1.717 component has practically the same buoyant density as the nuclear DNA, which was confirmed by examination of the DNA from isolated nuclei and also of the DNA in the 20,000$\times g$ supernatant of the homogenates. Besides these two

TABLE I. Content of DNA and chlorophyll in various fractions obtained from *Chlorella* cells.

	DNA		Chlorophyll		DNA/Chlorophyll		
Fractions[a] X	mg	X_I/X_O or (−DNase)/(+DNase)	mg	X_I/X_O or (−DNase)/(+DNase)	mg	X_I/X_O or (−DNase)/(+DNase)	X/H_O
H_O[b]	0.396	100	3.16	100	12.5	100	100
H_I[b]	0.369	93	2.95	93	12.5	100	100
P_O[b]	0.414	100	4.98	100	8.31	100	65
P_I[b]	0.393	95	4.69	94	8.38	102	67
L 2-1_O[b]	0.669	100	18.6	100	3.60	100	29
L 2-1_I[b]	0.618	92	16.0	86	3.86	107	31
L 2_1-2 (−DNase)[c]	0.330	100	19.6	100	1.68	100	13
L 2_1-2 (+DNase)[c]	0.130	40	17.9	91	0.743	42	6
L 2_2-2 (−DNase)[c]	0.602	100	46.2	100	1.30	100	10
L 2_2-2 (+DNase)[c]	0.466	77	41.9	91	1.11	85	9
S	0.813	100					
S (+Mg)[d]	0.800	98	—	—	—	—	—
S (+Mg, +DMase)[d]	0.084	10					

a. H: homogenates; P: 20,000$\times g$ pellet of homogenates; L 2-1: L 2 obtained by the 1st SDG-1; L 2_1-2 & L 2_2-2: L2_1 and L 2_2 obtained by the 2nd SDG-2, respectively; S: 20,000$\times g$ supernatant of homogenates.

b. Each fraction was divided into two equal portions; the one with the subscript O (e.g. H_O) was frozen at once with the addition of sodium dodecylsulphate, and the other with I (e.g. H_I) was incubated at 0°C till the whole fractionation run was finished.

c. Each fraction was divided into two equal portions; the one (−DNase) was that frozen at once with the addition of sodium dodecylsulphate, and the other (+DNase) was subjected to DNase-digestion. For the conditions of digestion see the text.

d. Mg^{++} was simply added to S at the final concentration of 0.02 *M*, since S contained 0.07 *M* K^+ and 0.002 *M* Ca^{++}. The other conditions for the DNase-digestion were the same as those for L 2_1 and L 2_2.

components, a DNA species having a buoyant density of 1.722 often became distinctly discernible in L 2. This component is thought to be derived from mitochondria, because it also appears distinctly in glucose-adapted cells and is quickly labeled with ^{32}P when the D-cells are grown for a short period in a 1% glucose medium in the dark, while μ_2 is labeled quickly with ^{32}P when the same D-cells are grown photoautotrophically.

It may be mentioned that, among the four DNA preparations obtained from L 2_1-2 and L 2_2-2, either DNase-treated or non-treated, a fibrous DNA material was observed after the isopropanol-treatment only in the case of L 2_1-2 untreated with DNase.

The results of quantitative analyses on the DNA and chlorophyll contents of the various fractions are shown in Table I. L 2_1 contains more broken chloroplasts than L 2_2, and unlike L 2_2 it loses a green material upon repeated density-gradient and wash centrifugations. Therefore, the difference between these two fractions in DNase-digestibility seems to reflect the intactness of the chloroplasts in these fractions. It may also be true that L 2_1-2 was contaminated somehow with the DNA of the nuclear fragments, which may be precipitated as a fibrous material and easily digested with DNase as is the S-DNA in Table I.

Thus, L 2_2-2 DNA may consist largely of chloroplast DNA, and its DNA-content after DNase digestion shown in Table I may be considered to represent the approximate content of DNA in the chloroplast. From the ratio of DNA to chlorophyll of L 2_2-2 (+DNase) compared with that of H_0 (the homogenates) in Table I, the DNA content of the chloroplast is calculated to be around 4×10^{-15}gm/chloroplast, which is of the same order of magnitude as the DNA content of bacterial cells.

The 1.692 component, μ_2, found in these chloroplast fractions was called a "metabolic DNA," because of its characteristic metabolic behavior (*cf.* ref. 7). The significance of the apparent turnover of the nucleotides in μ_2 in the chloroplast is still obscure. It could be involved in replication of the genetic DNA of the chloroplast or in the synthesis of a non-genetic DNA, or it could be a phenomenon related in some way to the photosynthetic process. At present we prefer the latter two possibilities, because it is tempting to surmise that the other component (1.717) is the genetic DNA of the chloroplast, in view of the fact that its content in a chloroplast is comparable to the content of DNA in a bacterial cell.

Acknowledgements

This study was supported by a grant from the U. S. Public Health

Service (GM-07601-06) and by a grant from the Ministry of Education, Japan. Thanks are due to Prof. H. Tamiya for his advice in the preparation of this manuscript.

REFERENCES

1. Iwamura, T., *Biochim. Biophys. Acta*, **42**, 161 (1960).
2. Iwamura, T., *Ibid.*, **61**, 472 (1962).
3. Iwamura, T., and S. Kuwashima, *Ibid.*, **80**, 678 (1964).
4. Iwamura, T., *J. Biochem.* (Tokyo), **42**, 575 (1955).
5. Tamiya, H., *J. Cellular Comp. Physiol.*, **62**, Suppl. 1, 157 (1963).
6. Marmur, J., *J. Mol. Biol.*, **3**, 208 (1961).
7. Iwamura, T., *Progr. Nucleic Acid Res. Mol. Biol.*, **5**, 133 (1966).

Chairman's Remarks

Major problems in the development of the photosynthetic apparatus in bacteria include the nature of controls over such development, and the question of the degree of its "autonomy" according to work reviewed by Dr. Jane Gibson. Although the presence of light *or* absence of oxygen is sufficient to induce photosynthetic lamellar development in purple bacteria (*R. rubrum*, *R. spheroides*), it may not be a coincidence that these are bacteria capable of respiration, and the respiratory lamellae have not been clearly separated from the photosynthetic ones. According to the work of Lascelles, oxygen or light affects the activity of delta-aminolevulinic acid synthetase; but the actual control mechanisms are obscure. It is also far from clear whether bacterial chromatophores contain information for their own continued synthesis, in any way resembling the situation for chloroplasts or mitochondria. Although a minor DNA component ("satellite" band in density gradient centrifugations) has been detected in some species of bacteria it is not apparent in others; and in no case has it yet been possible to demonstrate a physical association of DNA with isolated chromatophores.

In the subsequent discussion, G. Kikuchi emphasized some of the progress and remaining problems in current work on the induction of delta-aminolevulinic acid synthetase in a purple bacteria (*R. spheroides*) and on induction of bacteriocholorophyll synthesis. While some inhibitors (hydroxyurea, FUDR) inhibit the onset of both processes, others (mitomycin A, hemin) inhibit one or the other specifically. Although hydroxyurea and FUDR are usually thought to inhibit DNA synthesis, the evidence seems to show that this is *not* the basis for their inhibition of bacteriochlorophyll formation and ALA synthetase induction, in this instance.

R. Smillie reviewed some of the evidence for self-replication of chloroplasts, especially in algae. This included the presence of DNA, of several varieties of RNA, and of protein, lipid and carbohydrate. Complex enzymatic activities of isolated chloroplasts include synthesis of protein, RNA and DNA. Smillie was able to show that the DNA synthesized in vitro was specifically that of the "satellite band" previously associated with the occurrence of chloroplasts. A considerable point was made of the specific inhibition of chloroplast protein synthesis by chloramphenicol, a reagent which does not inhibit the growth of *Euglena* cells, or protein synthesis by higher plants, by fungi or by mammals, but which does inhibit the growth of bacteria and of blue-green algae. Actidione, on the other hand, inhibits cell growth in *Euglena*, but not the synthesis of at least some of the chloroplast enzymes ("Fraction I," or ribulose diphosphate carboxylase; or the TPN-specific triose phosphate dehydrogenase). By this criterion as well as that of size the chloroplast ribosomes are more primitive than those of the cytoplasm, and resemble those of bacteria. Finally, using the nucleic acid hybridization technique, Smillie presented some evidence that the RNA from chloroplast ribosomes hybridizes with chloroplast DNA, whereas the RNA from dark-grown heterotrophic cells does not.

T. Iwamura described some further experiments on the rapid turning-over of DNA of *Chlorella* cells. The evidence seems to indicate at this point that the rapid turnover is more closely related to electron transport and phosphorylation than it is to the ordinary events of DNA synthesis.

The intriguing destruction of the photosynthetic apparatus in *Chlorella protothecoides* due simply to the presence of glucose was described in further detail by Tamiko Oh-hama. Supporting data by S. Miyachi showed that the nature of the products of carbon dioxide fixation changed very early during de-greening, with a relative increase in lipids and decrease in carbohydrates. An entirely new field may have been opened up by the finding that light still affects the process of carbon dioxide fixation, after chlorophyll and chloroplasts seem to have disappeared.

Y. Chiba demonstrated in an elegant way the products of a chlorophyllase-catalyzed reaction involving exchange between the methyl and phytyl groups of chlorophyllide *a*. The transformation of protochlorophyllide *a* to chlorophyllide *a* in vitro was described further by L. Bogorad, who suggested a decrease in molecular weight of the chlorophyllide holochrome following conversion. This conclusion was challenged during the subsequent discussion, however. Finally, Bogorad described the inhibition of formation of new protochlorophyll by chloramphenicol, even though this poison has no effect on the other aspects of chlorophyll synthesis.

Although this session could not be described as complete in its coverage of developmental aspects, there was sufficient material included to give some picture of the scope of subtle interactions between chloroplasts or chromatophores, the cell in which they reside, and the external environment. For algae and higher plants the ancient speculation of "autonomy" of chloroplasts is much closer to precise experimental definition, while for the bacteria the evidence is still much more incomplete.

A. T. Jagendorf

PART IV

REGULATORY ASPECTS

Dynamic Metabolic Regulation of the Photosynthetic Carbon Reduction Cycle

J. A. Bassham and Martha Kirk

Lawrence Radiation Laboratory, University of California, Berkeley, California

The necessity for metabolic regulation of carbon reduction pathways in photosynthesis became apparent as the result of two discoveries which have altered earlier concepts of photosynthesis. The first of these was the recognition that direct products of photosynthesis include not only carbohydrates, but also amino acids and other products. The other discovery was that there is an exchange of molecules of some intermediate compounds of photosynthesis with molecules of the same compounds involved in glycolysis.

Some of the earliest studies of photosynthesis with $^{14}CO_2$ revealed labeling of amino acids and fats at times comparable to those required for the labeling of sugars.[1] Later quantitative tracer studies with $^{14}CO_2$ and photosynthesizing *Chlorella pyrenoidosa* showed that alanine and other amino acids are formed directly from intermediates of the basic carbon reduction cycle,[2] and that their rate of labeling with ^{14}C can account for as much as 30% of the total $^{14}CO_2$ incorporation during the first minutes of photosynthesis.[3] During that same period only about one-third as much ^{14}C was found in sucrose, a major carbohydrate product of photosynthesis in *Chlorella*. Benson and coworkers[4] have shown that galactolipids are rapidly labeled by *Chlorella*, with 40% of the ^{14}C label of the lipids appearing in the fatty acid moieties after 5 min photosynthesis in *Chlorella* with $^{14}CO_2$.

This direct formation of amino acids and fatty acids from intermediates of the photosynthetic carbon reduction cycle makes these compounds equivalent in some respects to free sugars as immediate products

---

Abbreviations: PGA, 3-phosphoglyceric acid; DHAP, dihydroxyacetone phosphate; Ga3P, glyceraldehyde-3-phosphate; FDP, fructose-1,6-diphosphate; F6P, fructose-6-phosphate; SDP, sedoheptulose-1,7-diphosphate; S7P, sedoheptulose-7-phosphate; E4P, erythrose-4-phosphate; Xu5P, xylulose-5-phosphate; R5P, ribose-5-phosphate; Ru5P, ribulose-5-phosphate; RuDP, ribulose-1,5-diphosphate; G6P, glucose-6-phosphate.

of carbon reduction in photosynthesis. It seems likely that some regulation of the flow of carbon from the basic cycle to the production of these compounds must be required in order to maintain a balance between their syntheses and their utilization in the further syntheses of macromolecules.

Green plant cells pass through various stages of development requiring synthesis of changing proportions of fats, proteins, carbohydrates and other macromolecules such as nucleic acids. Cells in the mature leaves of such plants as sugar beet may produce almost exclusively carbohydrate for translocation to other parts of the plant. Yet in young, rapidly growing leaves of these same plants, a considerable portion of the photosynthetically reduced carbon must be allocated to the production of proteins and lipids and other materials required for the building of the chloroplasts themselves and of the new cells.

To a large extent, of course, the regulation of these syntheses of macromolecules will be accomplished through the control of enzyme synthesis, according to mechanisms that are rapidly being discovered. Nevertheless, there would seem to be a need for a dynamic metabolic regulation of the pool sizes of these compounds. This regulation should be capable of rapid response to changing environmental conditions and should not be dependent upon synthesis of new enzymes. In short, the type of regulation needed to maintain reasonable pool sizes for small molecules produced directly from the photosynthetic carbon reduction cycle probably is dynamic regulation of key enzymes at or near the branch points in the cycle itself.

The second development which focused attention on the need for metabolic regulation was the finding that the intermediate compounds of photosynthesis and glycolysis in green cells are not completely isolated from one another. Earlier studies with *Chlorella* and $^{14}CO_2$, as well as ^{14}C-labeled glucose, had suggested that there is some separation between certain pools of intermediates of the photosynthetic carbon reduction cycle and of glycolysis and respiration.[5] However, it is now clear that there is an important exchange between some pools of photosynthesis and of glycolysis. Heber and Willenbrinck[6] reported that some labeled intermediates of the photosynthetic carbon cycle are transported from the chloroplast to the cytoplasm in *Elodea*. Based on his experiments with *Elodea* and *Spinacia*, Heber[7] concluded that the phosphoglyceric acid (PGA), dihydroxyacetone phosphate (DHAP), and fructose-1,6-diphosphate (FDP) "function as transport metabolites in photosynthesis." While those results were obtained by a post mortem "non-aqueous" isolation

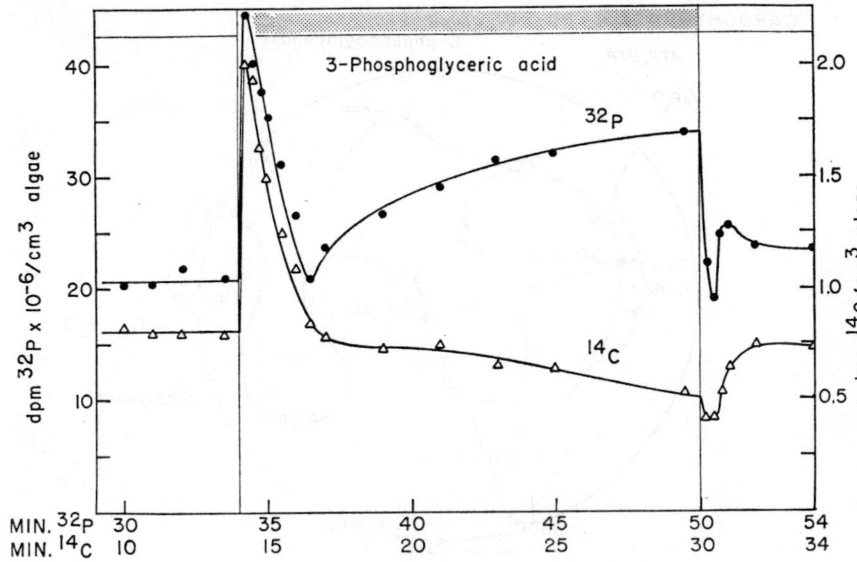

Fig. 1. Levels of ^{14}C- and ^{32}P-labeled PGA in *Chlorella pyrenoidosa* during photosynthesis and respiration.

of chloroplasts following photosynthesis with tracers by the intact leaf, the findings have in general been confirmed by other techniques described below.

In the meantime, studies of ^{14}C-and ^{32}P-labeling of metabolites by *Chlorella* photosynthesizing in light and respiring in the dark had demonstrated a clear interaction between the intermediate compounds of photosynthesis and glycolysis.[8] The results of a similar study are shown in Fig. 1. After some 14 min of photosynthesis with $^{14}CO_2$ and 34 min with ^{32}P-labeled phosphate, PGA was saturated with respect to labeling by each of these isotopes. When the light was turned off and the flow of electrons and ATP from the photochemical apparatus stopped, the reduction of PGA abruptly ceased, leading to a rapid rise in both ^{32}P and ^{14}C label. After about 30 sec the level of PGA declined, since the photosynthetic carboxylation reaction had by then stopped and PGA was no longer being formed. The decline in the level of labeled PGA is due to its conversion to other products such as alanine and its consumption by respiration via the tricarboxylic cycle (see Fig. 2).

After about 2 min the level of ^{32}P-labeled PGA rises to a steady-state level in the dark which is higher than the steady-state level in the light, while the ^{14}C-labeling of PGA continues to decline. This shows that the

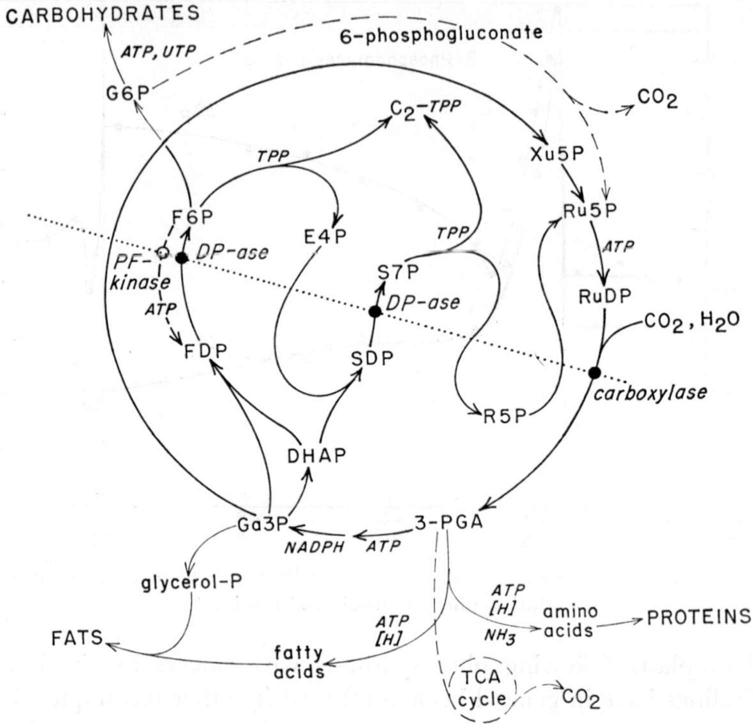

Fig. 2. Relations between photosynthesis, glycolysis, and synthesis of macromolecules, and regulatory control points. Compounds of the photosynthetic carbon reduction cycle above the dotted line are those which tend to be retained in the chloroplasts; those below the dotted line have been found to be readily transported from chloroplasts. Intermediate compounds in paths other than the photosynthetic carbon reduction cycle are not shown.

doubly-labeled PGA is being replaced by newly formed, singly-labeled PGA made from the glycolysis of unlabeled carbohydrate stores. Glycolysis of course introduces the same ^{32}P-label as photosynthesis, since in each case the phosphate group is derived from a rapidly turning-over pool of ATP, whether the latter is formed by photosynthetic phosphorylation or oxidative phosphorylation.

Up to this point the results could be interpreted as indicating isolated pools of PGA for photosynthesis and glycolysis. However, if this were the case, when the light is again turned on, the glycolytic pool of PGA should be unaffected and the ^{32}P-labeling should not drop significantly more than the ^{14}C label. The ^{32}P label of the PGA drops very much more than the ^{14}C label. This drop indicates that all of the ^{32}P-labeled PGA is immediately affected by the light. Thus, photosynthesis acts on the entire labeled pool of PGA. This data clearly established the fact that PGA, formed in the dark by glycolysis, can be immediately used in the photosynthetic carbon reduction cycle when the light is turned on.

With the interchangeability of PGA between photosynthesis and glycolysis complicating the kinetic studies, it became increasingly urgent that we perform parallel studies of photosynthesis in isolated chloroplasts. Chloroplasts capable of photosynthetic reduction of carbon dioxide at only 20% (usually much less)[9] of the rates to be expected *in vivo* did not appear to be suitable for quantitative studies, particularly of regulatory mechanisms. By modifying a number of conditions in isolation and incubation, it was possible to obtain rates of carbon dioxide assimilation by isolated spinach chloroplasts approaching *in vivo* rates for healthy leaves, at least for 10 or 15 min.[10] However, when the labeling of various products of this photosynthesis by isolated spinach chloroplasts was measured,[11] it was apparent that the levels of certain intermediates of the photosynthetic carbon reduction cycle, particularly DHAP, were abnormally high, as compared with levels seen with intact plants. The reason for this difference became clear when we studied the distribution of intermediate products between the chloroplasts themselves and the supernatant solution in which the chloroplasts had been suspended.[11,12] Over 80% of the carbon fixed appeared in the suspending medium. Of this, by far the greater portion was in PGA and DHAP, with substantial amounts appearing also in FDP, sedoheptulose-1,7-diphosphate (SDP), and glycolic acid. Ribulose-1,5-diphosphate (RuDP), fructose-6-phosphate (F6P), glucose-6-phosphate (G6P), and sedoheptulose-7-phosphate (S7P) were well retained in the chloroplasts.

Thus, there is a striking differential behavior in the migration of photosynthetic intermediates from chloroplast to suspending medium. In general, those compounds of the photosynthetic carbon reduction cycle following the carboxylation reaction and preceding the diphosphatase reaction (see Fig. 2) tend to diffuse from the chloroplast to the medium, while those compounds coming after the diphosphatase reaction and preceding the carboxylation reaction (except for pentose monophosphates) are well retained inside the chloroplasts.

Given the interaction between the intermediates of photosynthesis and glycolysis and the demonstrated movement of intermediates between chloroplasts and cytoplasm, it seemed clear that some form of light-dark dynamic metabolic regulation of key enzymes of photosynthesis and glycolysis is required.

Direct kinetic evidence for light-dark changes in the activities of enzymes of the carbon reduction cycle came from the same light-dark studies with *Chlorella* in the presence of ^{32}P and ^{14}C that were reported above.[8] The level of FDP plus SDP in light and dark is shown in Fig. 3. When the light is first turned off, these sugar diphosphates drop in concentration to nearly zero, as would be expected from the fact that light is no longer supplying cofactors for the reduction of PGA to triose phosphate, from which these sugar diphosphates are formed. However, the level of FDP plus SDP then rises, passing through a maximum to a new steady-state dark level. This result clearly suggests that the diphosphatase has become inactive and that a phosphofructokinase has been activated in the dark, and is utilizing ATP formed by oxidative phosphorylation. Note also that the sugar diphosphates formed in the light are made partly from endogenous sugars that are not fully labeled with ^{14}C.

The levels of ATP, as well as UTP and ADP, in light and dark are shown in Fig. 4. It can be seen that while the ATP initially drops rapidly when the light is turned off, it soon rises to a high dark steady-state level due to oxidative photophosphorylation. Thus, there is ample ATP for the PFkinase reaction.

The time required for activation of the diphosphatase reaction in light was revealed by a detailed kinetic experiment with *Chlorella* in which, following a period of steady-state photosynthesis with $^{14}CO_2$ and a period of darkness, the light was turned on again and samples were taken at 10-sec intervals for the next 2 min. In Fig. 5 it can be seen that the level of FDP rises very sharply for 30 sec. We attribute this rapid rise to the onset of reduction of PGA due to cofactors from the light reactions, together with the inactivity of the diphosphatase following a period of darkness. After 30 sec the diphosphatase has been activated and the level of fructose diphosphate then drops rapidly, since the photosynthetic carbon reduction cycle is not yet fully operative and the rate of reduction of PGA to triose phosphate has not reached the steady-state level. Later, the labeled fructose diphosphate level rises more slowly towards the steady-state level.

Fig. 5 also shows that the transient increase in DHAP is much smaller than that of FDP. This means that the transient peaks are not due merely to a wave of carbon coming from the sudden reduction of a large pool of PGA. If that were the case, the transient peak in DHAP, which precedes FDP in the cycle, would be higher than the peak in FDP. Instead, the smaller DHAP peak is a reflection of the higher FDP peak, and indicates an active, reversible aldolase reaction in light and dark.

Also, if the transient peaks were merely wave phenomena induced

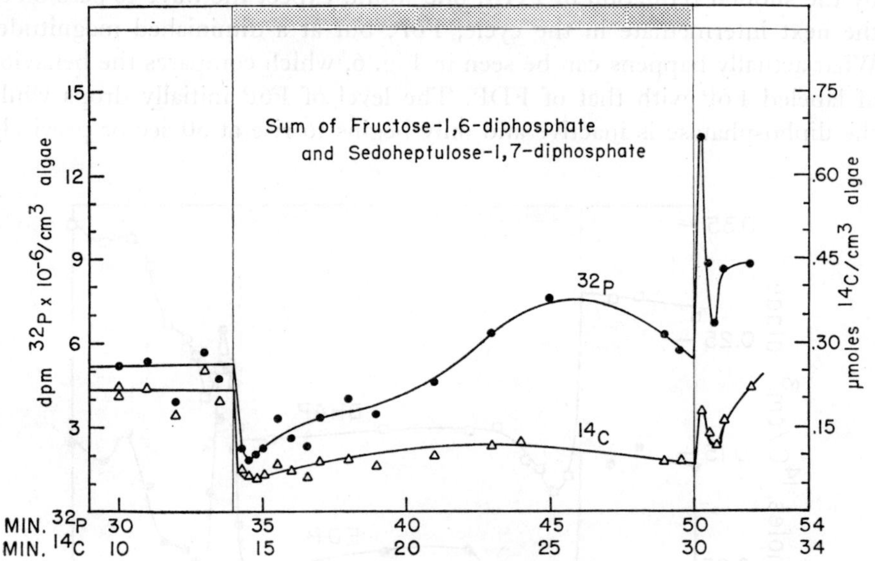

Fig. 3. Levels of ^{14}C-labeled FDP plus SDP in *Chlorella pyrenoidosa* during photosynthesis and respiration.

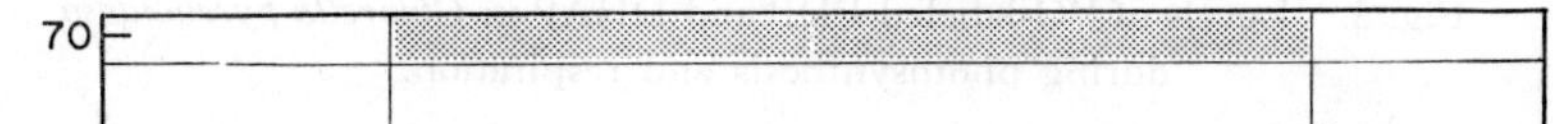

Fig. 4. Levels of ^{32}P-labeled ATP, UTP, and ADP in *Chlorella pyrenoidosa* during photosynthesis and respiration.

by the sudden reduction of PGA, one would expect the wave to pass on to the next intermediate in the cycle, F6P, but at a diminished magnitude. What actually happens can be seen in Fig. 6, which compares the behavior of labeled F6P with that of FDP. The level of F6P initially drops while the diphosphatase is inactive and only begins to rise at 30 sec or precisely

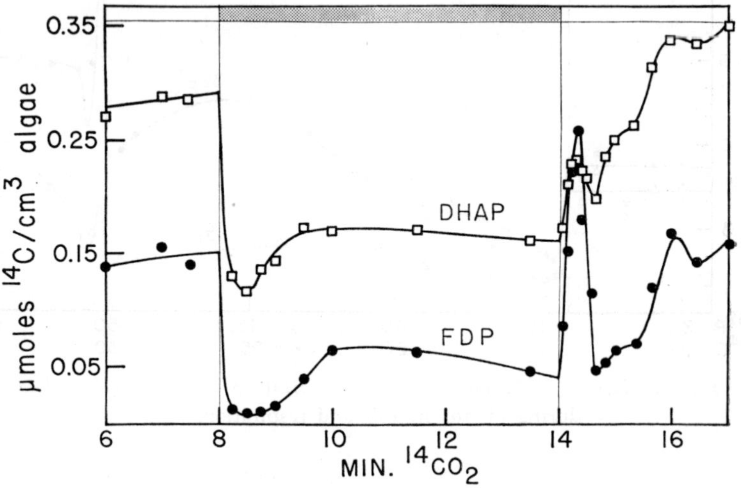

Fig. 5. Levels of ^{14}C-labeled FDP and DHAP in *Chlorella pyrenoidosa* during photosynthesis and respiration.

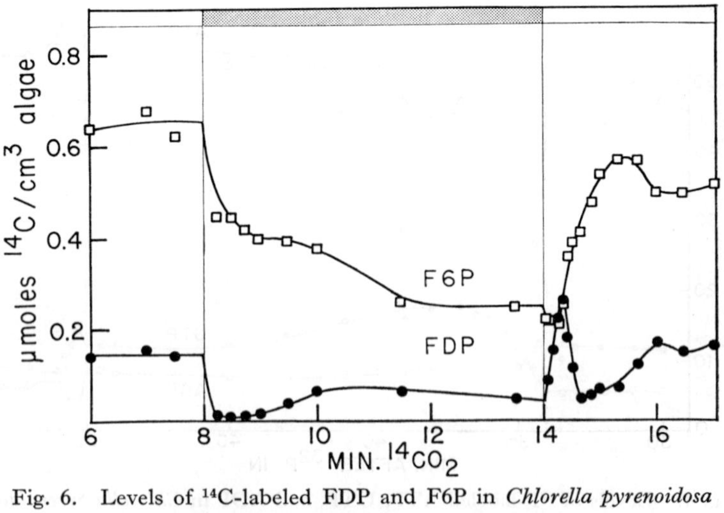

Fig. 6. Levels of ^{14}C-labeled FDP and F6P in *Chlorella pyrenoidosa* during photosynthesis and respiration.

the time when the diphosphatase has suddenly become activated. Once the diphosphatase is activated, there is a transient peak in the levels of F6P before steady state is achieved. This is due to the sudden removal of the diphosphatase bottleneck, allowing carbon from the accumulated FDP to flow into the sugar monophosphate pool. Precisely the same situation is seen with SDP and S7P (Fig. 7). The only difference is in the steady-state pool sizes of the compounds involved. This similarity in the transient peaks of FDP and SDP strongly suggests an identical mechanism of regulation and quite possibly that the same diphosphatase operates for both reactions.

The possibility that the transient peak in FDP and SDP could be caused by residual dark PFkinase activity seems unlikely. PFkinase activity during the first seconds of light should not be greater than the dark activity. The level of ATP does not rise more than 15% when the light is turned on (Fig. 4). The level of F6P, as just mentioned, initially drops.

Evidence for the light-dark regulation of the carboxylation reaction also came first from the light-dark studies of *Chlorella pyrenoidosa* with $^{14}CO_2$ and $H^{32}PO_4^{-2}$, described earlier.[8] When the light was turned off, the level of the carboxylation substrate, RuDP, fell rapidly at first. After 2 min darkness, the rate of fall in RuDP level was no longer proportional to the RuDP level, indicating that the activity of the enzyme for this reaction had diminished.

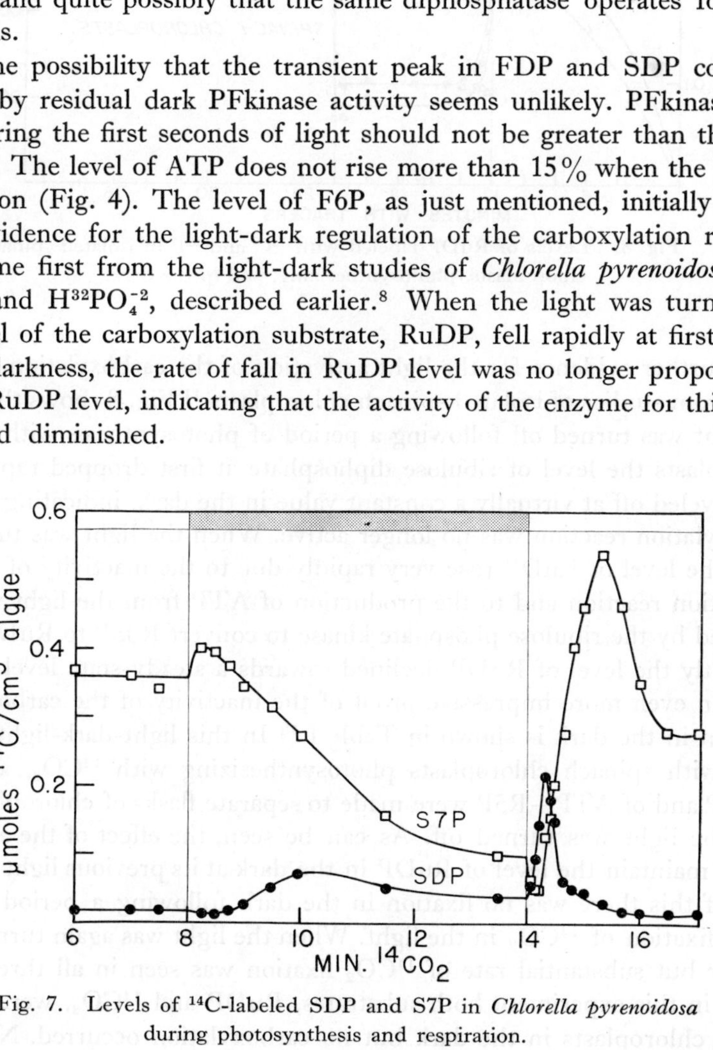

Fig. 7. Levels of ^{14}C-labeled SDP and S7P in *Chlorella pyrenoidosa* during photosynthesis and respiration.

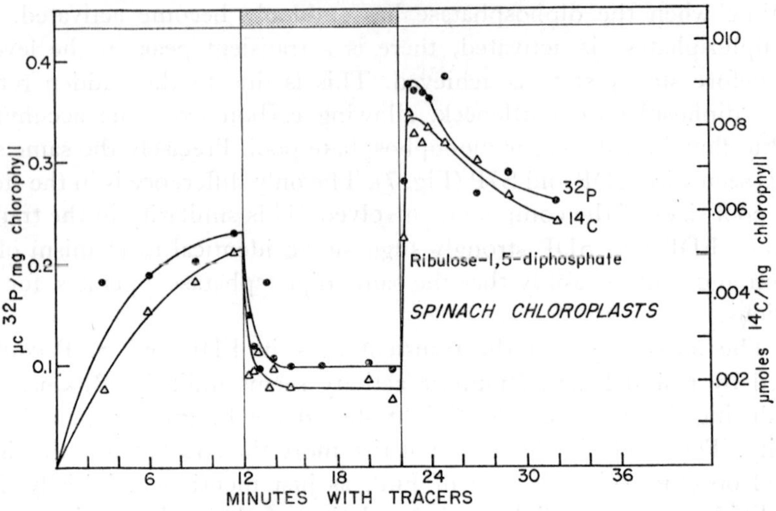

Fig. 8. Levels of RuDP labeled with ^{14}C and ^{32}P in isolated spinach chloroplasts, photosynthesizing, and in the dark.

Further evidence for the light activation of the carboxylation reaction came from studies of isolated spinach chloroplasts.[11] Fig. 8 shows that when the light was turned off following a period of photosynthesis with spinach chloroplasts the level of ribulose diphosphate at first dropped rapidly but then leveled off at virtually a constant value in the dark, indicating that the carboxylation reaction was no longer active. When the light was turned on again the level of RuDP rose very rapidly due to the inactivity of the carboxylation reaction and to the production of ATP from the light which is required by the ribulose phosphate kinase to convert Ru5P to RuDP. Subsequently the level of RuDP declined towards a steady-state level.

An even more impressive proof of the inactivity of the carboxylation reaction in the dark is shown in Table I.[13] In this light-dark-light experiment with spinach chloroplasts photosynthesizing with $^{14}CO_2$, additions of ATP and of ATP+R5P were made to separate flasks of chloroplasts just after the light was turned off. As can be seen, the effect of the additions was to maintain the level of RuDP in the dark at its previous light level. In spite of this there was no fixation in the dark following a period of very active fixation of $^{14}CO_2$ in the light. When the light was again turned on, a smaller but substantial rate of $^{14}CO_2$ fixation was seen in all three flasks. Thus, in this experiment both substrates, RuDP and $^{14}CO_2$, were present in the chloroplasts in the dark but no carboxylation occurred. Neverthe-

TABLE I.

	1st Light Period		Dark Period		2nd Light Period	
	Rate*	RuDP*	Rate	RuDP	Rate	RuDP
Control	143	0.08	− 3	0.02	82	0.23
+ATP	154	0.10	−10	0.07	104	0.37
+ATP, R5P	145	0.08	− 3	0.08	86	0.27

* Rates are given in μmoles $^{14}CO_2$ (mg Chl hr)$^{-1}$; RuDP concentration is given in μmoles ^{14}C(mg Chl)$^{-1}$.

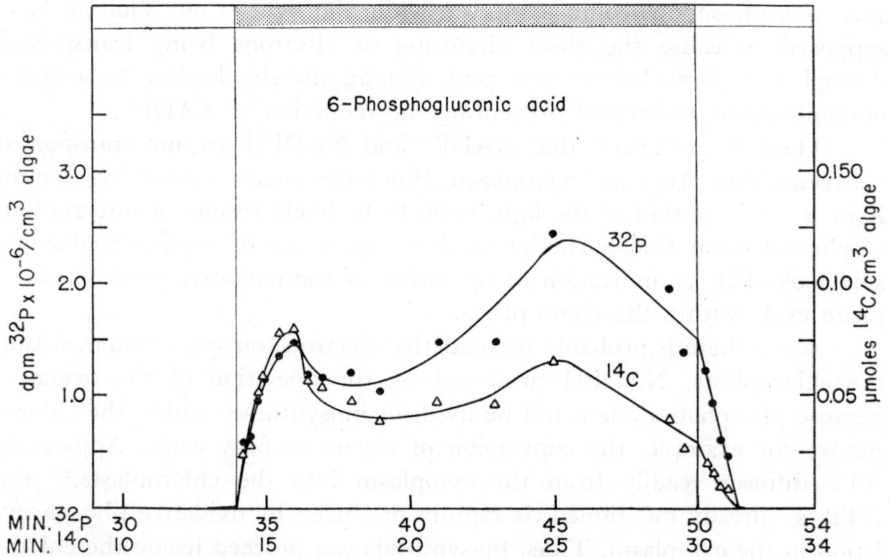

Fig. 9. Levels of 6-phosphogluconic acid labeled with ^{14}C and ^{32}P in *Chlorella pyrenoidosa* during photosynthesis and respiration.

less, the system was not permanently inactivated, for subsequent light was able to stimulate high fixation rates.

The reason for light-dark diphosphatase regulation seems obvious. Given the diffusion of ATP and of some sugar phosphates between chloroplasts and cytoplasm, phosphofructokinase and fructose diphosphatase should not be active at the same time, lest they together operate as an ATPase.

The reasons for regulation of the carboxylation reaction were not so apparent. Green cells contain a highly active oxidative pentose phosphate cycle. This cycle is immediately activated when the light is turned off, as indicated by the data shown in Fig. 9, obtained during the previously described steady-state experiment with photosynthesizing *Chlorella* in the presence of $^{14}CO_2$ and ^{32}P-labeled phosphate. As soon as the light is turned off both the ^{32}P and ^{14}C label of 6-phosphogluconic acid rise rapidly and are maintained at an appreciable level during the entire dark period. When the light is again turned on the formation of 6-phosphogluconic acid ceases.

The site of the operation of this oxidative pentose phosphate cycle may well be within the chloroplasts. It was found earlier[14] that the addition of vitamin K to photosynthesizing *Chlorella* causes the immediate appearance of 6-phosphogluconic acid even while the light is on. Vitamin K is supposed to cause the short circuiting of electrons being transported through the photoelectron transport system, thereby leading to a cyclic photophosphorylation and preventing the reduction of $NADP^+$.

There is evidence[15] that $NADP^+$ and NADPH are not transported between chloroplasts and cytoplasm. Since the effects caused by vitamin K or by interruption of the light seem to be likely results of interruption of photoelectron transport, the sudden appearance of 6-phosphogluconic acid looks like an indication of operation of the oxidative pentose phosphate cycle within the chloroplasts.

While there is probably no oxidative electron transport system within the chloroplasts, NADPH produced by the operation of the oxidative pentose phosphate cycle could be used for biosyntheses within the chloroplasts—for example, the conversion of sugars to fatty acids. Apparently ATP diffuses readily from the cytoplasm into the chloroplasts,[12] and ATP, as already mentioned, is rapidly produced by oxidative phosphorylation in the cytoplasm. Thus, biosynthesis can proceed inside the chloroplasts in the dark with ATP from the cytoplasm and NADPH from the operation of the oxidative pentose phosphate cycle in the chloroplasts.

If this hypothesis is correct, then there should be a regulatory mechanism which in the dark would block the transformation of pentose monophosphates to RuDP to PGA. Only one of these two steps needs be inactive in the dark, and it appears that it may be the carboxylation step from ribulose diphosphate to PGA. We have already demonstrated (Table I) that with added ATP, pentose monophosphates can be converted to ribulose diphosphate in the dark.

What is the mechanism of the light-dark regulation of the diphosphatase and ribulose diphosphate reactions? Isolated enzymes which catalyze these reactions are: D-fructose-1,6-diphosphate, 1-phosphohydrolase, 3.1.3.11 (hexose diphosphatase), and 3-phospho-D-glycerate carboxy-lyase (dimerizing), 4.1.1.39 (ribulosediphosphate carboxylase). Both enzymes are characterized by a pH optimum which is higher than normal physiological pH for chloroplasts and a requirement for a high level of magnesium ion.[16,17] Activation of these enzymes by light could be the result of movements of magnesium and hydrogen ions through membranes. Dilley and Vernon[18] reported a light-dependent uptake of H^+ and an efflux of K^+ and Mg^{++} ion by isolated spinach chloroplasts. This reported ion flow might appear at first to be contrary to that required for enzyme activation. However, the light-driven flow of ions is across the thylakoid membranes within the chloroplasts, and probably leads to a flow of hydrogen ions from the stroma region into the membrane-enclosed thylakoids, and a flow of magnesium ions into the stroma. Thus measurements of pH and of metal ions in the suspending medium may well reflect changes occurring within the stroma, particularly if the chloroplasts are broken or "leaky."

Another indication that light-induced ion flow may be responsible for light-dark regulation of the enzymes is to be found in the fact that certain fatty acids and fatty acid esters simultaneously inhibit photophosphorylation, the carboxylation reaction, and the diphosphatase reaction.[19] These inhibitions, which are reversible, are thought to be accomplished by some indirect effect of alteration in the properties of the chloroplast membranes. A likely alteration in membrane properties would be in the light-induced ion pumping capacity. Photophosphorylation is thought to require a pH gradient across the thylakoid membranes,[20] and, like the other inhibited enzymes, photophosphorylation requires magnesium ion.[21]

While light-induced ion pumping provides a plausible mechanism for the light-dark regulation of the diphosphatase and carboxylation reactions, we visualize a different, but as yet unknown, mechanism as providing a fine regulation of the diphosphatase reaction when plants are photosynthesizing with the lights on. At all times while photosynthesis is proceeding there would be sufficient diphosphatase activity to permit the cycle to regenerate pentose phosphates at a steady-state rate sufficient to provide for the operation of the basic photosynthetic carbon reduction cycle. However, additional diphosphatase activity would be so adjusted as to either pile up the excess fixed carbon at the level of PGA, DHAP, and FDP, which would be used for biosynthesis of proteins, fats, and other materials; or, with greater diphosphatase activity, the excess carbon would be accumulated in fructose-6-phosphate which could then be converted to carbohydrates (Fig. 2). This type of regulation would thus satisfy the changing requirements for biosynthesis in response to physiological change, such as growth and development.

We are inclined to make two predictions about this mechanism. First, it should be related to hormonal or genetic control and may well involve the adjustment of a regulatory factor by the activity of an inducible enzyme. Second, this regulatory factor may exert its effect through some quantitative change in the light activation. In this way it could provide the fine, or vernier, control needed to insure the precise diphosphatase activity to keep the basic carbon cycle running while at the same time adjusting the supply of carbon to biosynthetic pathways.

REFERENCES

1. Benson, A. A., Bassham, J. A., Calvin, M., Goodale, T. C., Haas, V. A., and Stepka, W., *J. Am. Chem. Soc.*, **72**, 1710 (1950).
2. Bassham, J. A., Benson, A. A., Kay, L. D., Harris, A. Z., Wilson, A. T., and Calvin, M., *Ibid.*, **76**, 1760 (1954).
3. Smith, D. C., Bassham, J. A., and Kirk, M., *Biochim. Biophys. Acta*, **48**, 299 (1961).
4. Benson, A. A., Wiser, R., Ferrari, R. A., and Miller, J. A., *J. Am. Chem. Soc.*, **80**, 4740 (1958).
5. Moses, V., Holm-Hansen, O., Bassham, J. A., and Calvin, M., *J. Mol. Biol.*, **1**, 21 (1959).
6. Heber, U., and Willenbrinck, J., *Biochim. Biophys. Acta*, **82**, 313 (1964).
7. Heber, U., in *Biochemistry of Chloroplasts* (T. W. Goodwin, ed.), Vol. II, pp. 71–87, Academic Press, London and New York (1967).
8. Pedersen, T. A., Kirk, M., and Bassham, J. A., *Physiol. Plant.*, **19**, 219 (1966).
9. Walker, D. A., in *Biochemistry of Chloroplasts* (T. W. Goodwin, ed.), Vol. II, pp. 53–69, Academic Press, London and New York (1967).
10. Jensen, R. G., and Bassham, J. A., *Proc. Natl. Acad. Sci.*, **56**, 1095 (1966).
11. Bassham, J. A., and Jensen, R. G., in *Photosynthesis in Plant Life* (A. San Pietro, F. A. Greer, and T. J. Army, eds.), pp. 79–110, Academic Press, New York (1967).
12. Bassham, J. A., Kirk, M., and Jensen, R. G., *Biochim. Biophys. Acta*, **253,** 211 (1968).
13. Jensen, R. G., and Bassham, J. A., *Biochim. Biophys. Acta*, **153,** 227 (1968).
14. Gould, E. S., and Bassham, J. A., *Biochim. Biophys. Acta*, **102**, 9 (1965).
15. Heber, U., and Santarius, K. A., *Biochim. Biophys. Acta*, **109**, 391 (1965).
16. Weissbach, A., Horecker, B. L., and Hurwitz, J., *J. Biol. Chem.*, **218**, 795 (1956).
17. Preiss, J., Biggs, M. L., and Greenberg, E., *J. Biol. Chem.*, **242**, 2292 (1967).
18. Dilley, R. A., and Vernon, L. P., *Arch. Biochem. Biophys.*, **111**, 365 (1965).
19. Pedersen, T. A., Kirk, M., and Bassham, J. A., *Biochim. Biophys. Acta*, **112**, 189 (1966).
20. Jagendorf, A. T., and Uribe, F., *Brookhaven Symposia in Biology*, No. 19, p. 215 (1967).
21. Arnon, D. I., Whatley, F. R., and Allen, M. B., *J. Am. Chem. Soc.*, **76**, 6324 (1954).

The Role of Ribose-5-P Isomerase in Regulation of the Calvin Cycle in *Rhodospirillum Rubrum**

LOUISE ANDERSON‡, L. E. WORTHEN†, AND R. C. FULLER‡

Department of Microbiology, Dartmouth Medical School, Hanover, N.H.

Ribose 5-P isomerase (D-ribose-5-P ketol-isomerase, E.C.5.3.1.6) is a key enzyme in heterotrophic carbohydrate metabolism and the first of three enzymes of autotrophic carbohydrate metabolism catalyzing the formation of 3-P-glycerate from ribose-5-P and CO_2 through the following series of reactions:

$$\text{ribose-5-P} \xrightarrow{\text{isomerase}} \text{ribulose-5-P} \xrightarrow{\text{kinase}}$$

$$\text{ribulose-1,5-dip} \xrightarrow[\text{CO}_2]{\text{carboxylase}} \text{3-P-glycerate}$$

Although several compounds (AMP, NAD, NADH, ATP) are known to affect ribose-5-P dependent $^{14}CO_2$ fixation in extracts of Chromatium,[1] *Hydrogenomonas facilis*,[2] and several Thiobacilli,[3–5] the effect of these and other metabolic effectors on the separate reactions leading from ribose-5-P to 3-P-glyceric has not been determined. Ribose-5-P isomerase has been partially purified and partially characterized from a variety of plant, animal and bacterial sources.[6–13]

The purpose of the present study was to isolate and characterize the isomerase from the purple non-sulfur photosynthetic bacterium *Rhodospirillum rubrum* and to investigate the effect of several possible activators and inhibitors on the enzyme-catalyzed reaction as part of a broader study of regulation of the activity of the reductive pentose phosphate (Calvin) cycle in this photosynthetic bacterium.

Phosphoriboisomerase was measured by the colorimetric method of Axelrod and Jang.[6] Na ribose-5-P was used as substrate.

The method used for the purification of the enzyme is given in Table

* Supported in part by a grant, GM10765–05, from the National Institute of General Medical Sciences.

† Present address: University of Maine, Orono, Maine.

‡ Present address: Graduate School of Biomedical Sciences, Biology Division, P.O. Box Y, Oak Ridge National Laboratory, Oak Ridge, Tennessee, 37830.

I. *R. rubrum*, strain S-1, was cultured on 100 m*M* L-malate, 5 m*M* $(NH_4)_2$-SO_4 as previously described.[14] Cells were suspended in potassium phosphate or Tris buffer and released from a French pressure cell. The extract was centrifuged 45 minutes at 100,000×*g*. Protamine sulfate was added

TABLE I. Purification of ribose-5-P isomerase from *R. rubrum*.

	Total Units*	Specific Activity†	Purification
100,000×*g* Supernatant	44	0.18	
Protamine Sulfate 35–50% $(NH_4)_2SO_4$		0.28	1.5
G-100 Sephadex	1.8	1.6	8.7
Hydroxyapatite	14	1.9	11
DEAE	1.5	4.2	23

* μmoles Ru5P formed/min.

† μmoles Ru5P formed/min/mg protein.

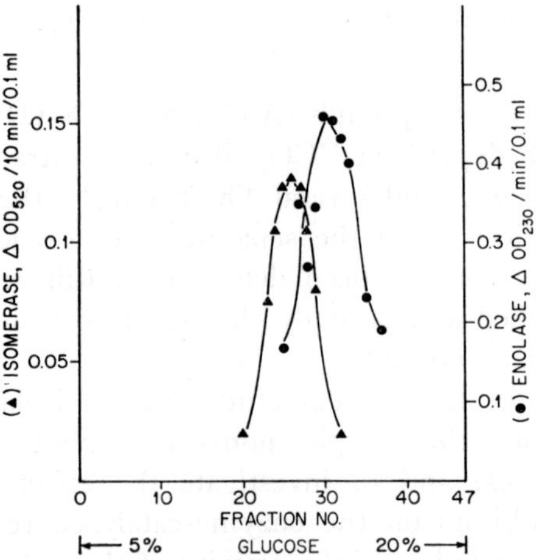

Fig. 1. Determination of sedimentation coefficient of ribose-5-P isomerase by glucose density gradient analysis. 0.6 units of isomerase, purified through the $(NH)_4)_2SO_4$ step, and 300 μg pure yeast enolase (generously donated by Dr. E. W. Westhead) were dissolved in 100 μl 2.5% buffered glucose and layered onto a linear 5 to 20% gradient of glucose in 10 m*M* potassium phosphate, 10 m*M* $MgCl_2$, 5 m*M* mercaptoethanol, pH 7.6. After 15 hours centrifugation at 38,000 rpm and 4°C in an SW 39 rotor, 47 fractions were collected and diluted with 1 ml 100m*M* Tris-HCl, pH 7.6. The isomerase was assayed by the usual method. Enolase was assayed by the method of Westhead and Malmström (1957). The $S_{20,w}$ of enolase is 5.9.[22] Assuming direct proportionality between sedimentation coefficient and distance traveled on the gradient[23] the $S_{20,w}$ for isomerase is 5.1 (average result from 4 determinations).

in order to precipitate nucleic acids. $(NH_4)_2SO_4$ was added to the supernatant and the fraction precipitating between 35 and 50% saturation was dissolved in 2 ml buffer and passed through a G-100 Sephadex column. 2 g washed hydroxyapatite powder was added to the eluate, the slurry centrifuged, the apatite washed once, and the combined supernatants placed on a DEAE column. A linear NaCl gradient in 10 m*M* Tris, 5 m*M* mercaptoethanol, pH 7.0, was used to elute isomerase. The total purification effected by this procedure was approximately 50-fold over the original crude (chromatophore-containing) extract.

The $S_{20,w}$ value of the isomerase was estimated by glucose density gradient analysis to be 5.1 (Fig. 1). Since exclusion from Sephadex correlates with Stokes radii,[15] the exact coincidence of exclusion of isomerase and enolase from Sephadex G-100 indicates that the Stokes radius of ribose-5-P isomerase is 27 Å (Fig. 2). Assuming a partial specific volume of 0.725 cm^3/g, the molecular weight of the isomerase can be estimated (Siegel and Monty, 1966) to be 57,000 and the frictional ratio 1.47.

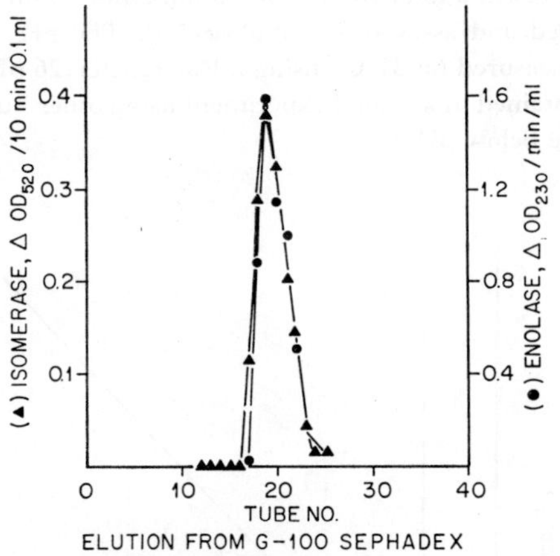

Fig. 2. Determination of Stokes radius of ribose-5-P isomerase by exclusion from Sephadex. 8 units of isomerase, purified through the $(NH_4)_2$-SO_4 fractionation step, and 500 μg pure yeast enolase in 1 ml 7% saturated $(NH_4)_2SO_4$ in buffer were layered onto a 2.5 × 35 cm coarse (40–120 μ) G-100 Sephadex column and eluted with 10 m*M* potassium phosphate, 10 m*M* $MgCl_2$, 5 m*M* mercaptoethanol, pH 7.6. 50 drop (4.5 ml) fractions were collected. The void volume for the column was 650 drops (fraction 13). The Stokes radius of enolase was calculated to be 27 Å from the amino acid molecular weight and the known partial specific volume[22] using the conventional formula.[15] Since exclusion from Sephadex correlates with Stokes radii,[15] the Stokes radius of ribose-5-P isomerase is 27 Å. (Substantiated by 3 separate determinations.)

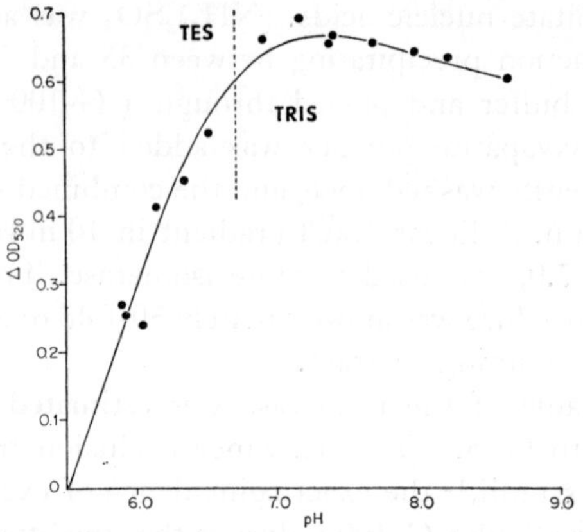

Fig. 3. Dependence of ribose-5-P isomerase on pH. The reaction mixture contained enzyme, purified through the DEAE step, 4 μmoles sodium ribose-5-P, 150 μmoles N-Tris (hydroxymethyl)-methyl-2-amino-ethane-sulfonic acid (TES) buffer and 1 μmole Tris below pH 6.9 and 151 μmoles Tris above pH 6.9 in a total volume of 1.8 ml. After 10 minutes at 37°C, 0.6 ml was removed and assayed for ribulose-5-P. The pH of the remaining mixture was measured (at 37°C) using a Radiometer 26 pH meter. Similar results were obtained in a second experiment using other buffers. No activity was measurable below pH 5.5.

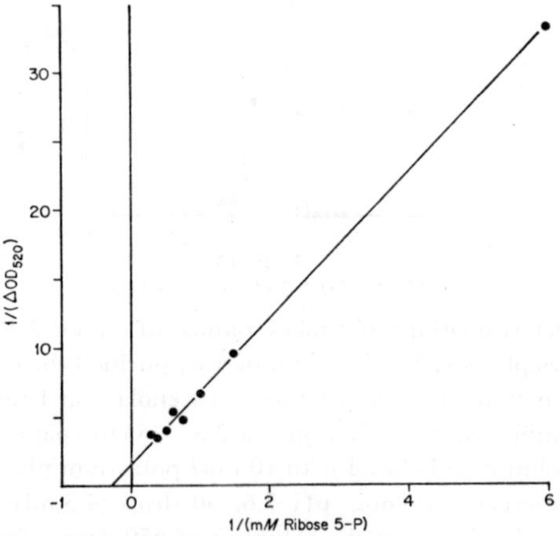

Fig. 4. Determination of Michaelis constant by the method of Lineweaver and Burk. The reaction mixture contained enzyme purified through the DEAE step, 50 μmoles Tris buffer, and sodium ribose-5-P, as indicated, in 0.6 ml. Isomerase activity was measured at 37°C by the usual procedure. The Km for ribose-5-P is $4 \times 10^{-3} M$ (average of 4 determinations).

The enzyme-catalyzed isomerization was linear with time, falling off as equilibrium was approached, and linear with respect to enzyme concentration. There is a broad pH optima between pH 7 and 7.4 (Fig. 3). The Km determined by the method of Lineweaver and Burk is 4×10^{-3} *M* for ribose-5-P (Fig. 4).

Autotrophic growth conditions result in high levels of ribulose-1,5-diP carboxylase activity, but not of isomerase activity, in extracts of *R. rubrum*.[16] The carboxylase from this bacteria is smaller (m. wt., 112,000) than the plant carboxylase (m. wt., 557,000), has Km's for CO_2 and ribulose-1,5-diP of the same order as the plant enzyme[17] but does not require the addition of a divalent cation for activity.[18]

A number of compounds have been tested for their effect on the reactions catalyzed by the isomerase and by the carboxylase (Table II). The results given here are the average from at least two separate experiments. Isomerase inhibition was measured using sufficient amounts of enzyme, which had been purified through the hydroxyapatite or DEAE step, to allow measurement of activity while the reaction was linear with time. The units of enzyme activity used varied somewhat in separate determinations. Details of the purification and characterization of ribulose-1,5-diP carboxylase from *R. rubrum* will be given elsewhere (manuscript in preparation). Inhibition of the carboxylase was measured with purified carboxylase, using a $^{14}CO_2$ assay similar to that described by Paulsen and Lane[19] when the spectrophotometric system of Racker[20] was not applicable.

Phosphate inhibits both isomerase and carboxylase. Arsenate inhibits only carboxylase. Arsenite inhibits the isomerase, but not the carboxylase. Both enzymes, when partially purified, retain activity best in the presence of mercaptoethanol. Ammonium sulfate inhibits neither the isomerase nor the carboxylase from *R. rubrum* (Table II). D-3-P-Glyceric acid inhibits the carboxylase but has no effect on the isomerase. Citrate inhibits both the isomerase and carboxylase; unlike the plant enzyme the carboxylase from *R. rubrum* does not require the addition of a divalent cation for activity. High substrate levels have no effect on either the carboxylase or the isomerase from *R. rubrum*. Ribulose-1,5-diP, however, inhibits the isomerase. The kinetics of this inhibition appear to be complex and the variable results may be due to inhibition by the product of the reaction, ribulose-5-P, in combination with the added inhibitor.

None of the energetically related nucleotides tested (Table II) inhibited the carboxylase. The isomerase, however, was inhibited by AMP and by NAD. Like the inhibition by ribulose-1,5-diP the inhibition by

TABLE II. Inhibition of ribose-5-P isomerase and ribulose-1,5-diP carboxylase from *R. rubrum*.

Inhibitor	% Inhibition	
	Isomerase	Carboxylase
10 m*M* PO_4^{3-} (K^+)	20	40
10 m*M* AsO_4^{3-} (Na^+)	0	50
10 m*M* AsO_2^{2-} (Na^+)	22	0
10 m*M* $(NH_4)_2SO_4$	0	0
10 m*M* pyruvate	0	0
10 m*M* P-enolpyruvate	0	0
10 m*M* glucose-6-P	0	0
10 m*M* D-3-P-glycerate	0	30
10 m*M* citrate	44	
1 m*M* citrate	2	75
10 m*M* ribose-5-P	0	0
1 m*M* ribulose-1, 5-diP	32*	0
10 m*M* NAD	70†	0
10 m*M* NADP	0	0
10 m*M* NADH	—	0
1 m*M* NADH	0	0
10 m*M* AMP	18‡	0
10 m*M* ADP	0	0
10 m*M* ATP	0	0

* values obtained in 3 different experiments were 0, 34 and 62.
† values obtained in 3 different experiments were 45, 66 and 100.
‡ values obtained in 3 different experiments were 0, 23 and 32.
Preliminary kinetic experiments indicate that the amount of inhibition brought about by either ribulose-1,5-diP, NAD, or AMP may be influenced by the concentration of the product of the reaction, ribulose-5-P. This would account for the variation observed in this case.

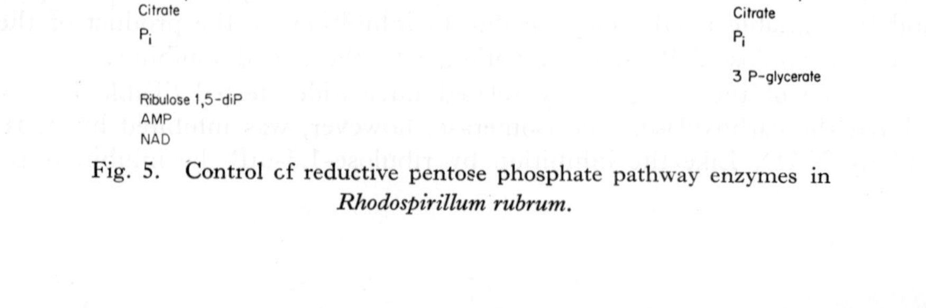

Fig. 5. Control of reductive pentose phosphate pathway enzymes in *Rhodospirillum rubrum*.

AMP and NAD was variable. The kinetics of inhibition were similar to those obtained for ribulose-1,5-diP. Fig. 5 summarizes the effect of various inhibitors on both the isomerase and carboxylase reactions in *R. rubrum.*

Clearly ribose-5-P isomerase in this organism is an important enzyme from the standpoint of control of the reductive pentose phosphate cycle. High levels of AMP and oxidized NAD signal a lack of available energy for the operation of the reductive pentose phosphate cycle. These compounds could directly inhibit the operation of the cycle by suppressing isomerase activity. Ribulose-1,5-diP is also a negative modifier of the isomerase. This would appear to be simple feedback inhibition with complex kinetics.

Citrate inhibits both the isomerase and the carboxylase. *Rhodospirillum rubrum* is a facultative phototroph capable of existing either autotrophically or heterotrophically. It assimilates acetate, ethanol, malate, and other 4-carbon dicarboxylic acids. The 4-carbon dicarboxylic acids, in particular, should lead to high steady state levels of citrate, relative to the levels which might be expected when the organism is solely dependent on the reductive assimilation of CO_2 for carbon. It is to the advantage of the organism to assimilate preformed carbon units rather than to reduce CO_2. High intracellular levels of citrate apparently suppress the operation of the reductive pentose phosphate cycle, allowing the utilization of the energy rich compounds ATP and NADH for other purposes.

We have previously shown that the nature of the carbon source utilized for growth influences the levels of several enzymes in this organism.[16] In those studies it was shown that during growth light, per se, had no effect on the demonstrable activity levels of any of the enzymes of the tricarboxylic acid cycle or of the reductive pentose phosphate pathway measured. The control of ribose-5-P isomerase activity, however, but not of ribulose-1,5-diP carboxylase activity, appears to be intimately linked to compounds involved in energy metabolism. Light, through the formation of ATP and NADH, is apparently involved in at least two ways: one, directly in the operation of the Calvin cycle in *R. rubrum* through the cyclic regeneration of ribulose-1,5-diP; and two, indirectly through phosphorylation and reduction, respectively, of the pentose phosphate isomerase inhibitors, AMP and NAD.

REFERENCES

1. Johnson, E. J., *Arch. Biochem. Biophys.*, **114**, 178 (1966).
2. McFadden, B. A., and Tu, C. L., *Biochem. Biophys. Res. Comm.*, **19**, 728 (1965).
3. Johnson, E. J., and Peck, H. D., Jr., *J. Bacteriol.*, **89**, 1041 (1965).

4. Aleem, M. I. H., and Huang, E., *Biochem. Biophys. Res. Comm.*, **20**, 515 (1965).
5. Gale, N. L., and Beck, J. V., *Ibid.*, **24**, 792 (1966).
6. Axelrod, B., and Jang, R., *J. Biol. Chem.*, **209**, 847 (1954).
7. Hurwitz, J., Weissbach, A., Horecker, B. L., and Smyrniotis, P. Z., *J. Biol. Chem.*, **218**, 769 (1956).
8. Tabachnick, M., Srere, P. A., Cooper, J., and Racker, E., *Arch. Biochem. Biophys.*, **74**, 315 (1958).
9. Dickens, F., and Williamson, D. H., *Biochem. J.*, **64**, 567 (1956).
10. Agosin, M., and Aravena, L., *Enzymologia*, **22**, 281 (1960).
11. Dobrogosz, W., J., and Demoss, R. D., *Biochim. Biophys. Acta*, **77**, 629 (1963).
12. Urivetsky, M., and Tsuboi, K. K., *Arch. Biochem. Biophys.*, **103**, 1 (1963).
13. Rao, P. S., and Nicholas, D. J. D., *Biochim. Biophys. Acta*, **124**, 221 (1966).
14. Anderson, L., and Fuller, R. C., *Plant Physiol.*, **42**, 491 (1967).
15. Siegel, L. M., and Monty, K. J., *Biochim. Biophys. Acta*, **112**, 346 (1966).
16. Anderson, L., and Fuller, R. C., *Plant Physiol.*, **42**, 497 (1967).
17. Anderson, L., Worthen, L. E., and Fuller, R. C., *Fed. Proc.*, **26**, 725 (1967).
18. Anderson, L., unpublished data.
19. Paulsen, J. M., and Lane, M. D., *Biochem.*, **5**, 2350 (1966).
20. Racker, E., *Methods in Enzymol.*, **V**, 266 (1962).
21. Westhead, E. W., and Malmström, B. G., *J. Biol. Chem.*, **228**, 655 (1957).
22. Malmström, B. G., *The Enzymes*, **5**, 478 (1961).
23. Martin, R. G., and Ames, B. N., *J. Biol. Chem.*, **236**, 1372 (1961).

Warburg Effect: Control of Photosynthesis by Oxygen

MARTIN GIBBS, PETER W. ELLYARD* AND ERWIN LATZKO†

Department of Biology, Brandeis University,
Waltham, Massachusetts

INTRODUCTION

In 1920, Warburg[1] made the observation that the rate of release of O_2 by illuminated *Chlorella* was inhibited by molecular oxygen. This effect on photosynthesis is now commonly referred to as the "Warburg effect." This effect has been observed in a wide variety of photosynthetic organisms, including algae, bryophytes, ferns, and angiosperms. Turner and Brittain[2] have summarized the results of the considerable amount of work on the Warburg effect.

Some characteristics of the Warburg effect

Tamiya and Huzisige[3] were the first to note that the effect was completely and rapidly reversible when the plant was transferred from an aerobic to an anaerobic atmosphere. They also presented evidence demonstrating that the magnitude of the inhibition of O_2 is strikingly affected by CO_2 concentration. On the basis of their report[3] and those of others,[4,5] it may be concluded that when photosynthesis is limited by CO_2, the hindrance caused by O_2 is maximal.

Benson and Calvin[6] found that the incorporation of $^{14}CO_2$ into glycolic acid was reduced 10-fold by a reduction of O_2 partial pressure from 20% to 1% in *Scenedesmus*. More recently, Coombs and Whittingham[7] have reported similar data and extended the work of Benson and Calvin[6] by noting a sharp increase in sugar mono and diphosphates coupled to a decrease in glycolate.

Two different mechanisms have been suggested for the synthesis of glycolate. One group envisages that glycolate is formed from the upper two carbons of a pentulose, hexulose or heptulose mono or diphosphate.[8,9] The source of the oxidizing power is not evident. Coombs and Whittingham[7] have suggested hydrogen peroxide while Hess and Tolbert[10] have favored an oxidant produced by photosystem II. The second proposed mechanism

* Present address. Charles F. Kettering Research Laboratory, Yellow Springs, Ohio.
† Present address. Chemisches Institut, Freising-Weihenstephan, Germany.

is that glycolate results from a direct condensation of two C_1 units with the help of reducing power generated in the light. It was first proposed by Tanner *et al.*[11] and later supported by Warburg and Krippahl[12] and Zelitch.[13]

Theories advanced to account for the Warburg effect

Warburg[1] suggested that the role of O_2 might be to reoxidize the primary photochemical product, thus competing for reducing power with O_2. This view has been favored by a number of workers.[3,14]

Oxygen could inhibit by hindering the action of one or more enzymes in the photosynthetic carbon reduction cycle. Some evidence for this point of view has been presented by Turner *et al.*[15] Finally Bassham and Kirk[16] suggested that O_2 caused inhibition of photosynthesis by increasing the loss of carbon (glycolic acid) from the photosynthetic carbon cycle, thereby decreasing the level of ribulose 1,5-diphosphate. However, in their experiments, the level of ribulose 1,5-diphosphate remained high even when CO_2-free air was substituted for air and very little glycolic acid was formed. Coombs and Whittingham[7] have presented similar evidence.

A wide variety of hypotheses have been advanced to explain the Warburg effect. Essentially all work reported in the literature has been performed with the intact cell. It is our purpose to define the Warburg effect in the chloroplast and to attempt an explanation using the isolated organelle.

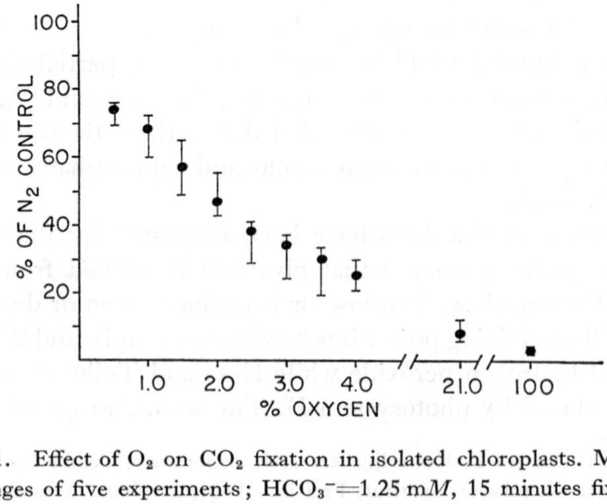

Fig. 1. Effect of O_2 on CO_2 fixation in isolated chloroplasts. Means and ranges of five experiments; $HCO_3^- = 1.25$ m*M*, 15 minutes fixation.

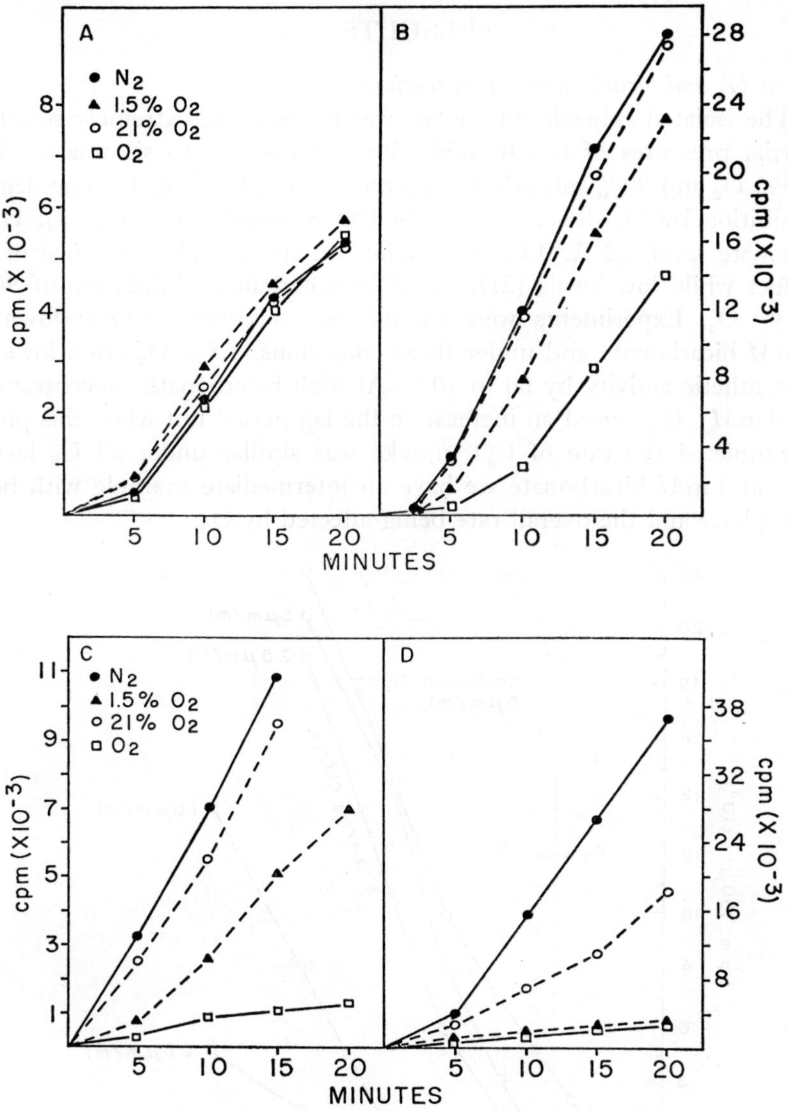

Fig. 2. Effect of O_2 and HCO_3^- concentration on CO_2 fixation in isolated chloroplasts.

A)	13 m*M* HCO_3^-	Max. rate=42 μmoles mg chl^{-1} hr^{-1}
B)	10 m*M* HCO_3^-	Max. rate=94 ,, ,, ,,
C)	4 m*M* HCO_3^-	Max. rate=35 ,, ,, ,,
D)	1 m*M* HCO_3^-	Max. rate=31 ,, ,, ,,

RESULTS

Effect of O_2 and bicarbonate concentrations

The isolated spinach chloroplast was found to be extremely sensitive to partial pressures of O_2. Roughly 50% of photosynthesis was blocked by 1.5% O_2 and 90% inhibition occurred in 21% O_2 (Fig. 1). The degree of inhibition by O_2 depended on the CO_2 concentration (Fig. 2). High bicarbonate levels (2 A, 13m*M*) resulted in the complete absence of an O_2 effect while low levels (2D, 1m*M*) resulted in an inhibition of 90% by 21% O_2. Experiments were carried out at concentrations down to 0.25 m*M* bicarbonate and under these conditions, 1.5% O_2 often lowered photosynthetic activity by 60 to 70%. At high bicarbonate concentrations (2B, 10 m*M*), O_2 caused an increase in the lag period but when this phase was completed the rate of CO_2 uptake was similar under all O_2 levels. In 2C, at 4 m*M* bicarbonate we have an intermediate example with both the lag phase and the overall rate being affected by O_2.

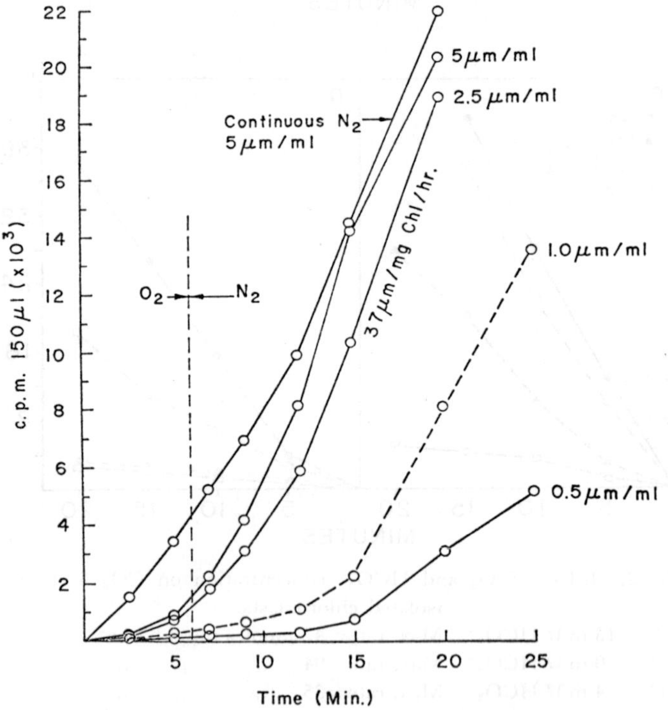

Fig. 3. The influence of changing from 100% O_2 to N_2 on CO_2 fixation at several bicarbonate levels.

Reversibility

The Warburg effect in isolated chloroplasts was reversible (Fig. 3). The rapidity of recovery depended upon the bicarbonate concentration. At the higher levels (5 m*M* and 2.5 m*M*), recovery was observed even in 100% O_2 with a slightly increased lag. The lower levels (1.0 m*M* and 0.5 m*M*) did not recover unless N_2 was introduced.

Influence of sugar phosphate

Bamberger and Gibbs[17,18] demonstrated that the lag period in photosynthesis could be shortened or completely removed by the addition of intermediates of the photosynthetic carbon reduction cycle such as fructose 1,6-diP, glycerate 3-P, ribose 5-P, and triose-P (glyceraldehyde 3-P, dihydroxyacetone-P). Experiments were done to test whether these phosphorylated compounds influenced the inhibition due to O_2. Such an experiment is illustrated in Fig. 4. Two dark periods of 2.5 minutes each were interjected between the 5 minute light periods. The major differences between N_2 and O_2 existed at the end of the 5 minute light period. In the second light period the chloroplast in O_2 had a slight lag but attained a

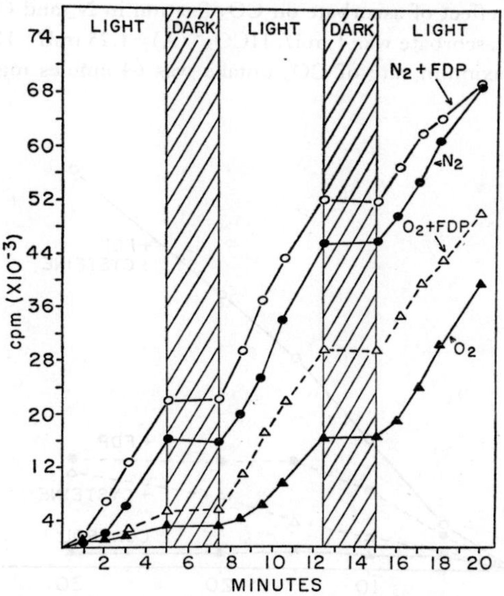

Fig. 4. The influence of 1 m*M* fructose 1,6-diP (FDP) on CO_2 fixation in N_2 and O_2. HCO_3^- was 10 m*M*. Maximum rate was 85 μmoles mg chl^{-1} hr^{-1}.

photosynthetic rate similar to the N_2 treatment in the third light period. The addition of 1m*M* fructose 1,6-diP or ribose 5-P shortened the lag periods in both N_2 and O_2, allowing linear fixation in the first light period in the former and in the second light period in the latter. In the third light period, a similar rate was attained by all treatments.

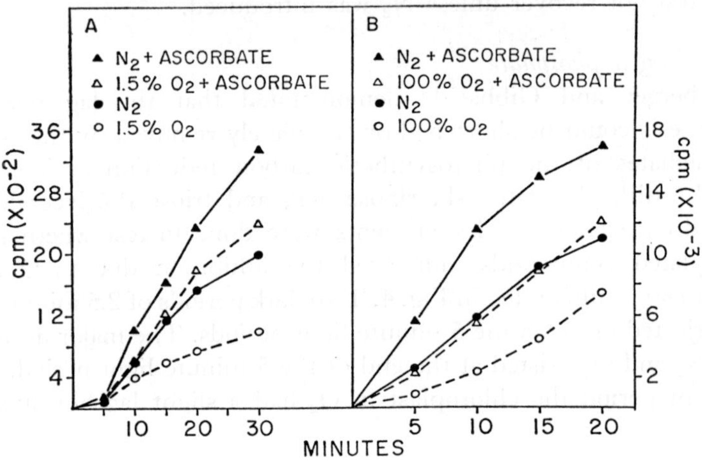

Fig. 5. The effect of ascorbate on CO_2 fixation in N_2 and O_2. Final concentration of ascorbate was 1 m*M*. HCO_3^- (A)=1.25 m*M*; HCO_3^- (B)= 10 m*M*. The maximum rate of CO_2 uptake was 64 μmoles mg chl^{-1} hr^{-1}.

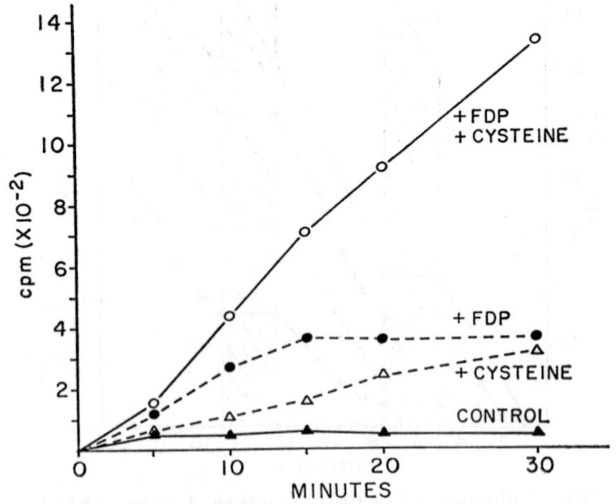

Fig. 6. Effect of additions of 1 m*M* fructose 1,6-diP (FDP) and 1 m*M* cysteine on $^{14}CO_2$ fixation in air.

Effect of ascorbate and reduced thiol compounds

Carbon dioxide fixation of isolated chloroplasts in air was stimulated by ascorbate, cysteine, reduced glutathione and BAL (2,3-dimercapto-propanol). These compounds, of which ascorbate is a representative example, were effective in overcoming inhibition to some extent (Fig. 5). It was found that reduced lipoic acid and dithiothreitol were also effective. The dithiols were more effective than the monothiols. Although all compounds were effective in O_2 concentrations up to 21%, the reduced thiol

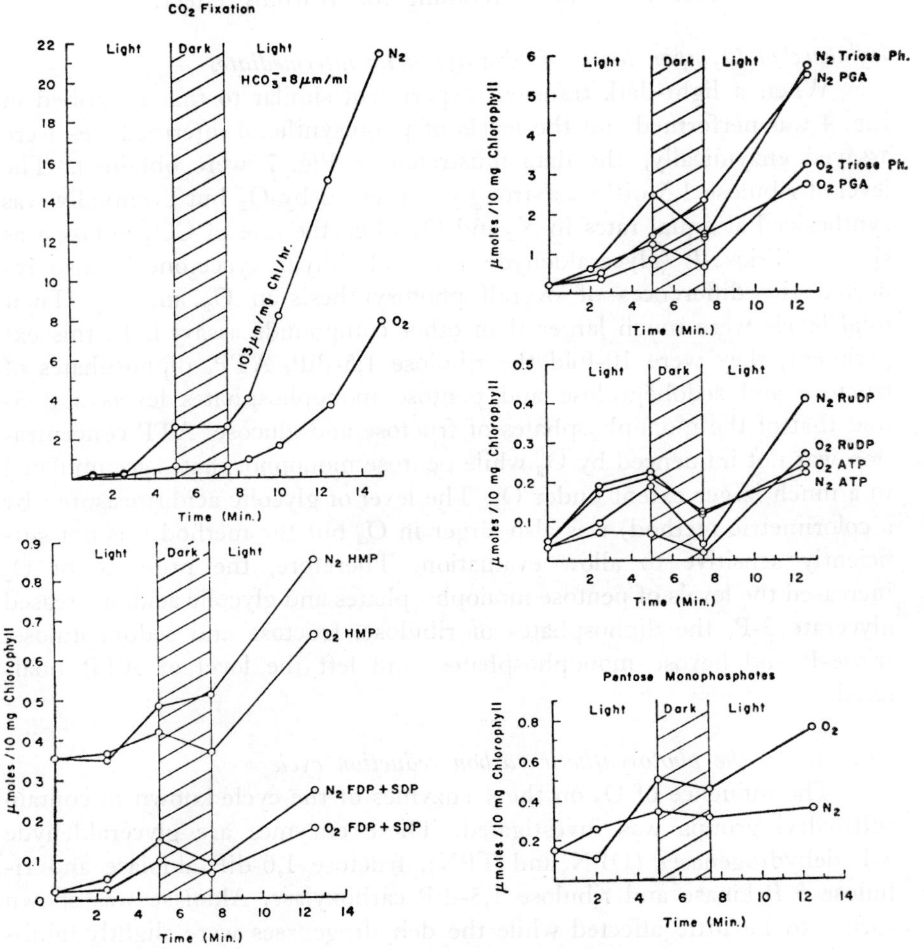

Fig. 7. The influence of N_2 and O_2 on CO_2 fixation and the levels of intermediates (assayed enzymatically) of the carbon reduction cycle. $HCO_3^- = 8$ m*M*.

compounds were not effective in 100% O_2. This may be due to the oxidation of the thiols in 100% O_2 to a form which inhibited photosynthesis. Oxidized glutathione, oxidized lipoate and cystine were found to inhibit photosynthesis.

Unlike the sugar phosphates, the reduced thiols did not significantly affect the length of the lag period. The effect of additions of both fructose 1,6-diP and cysteine to a system already inhibited 90% by 21% O_2 is shown in Fig. 6. The fact that their stimulatory effects were additive indicated that each had a different role in overcoming the Warburg effect.

Influence of O_2 upon levels of photosynthetic intermediates

When a light-dark transient experiment similar to that described in Fig. 4 was performed and the levels of photosynthetic intermediates were assayed enzymically, the data illustrated in Fig. 7 were obtained. The level of ribulose 1,5-diP was strongly depressed by O_2 but eventually was synthesized at equal rates in N_2 and O_2 when the rate of CO_2 uptake was similar. Triose-P (glyceraldehyde 3-P and dihydroxyacetone-P) also reflected the differences of overall photosynthesis in O_2 and N_2. Their total levels were much larger than other compounds assayed. In this experiment, they were 10-fold the ribulose 1,5-diP, ATP, diphosphates of fructose and sedoheptulose and pentose monophosphates levels and 5-fold that of the monophosphates of fructose and glucose. ATP concentration was not influenced by O_2 while pentose monophosphates accumulated to a much larger extent under O_2. The level of glycolic acid (measured by a colorimetric method) was also larger in O_2 but the method was not sufficiently sensitive to allow evaluation. Therefore, the presence of O_2 increased the levels of pentose monophosphates and glycolic acid, decreased glycerate 3-P, the diphosphates of ribulose, fructose and sedoheptulose, triose-P and hexose monophosphates, and left the level of ATP unaltered.

Enzymes of the photosynthetic carbon reduction cycle

The influence of O_2 on the 4 enzymes of the cycle known to contain sulfhydryl groups was investigated. These enzymes are glyceraldehyde 3-P dehydrogenases (DPN and TPN), fructose 1,6-diP aldolase and ribulose 5-P kinase and ribulose 1,5-diP carboxylase. Aldolase was shown earlier to be little affected while the dehydrogenases were slightly inhibited.[19] In the present study, the carboxylase was also demonstrated to be unaffected by O_2. In contrast, phosphoribulokinase was extremely sensitive to O_2 (Table I). The chloroplasts from which the kinase was extracted

TABLE I. Effect of O_2, light, mono and dithiols on activity of phosphoribulokinase in chloroplast preparations.

Treatment	O_2 conc.	HCO_3^- m*M*	Addition final conc.	CO_2 fixation μmoles mg chl^{-1} hr^{-1}	Phosphoribulokinase μmoles mg chl^{-1} hr^{-1}
10 min, light	0(N_2)	10	—	64	134
10 min, light	100%	10	—	44	53
10 min, dark	0(N_2)	5	—	—	99
10 min, dark	100%	5	—	—	21
10 min, dark	100%	5	R5P, 1m*M*	—	42
10 min, light	100%	5	R5P, 1m*M*	—	63
14 min, light	0(N_2)	3	—	25	54
14 min, light	0(N_2)	3	GSH, 1m*M*	46	97
14 min, light	0(N_2)	3	BAL, 0.5m*M*	52	350

The enzyme was extracted after 7.5 min of photosythesis.
GSH and BAL, where indicated, were added to the incubation mixture.

TABLE II. Effect of TPN concentration on the rate of photoreduction in nitrogen and oxygen.

Oxygen Concentration	Added TPN (μ*M*)	*V*/*Vm** (×100)
0(N_2)	0	6
"	0.5	17
"	1.0	25
"	5.0	57
"	10.0	67
"	100.0	100*
100%	0	6
"	0.5	19
"	1.0	29
"	5.0	46
"	10.0	57
"	100.0	86

The concentration of ferredoxin was 4 units/ml. The endogenous TPN concentration was 0.5 μ*M*. The total volume was 3 ml.
* Rate in 100μ*M* TPN under N_2 was 68 μmoles mg chl^{-1} hr^{-1}

were allowed to carry out photosynthesis for 15 minutes. In no case was the specific activity (μmoles ribulose 5-P metabolized per mg chlorophyll per hr) of the kinase or of the other 2 enzymes the rate-limiting step in the overall process. It is clear that O_2 can lower the activity of at least 2 enzymes of the photosynthetic carbon reduction cycle, but from a consideration of specific activities it is not certain how significant this is.

Photoreduction of TPN and photophosphorylation

O_2 had no effect on the rate of TPN reduction or of coupled photophosphorylation. The experiments were performed in 100% O_2 and at levels of TPN down to so-called "endogenous" levels (Table II). The "endogenous" level of TPN under our conditions (10 μg chlorophyll) was roughly 0.5 μM.

DISCUSSION

The isolated chloroplast manifested the characteristics of the Warburg effect previously observed in intact cells. The most notable of these were rapid reversibility and an increased inhibition at lower bicarbonate concentrations. Clearly the primary aspects of the Warburg effect are associated with the chloroplast and these aspects can be studied in an isolated chloroplast preparation.

Warburg effect and the electron transport system

To account for the Warburg effect, some investigators have envisaged that one or more intermediates of the electron transport system may be produced in diminished quantities under O_2. The only known autoxidizable substance of the electron transport system is ferredoxin. However, the preferred oxidant of reduced ferredoxin is TPN rather than O_2 (Table II). When TPN is *not* available, O_2 can serve as a Hill oxidant in a Mehler type reaction.[20]

Our data are difficult to resolve with the findings of the Tokyo group.[3,14] A major difficulty is the recognition of their "R" which they do not consider to be TPNH. Even if TPNH were "R," or a substantial part of "R," there is not sufficient TPNH oxidase present in chloroplasts to account for a rapid oxidation of reduced pyridine nucleotide. An oxidase for TPN was generally absent from our preparations and if present, had an activity of less than 10 μmoles TPNH oxidized per mg chlorophyll per hour.

The fact that the photoreduction of TPN by chloroplasts is insensitive to O_2 would seem to weaken theories which tie O_2 inhibition to the formation of "reducing power." If CO_2 were limiting and TPN was not available, pseudocyclic photophosphorylation could occur. However, O_2 did not alter the level of ATP in the chloroplast (Fig. 7).

Warburg effect and carbon metabolism

During photosynthesis, there is fairly convincing evidence that O_2 has its greatest influence on chloroplast carbon metabolism. It changes the level of intermediates of the photosynthetic carbon reduction cycle and particularly of glycolic acid (Fig. 7) and affects the activity of at least 2 enzymes of the cycle (Table I).

Some theories attempt to account for the Warburg effect in terms of a drainage of carbon from the carbon cycle in the form of glycolic acid, resulting in a limiting amount of CO_2 acceptor.[16] Our experiments tend to provide some evidence for this point of view. Any means, whether it be high O_2 or low CO_2, that favors the production of glycolic acid inhibits photosynthesis. Coupled to an increase in glycolic acid is a decrease in essentially all other intermediates of the photosynthetic carbon reduction cycle. However, there is apparently no system present in the chloroplast which is capable of metabolizing glycolic acid. Glycolate 1-^{14}C added to illuminated chloroplasts was not oxidized. These results agree with the finding of Chan and Bassham[21] who reported only 0.27% of glycolate added to chloroplasts was metabolized. In the chloroplast, glycolic acid, like starch, must be considered an end product of carbon metabolism. Sucrose synthesis by chloroplasts is variable and only infrequently encountered.[22] In the intact plant, glycolic acid may reenter the carbon reduction pathway by way of glyoxylic acid, serine, and glycerate 3-P and contribute to the level of intermediates of the carbon reduction cycle.[23,24] This difference in further metabolism of glycolic acid by the intact cell could account for the striking depletion of these intermediates in the chloroplast in contrast to a smaller effect found in the intact cell.

We were able to confirm the results of Bradbeer and Anderson,[8] who found low rates of glycolate synthesis from sugar phosphates. The best rate for illuminated chloroplast fragments obtained by them was 220 μmoles of glycolic acid per mg chlorophyll per hour while we have achieved a value of 178. The intact chloroplast can synthesize glycolic acid from CO_2 at 100 times this rate. Some speculation is in order. Either the sugar phosphates are not the precursors of glycolic acid or these compounds cannot enter the intact chloroplast. With respect to the low rates with fragmented chloroplasts, the possibility must be considered that the 2-carbon precursor derived from the sugar phosphates is oxidized to glycolic acid by a substance produced during electron transport. Fractionation of the chloroplast modifies the required spatial arrangement (see also ref. 25).

Finally, it is of interest to note that in our experiments (data not shown here) the addition of DCMU resulted in a marked decrease in the percentage of $^{14}CO_2$ incorporated into glycolate at the 1mM bicarbonate level. This is also true when ascorbate is added as an alternative electron source for photosystem I. We consider these data as evidence that cooperation between the carbon cycle and photosystem II is required for glycolic acid synthesis (see also Hess and Tolbert).[26]

The possibility must be considered that glycolic acid is synthesized not only from sugar phosphates but also directly from CO_2 by a C_1+C_1 condensation. The formation of glycolic acid from sugar mono or diphosphates represents a net carbon loss from the carbon reduction cycle and this may be reflected as a shortage of acceptor, ribulose 1,5-diP. If glycolic acid were made by a C_1+C_1 condensation, there would be less CO_2 available for carboxylation by ribulose 1,5-diP carboxylase. At low bicarbonate, glycolic acid is the dominant compound and pyridine nucleotides would be reduced. Electrons could flow from the reduced pyridine nucleotides or reduced intermediates of the electron transport system directly to CO_2 or its reduced products. Oxygen inhibits the photosynthetic carbon reduction cycle by lowering the activity of at least 2 enzymes, and this would further lower the capacity of the cycle to compete with the C_1+C_1 pathway for the limiting CO_2 available. The high K_m (20 mM) of ribulose 1,5-diP carboxylase may also play a role in directing the traffic of CO_2 into one or the other pathway.

The Warburg effect is as complex as its counterpart in carbohydrate breakdown, the Pasteur effect. It is certainly as much an inhibition by low CO_2 as by high O_2, for these compounds are inseparable in their effects. Clearly the effect of O_2 on a system as involved as photosynthesis cannot be defined in narrow terms but must eventually be considered in terms of its influence on the entire process.

Acknowledgement

This investigation was aided by grant AT (30-1)-3447 from the United States Atomic Energy Commission and the National Science Foundation.

REFERENCES

1. Warburg, O., *Biochem. Z.*, **103**, 188 (1920).
2. Turner, J. S., and Brittain, E. G., *Biol. Rev.*, **37**, 130 (1962).
3. Tamiya, H., and Huzisige, H., *Stud. Tokugawa Inst.*, **6**, 83 (1949).
4. Turner, J. S., Todd, M., and Brittain, E. G., *Aust. J. Biol. Sci.*, **9**, 494 (1956).
5. Coombs, J., and Whittingham, C. P., *Phytochemistry*, **5**, 643 (1966).
6. Benson, A. A., and Calvin, M., *J. Exptl. Bot.*, **1**, 63 (1950).
7. Coombs, J. and Whittingham, C. P., *Proc. Roy. Soc. B.*, **164**, 511 (1966).
8. Bradbeer, J. W., and Anderson, C. M. A., in *Biochemistry of Chloroplasts* (T. W. Goodwin, ed.), Vol. II, pp. 175–179, Academic Press, New York, 1967.
9. Hess, J. L., and Tolbert, N. E., *J. Biol. Chem.*, **241**, 5705 (1966).

10. Hess J. L., and Tolbert, N. E., *Plant Physiol.*, **42**, 371 (1967).
11. Tanner, H. A., Brown, T. E., Eyster, C., and Treharne, R. W., *Biochem. Biophys. Res. Comm.*, **3**, 205 (1960).
12. Warburg, O., and Krippahl, G., *Z. Naturforsch.*, **15b**, 197 (1960).
13. Zelitch, I., *J. Biol. Chem.*, **240**, 1869 (1965).
14. Miyachi, S., Izawa, S., and Tamiya, H., *J. Biochem.*, **42**, 221 (1955).
15. Turner, J. S., Turner, J. F., Shortman, K. D., and King, J. E., *Aust. J. Biol. Sci.*, **11**, 336 (1958).
16. Bassham, J. A., and Kirk, M., *Biochem. Biophys. Res. Comm.*, **9**, 376 (1962).
17. Bamberger, E. S., and Gibbs, M., *Plant Physiol.*, **40**, 919 (1965).
18. Bamberger, E. S., *Ph. D. Thesis*, Cornell University, 1964.
19. Gibbs, M., Bamberger, E. S., Ellyard, P. W., and Everson, R. G., in *Biochemistry of Chloroplasts* (T. W. Goodwin, ed.), Vol. II, pp. 1–37, Academic Press, New York, 1967.
20. Fewson, C. A., Black, C. C., and Gibbs, M., *Plant Physiol.*, **38**, 680 (1963).
21. Chan, H. W. S., and Bassham, J. A., *Biochim. Biophys. Acta*, **141**, 429 (1967).
22. Everson, R. G., Cockburn, W., and Gibbs, M., *Plant Physiol.*, **42**, 840 (1967).
23. Wang, D., and Waygood, E. R., *Plant Physiol.*, **37**, 826 (1962).
24. Rabson, R., Tolbert, N. E., and Kearney, P. C., *Arch. Biochem. Biophys.*, **98**, 154 (1962).
25. Havir, E. A., and Gibbs, M., *J. Biol. Chem.*, **238**, 3183 (1963).
26. Hess, J. L., and Tolbert, N. E., *Plant Physiol.*, **42**, 1123 (1967).

Effects of Flash Lights of Different Wavelengths on Dark $^{14}CO_2$-Fixation in *Chlorella* Cells

SHIGETOH MIYACHI, TAMIKO OH-HAMA AND NORIKO OGASAWARA

Institute of Applied Microbiology, University of Tokyo
and
The Tokugawa Institute for Biological Research, Tokyo

INTRODUCTION

In a previous report it was shown that a part of ^{14}C fixed by *Chlorella* cells in the dark is removed by the effect of brief illumination, indicating the photochemical decarboxylation of the dark ^{14}C-fixation products.[1] The "photoinduced decarboxylation" of the dark fixation products was observed under anaerobic conditions, but no such effect was detected when the algal cells were exposed to a flash light during the course of aerobic dark fixation. The intensity of flash light applied in the previous experiments has been relatively high ($2 \times 10{,}000$ lux).

It was found in the present study that photochemical decarboxylation is observed also under aerobic conditions when light is applied to the algal cells at a proper intensity for a short period.

The action spectrum for the effects of flash light on nonsynchronized algal cells showed that red light (max. 600 mμ) is effective for photochemical decarboxylation. It was also found that $^{14}CO_2$-fixation is strongly enhanced by flash light of near far red (734 mμ), red (679 mμ), green (512 mμ) and blue (453 mμ) color. These responses are insensitive to CMU (3-(4′-chlorophenyl)-1,1-dimethylurea). Using the technique of synchronous culture it was found that the algal cells at D_a-stage, smaller in size and stronger in photosynthetic activity, show strong responses to flash lights, whereas those at the early L_3-stage, larger in size and weaker in photosynthetic activity, show only sluggish responses. It was also found that the action spectrum for the response of algal cells changes according to their developmental stages.

RESULTS AND DISCUSSION

Effects of brief illumination at different light intensities

Cells of *Chlorella ellipsoidea* were grown at 25° in an oblong flat vessel containing 1 liter of inorganic medium[2] with constant bubbling of air containing ca. 2% CO_2. The light intensity applied was 750 lux during the first two days, 2,500 lux during the 3rd through 4th day and 6,000 lux during the 5th through 6th day. The light intensity was lowered again to 750 lux at the end of the 6th day and the cells were harvested on the 7th day. The harvested cells were washed twice with $M/500$ K_2SO_4-solution and kept at 0°. Immediately before the experiment, the cells were sus-

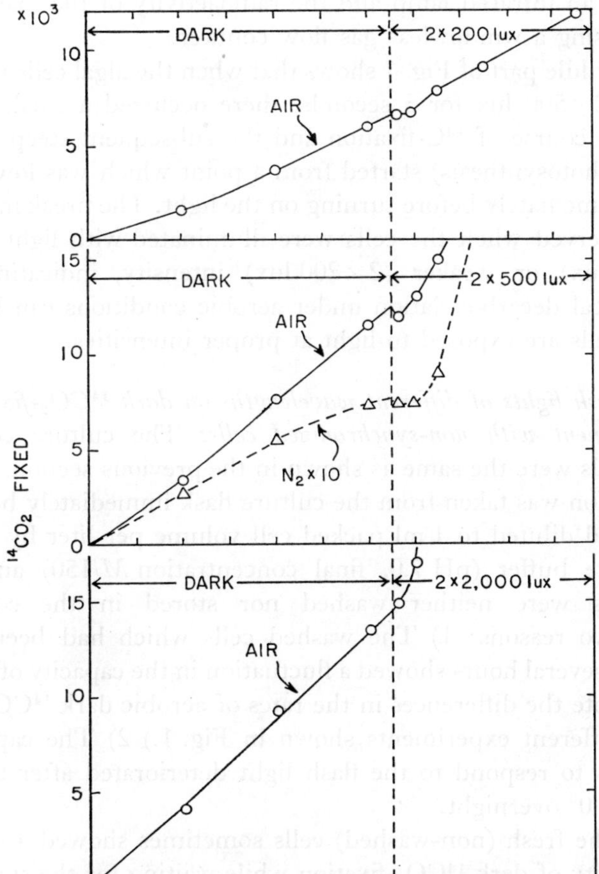

Fig. 1. **$^{14}CO_2$ outburst from the product(s) of aerobic dark $^{14}CO_2$-fixation immediately after turning on the light. (The curve indicated as $N_2 \times 10$ shows the course of anaerobic $^{14}CO_2$-fixation. The values shown here are ten times higher than those actually obtained.)**

pended in $M/100$ phosphate buffer (pH 7.0), the cellular concentration adjusted to 25 ml (packed cell volume) per liter and the suspension was placed in the "lollipop."[3] Then CO_2-free air was bubbled in the dark and $NaH^{14}CO_3$-solution (final concentration, 2 mM) was injected after 10 minutes. Portions of cell suspension were taken out at intervals and immediately transferred to a vessel containing acetic acid (final concentration, 10%). At 6.5 minutes after the addition of $NaH^{14}CO_3$, the light was turned on from both sides of the lollipop. The acid-treated cell suspensions were dried under an infrared lamp and the radioactivity of the fixed ^{14}C determined by using a windowless gas flow counter.

The middle part of Fig. 1 shows that when the algal cells were exposed to light of 2×500 lux for 5 seconds, there occurred a small but distinct break in the course of ^{14}C-fixation and the subsequent steep ascension of the curve (photosynthesis) started from a point which was lower than that observed immediately before turning on the light. The break in ^{14}C-fixation was not observed when the cells were illuminated with lights of stronger ($2\times2{,}000$ lux) or weaker (2×200 lux) intensity, indicating that the photochemical decarboxylation under aerobic conditions can be seen only when the cells are exposed to light at proper intensities.

Effects of flash lights of different wavelengths on dark $^{14}CO_2$-fixation

Experiment with non-synchronized cells: The culture conditions of *Chlorella* cells were the same as shown in the previous section. A 6-day-old cell suspension was taken from the culture flask immediately before the experiment and diluted to 1 ml packed cell volume per liter by the addition of phosphate buffer (pH 7.0, final concentration $M/150$) and deionized water. Cells were neither washed nor stored in the cold, for the following two reasons: 1) The washed cells which had been kept at 0° longer than several hours showed a fluctuation in the capacity of dark $^{14}CO_2$-fixation. (Note the differences in the rates of aerobic dark $^{14}CO_2$-fixation in the three different experiments shown in Fig. 1.) 2) The capacity of the washed cells to respond to the flash light deteriorated after the cells had been kept at 0° overnight.

Even the fresh (non-washed) cells sometimes showed a small change in the capacity of dark $^{14}CO_2$-fixation while waiting for the start of the experiment. As will be shown below, therefore, the time course of the change was always followed during each experiment.

Small test tubes covered with aluminum foil were provided with 1 ml each of cell suspension in a dark room. Then 0.1 ml of $NaH^{14}CO_3$ solution (final concentration, 2.5 mM) was added to each test tube at an interval

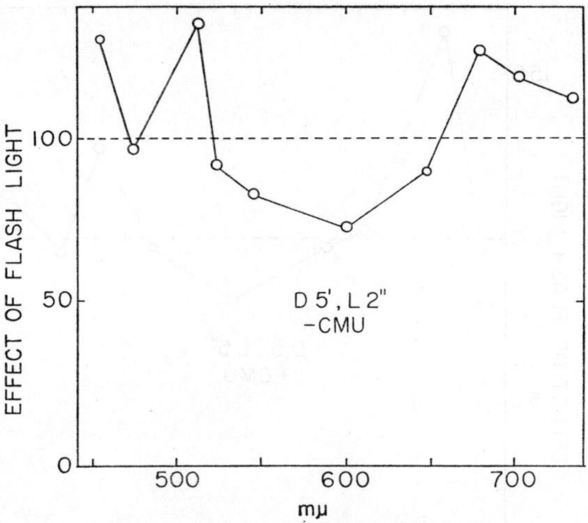

Fig. 2. Effects of flash lights of different wavelengths on dark $^{14}CO_2$-fixation. Flash light (2 sec) was given at the end of dark $^{14}CO_2$-fixation which lasted for 5 minutes. The effect was expressed taking the amount of ^{14}C fixed in the dark (no flash light) as 100.

of 1.5 minutes. Each test tube was shaken for 10 seconds by using a thermo-mixer immediately and $3\frac{2}{3}$ minutes after the addition of $NaH^{14}CO_3$. Five minutes after the addition of $NaH^{14}CO_3$, acetic acid was injected into the first and the last test tubes, as well as into every third tube, in order to follow the time course of the change in $^{14}CO_2$-fixation in total darkness during the course of the experimental period. The aluminum foil of other test tubes was taken off after the second shaking and the test tubes were exposed to monochromatic flash light (2 sec or 5 sec) five minutes after the addition of $NaH^{14}CO_3$. Immediately after the light exposure acetic acid was added to stop the reaction. The level of the dark $^{14}CO_2$-fixation of the cells which were exposed to flash light was estimated according to the time course curve obtained with the cells incubated in total darkness, and the difference between the radioactivity obtained with the flash light-exposed cells and that estimated for the dark $^{14}CO_2$-fixation was assumed to be due to the effect of flash light. The monochromatic light was obtained by using an interference filter and each light intensity was adjusted to 800 erg/cm^2·sec.

Fig. 2, a typical result, shows that red light (max. at 600 mμ) is effective for light-induced decarboxylation. It was also found that $^{14}CO_2$-fixation is greatly increased by brief exposure of cells to near far red (702-743 mμ), red (679 mμ), green (512 mμ) and blue (453 mμ) light. Fig. 3 shows that none of the effects disappears in the presence of CMU ($5\times10^{-5}M$). Although

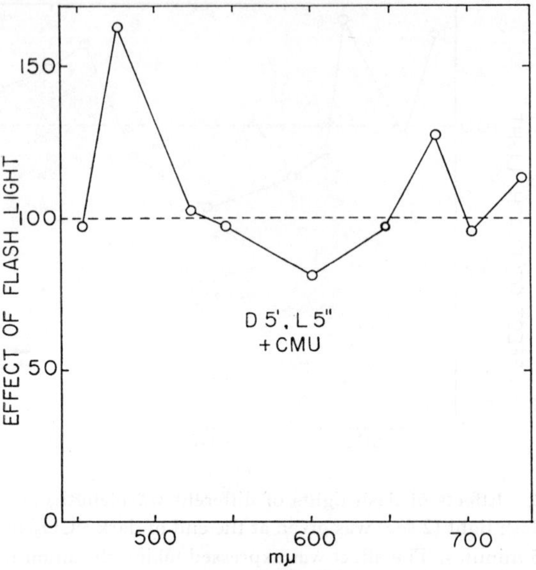

Fig. 3. Effects of flash lights of different wavelengths on dark $^{14}CO_2$-fixation in the presence of CMU. Flash light (5 sec) was given at the end of dark $^{14}CO_2$-fixation. Other experimental conditions were the same as described for Fig. 2.

TABLE I. Effects of flash lights of different wavelengths on dark $^{14}CO_2$-fixation.

Exp.	CMU	Illumination	Wavelength (mμ)				
			453	512	600	679	734
1	−	5 sec	115.0	137.4	81.4	132.1	113.9
2	−	2 sec	136.5	107.3	—	150.8	118.4
3	−	2 sec	129.7	134.8	72.8	126.0	112.1
4	+	5 sec	97.4	—	81.1	127.3	113.6
5	+	5 sec	110.2	103.9	88.8	112.8	94.7
6	+	5 sec	112.7	117.2	—	103.5	103.4
7	+	5 sec	95.7	117.8	105.5	319.2	166.6

The effect is expressed by taking the amount of ^{14}C fixed in the dark (no flash light) as 100.

the effect was not tested in the experiment shown in Fig. 3, the enhancement due to 512 mμ flash light was observed in other experiments, the results of which are summarized in Table I. As will be seen in the table, the degree of the response of algal cells to flash light of the same wavelength differed from harvest to harvest. In some cases there was no response to flash light of certain specific wavelengths (e.g. 734 mμ flash light had no effect in Exp. 5).

As mentioned earlier, all the results shown in Table 1 were obtained by using cells harvested on the 6th day of culture. Somewhat different action spectra were obtained when cells of different culture age were exposed to flash light: cells harvested on the 7th day (light intensity applied during the last three days: 6,000 lux) frequently showed no increase in CO_2-fixation with 512 mμ flash light, and the wavelength for maximum photodecarboxylation moved to a shorter wavelength (546 mμ). Enhancement of $^{14}CO_2$-fixation was also observed with 600 mμ flash light. One possible explanation for the differences in action spectra obtained with cells harvested on different days may be that the responses of *Chlorella* cells to flash lights of different wavelengths vary according to the developmental stage.

Experiment with synchronized cells: Using the technique of synchronous culture (D_8LD-cycle according to Tamiya *et al.*[4]) large cells at the developmental stage just before cell division (early L_3-cells) and small cells (D_a-cells), which were obtained by incubating the divided cells in the light for 3 hours, were prepared, and the modes of response of these cells to flash lights of different wavelengths were compared.

It was found that the response of early L_3-cells to flash light was sluggish at all wavelengths tested (Fig. 4). On the other hand, a very strong response was observed in D_a-cells, which showed twice as high photosynthetic activity (measured by continuous illumination of 679 mμ

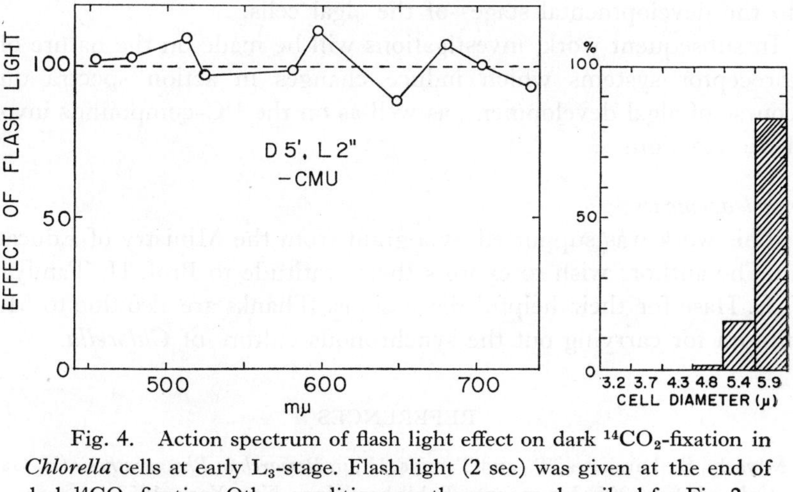

Fig. 4. Action spectrum of flash light effect on dark $^{14}CO_2$-fixation in *Chlorella* cells at early L_3-stage. Flash light (2 sec) was given at the end of dark $^{14}CO_2$-fixation. Other conditions are the same as described for Fig. 2.

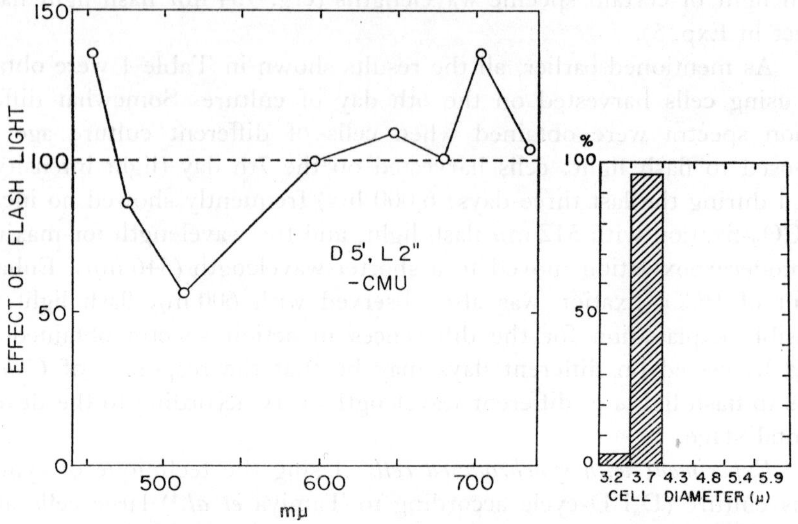

Fig. 5. Action spectrum of flash light effect on dark $^{14}CO_2$-fixation in *Chlorella* cells at D_a-stage. Experimental conditions are the same as described for Fig. 4.

light, 800 erg/cm²·sec) as the early L_3 cells (Fig. 5). An interesting finding is that the most pronounced photodecarboxylation was observed at 512 mμ which induced an increment in $^{14}CO_2$-fixation in non-synchronized cells as well as in cells at the early L_3-stage. These results strongly indicate that the action spectra of the flash light effect vary greatly according to the developmental stages of the algal cells.

In subsequent work, investigations will be made on the nature of the photoreceptor systems which induce changes in action spectra during the course of algal development, as well as on the ^{14}C-compounds involved in these reactions

Acknowledgements

This work was supported by a grant from the Ministry of Education. Japan.The authors wish to express their gratitude to Prof. H. Tamiya and Prof. E. Hase for their helpful discussions. Thanks are also due to Mr. T. Nishimura for carrying out the synchronous culture of *Chlorella*.

REFERENCES

1. Miyachi, S., Hirokawa, T., and Tamiya, H., in *Research in Photosynthesis* (H. Gaffron *et al.*, eds.), p. 205, Interscience Publishers, Inc., New York, 1957.

2. Tamiya, H., Iwamura, T., Shibata, K., Hase, E., and Nihei, T., *Biochim. Biophys. Acta*, **12**, 23 (1953).
3. Miyachi, S., Izawa, S., and Tamiya, H., *J. Biochem.*, **42**, 221 (1955).
4. Tamiya, H., Morimura, Y., Yokota, M., and Kunieda, R., *Plant and Cell Physiol.*, **2**, 383 (1961).

Mode of Coupling of the Phosphorylation System to the Electron Transport System in *Rhodospirillum Rubrum* Chromatophores*

T. Horio, K. Nishikawa, Y. Horiuti and T. Kakuno

Division of Enzymology, Institute for Protein Research, Osaka University, Osaka

INTRODUCTION

It is well known that chromatophores prepared from *Rhodospirillum rubrum*, when illuminated, can synthesize ATP from ADP and Pi (orthophosphate).[1] Horio and Kamen[2] have found that photosynthetic ATP formation takes place only when chromatophores are at an appropriate oxidation-reduction potential (E_h): the optimum E_h value for the reaction medium is around 0 volt at pH 8, and when the E_h value is either too negative or positive, the activity for photosynthetic ATP formation becomes insignificant.[3] This phenomenon is called "maximum coupling."

The present paper deals with the process of photosynthetic ATP formation in bacterial chromatophores as it is related to various partial reactions,[4] especially the E_h-dependent ones, and the cytochromes bound in chromatophores. The mechanism for coupling of the energy conversion system to the electron transport system will be discussed.

MATERIALS AND METHODS

A carotenoid-less mutant of *R. rubrum*, the so-called blue-green mutant (G-9), obtained from Dr. J. W. Newton, was used in most cases. The results obtained with the mutant were essentially the same as those obtained with the wild type strain which had been used in this series of studies. Chromatophores were prepared by disrupting light-grown cells by a grinding method[5,6] or by extrusion from a French pressure cell.[3,7] Experimental conditions for assays of ATP formation, ATPase, pyrophosphatase (PPi-ase), ATP-Pi exchange, ADP-ATP exchange, adenylate

---

* Presented in part at the Symposium on Electron Transport (organized by A. F. Brodie) at the Meeting of the American Society of Microbiology, Los Angeles, California, May 1966.

kinase, pyrophosphate (PPi) formation, etc. were the same as those described previously.[2-8]

RESULTS AND DISCUSSION

Effect of oxidation-reduction potential on photosynthetic ATP formation

With chromatophores prepared by the grinding method,[2,4] the rate of photosynthetic ATP formation increases with increasing concentrations of ascorbate (Fig. 1). The rate is maximal at approximately 6.7×10^{-2} *M* ascorbate and decreases at higher concentrations. When the chromatophores are sonicated for a short period (a few minutes) and then washed, or simply aerated, photosynthetic ATP formation without addition of ascorbate becomes insignificant. With the "electron-depleted" chromatophores thus prepared, the addition of an appropriate electron donor such

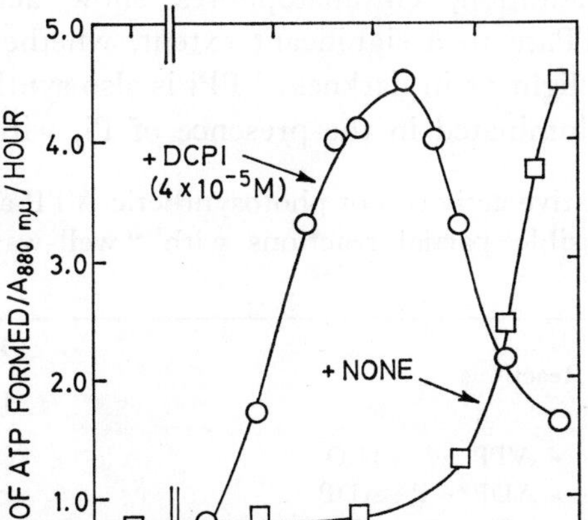

Fig. 1. Effect of ascorbate concentration on photosynthetic ATP formation brought about by chromatophores.

Chromatophores were Prepared from a blue-green mutant (G-9). The standard components of the reaction mixture were as follows: 0.5 ml 0.2 *M* glycylglycine-10% sucrose buffer (pH 8.0), 0.1 ml 0.1 *M* $MgCl_2$, 0.1 ml 0.1 *M*[^{32}P]Pi solution (pH 8) (approx. 10^6 counts/min. excess), 0.1 ml 0.05 *M* ADP, 0.1 ml chromatophores ($A_{873m\mu}$/ml, approx. 50). The total volume was adjusted to 1.5 ml with water. Reactions were carried out at 30°C for 3 min. in the light. Methods for ATP determination were described previously.[2-4]

as ascorbate is required for the induction of photosynthetic ATP formation.

If cytochrome c_2 purified from cells of *R. rubrum*[9] is added in its oxidized form to the reaction mixture, the ascorbate concentration required for the maximum coupling is markedly lowered.[2] Cytochrome *cc′* (previously called "RHP")[9] has almost the same influence as cytochrome c_2, provided that it is added at low concentrations ($\leqq 1\times10^{-6}\,M$).[2] No doubt, the added cytochromes mediate the flow of electrons from ascorbate to the electron transport system associated with chromatophores. The redox-dye, 2,6-dichlorophenol indophenol (DCPI), can substitute for the added cytochromes (Fig. 1).

Partial reactions of energy-conversion system leading to ATP formation

Besides photosynthetic ATP formation, chromatophores can catalyze the oxidation of NADH by molecular oxygen which couples ATP formation.[10] In addition, chromatophores show activity for ATP-Pi exchange and ATPase to a significant extent, whether the reactions are carried out in the light or in darkness.[4] PPi is also synthesized when chromatophores are illuminated in the presence of Pi, with or without ADP,

TABLE I. Relative activities of photosynthetic ATP and PPi formation and their possible partial reactions with "well-washed" chromatophores.

Reactions	Percentage activities: Light	Percentage activities: Darkness
ADP + Pi ⟶ ATP + H_2O	(100)	0
ADP* + ADP∼P ⟶ ADP*∼P + ADP	0	0[a]
2ADP ⟶ ADP∼P + AMP	0	0[b]
Pi* + ADP∼P ⟶ ADP∼P* + Pi	50	10
ATP + H_2O ⟶ ADP + Pi	15	20
2Pi ⟶ PPi + H_2O	20	0
PPi + H_2O ⟶ 2Pi	15	20
PPi + Pi* ⟶ PPi* + Pi	—	3
PPi* + ADP ⟶ Pi* + ADP∼P*	0	0

Chromatophores which were washed until both "ready-made" and "latent" adenylate kinase activities (see text and Fig. 3) became negligible were used. Those prepared from a wild type strain and from a blue-green mutant (G-9) produced essentially the same results. All activities were measured in terms of molar rate. The rates of various reactions are expressed as percentages of that of photosynthetic ATP formation. See ref. 4, 7, 8 for details.

a. 5 to 15 after sonication for 30 minutes (see Fig. 2).

b. See text.

but at an initial rate only 1/5 to 1/10 as fast as that of photosynthetic ATP formation.[8, 11] A strong activity for PPi-ase is observed with chromatophores repeatedly washed with the buffer solutions which have been currently used in the preparation. PPi-Pi exchange in darkness is detectable in chromatophores, but the level of activity is relatively low, only 1/4 to 1/5 as high in rate as that for ATP-Pi exchange in darkness.[8] When chromatophores are incubated in darkness with ADP and PPi, ATP formation is hardly detectable.[8] Therefore, it seems unlikely that PPi is converted to a high-energy phosphoryl intermediate leading to ATP formation with ADP in darkness.

The relative activities of these reactions are summarized in Table I.

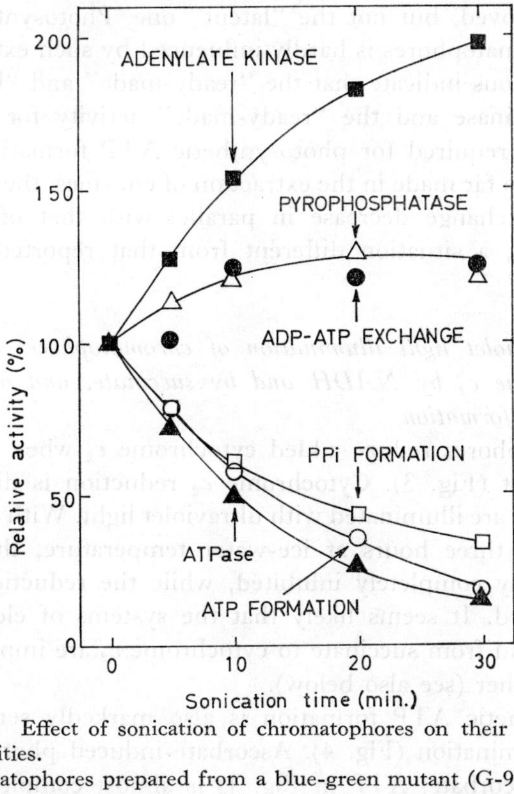

Fig. 2. Effect of sonication of chromatophores on their various enzymic activities.

Chromatophores prepared from a blue-green mutant (G-9) were used. See Fig. 1 for ATP formation. Other activities were assayed as follows: ATPase, by [^{32}P] Pi from γ-[^{32}P] ATP; PPi formation, by PPi from [^{32}P] Pi; adenylate kinase, by ATP from ADP; PPi-ase, by Pi from [^{32}P] PPi; ADP-ATP exchange, by [^{14}C] ATP converted from [^{14}C] ADP in the presence of ATP. Rate of PPi formation was measured in terms of moles of PPi formed. Reactions for ATP and PPi formation were carried out in the light, the others in the dark. See ref. 4, 7, 8 for details.

Effect of sonication of chromatophores on various enzyme activities

When chromatophores are sonicated, all the activities shown in Table I are influenced. As the sonication time is prolonged, the rates of photosynthetic ATP and PPi formation, as well as those of ATPase and of ATP-Pi exchange, are decreased, while those of adenylate kinase, PPi-ase and ADP-ATP exchange are increased (Fig. 2). Among these activities associated with chromatophores prepared by conventional methods, both adenylate kinase activities ("ready-made" and "latent", detectable before and after the sonication, respectively) are removed from the chromatophores by repeated washing with the buffer solutions. By the same washing procedure, the "ready-made" activity for ADP-ATP exchange is completely removed, but not the "latent" one. Photosynthetic ATP formation by chromatophores is hardly influenced by such extensive washing. These observations indicate that the "ready-made" and "latent" activities for adenylate kinase and the "ready-made" activity for ADP-ATP exchange are not required for photosynthetic ATP formation. In contrast, in all attempts so far made in the extraction of enzymes, the rates of ATPase and ATP-Pi exchange decrease in parallel with that of photosynthetic ATP formation, a situation different from that reported for mitochondria.[12,13]

Effect of ultraviolet light illumination of chromatophores on reductions of added cytochrome c_2 by NADH and by succinate, and on photosynthetic ATP and PPi formation

Chromatophores reduce added cytochrome c_2 when NADH or succinate is present (Fig. 3). Cytochrome c_2 reduction is diminished when chromatophores are illuminated with ultraviolet light. With chromatophores illuminated for three hours at ice-water temperature, the reduction by NADH is nearly completely inhibited, while the reduction by succinate is less influenced. It seems likely that the systems of electron transport from NADH and from succinate to cytochrome c_2 are impaired independently of each other (see also below).

Photosynthetic ATP formation is also markedly sensitive to ultraviolet light illumination (Fig. 4). Ascorbate-induced photosynthetic ATP formation (+ascorbate, ATP in Fig. 4) is almost completely annihilated by one hour illumination, but ascorbate-induced photosynthetic PPi formation (+ascorbate, PPi) is impaired to only one half even after three hours' illumination, indicating that the flow of electrons in the photosynthetic, cyclic electron transport system, which couples PPi formation, is

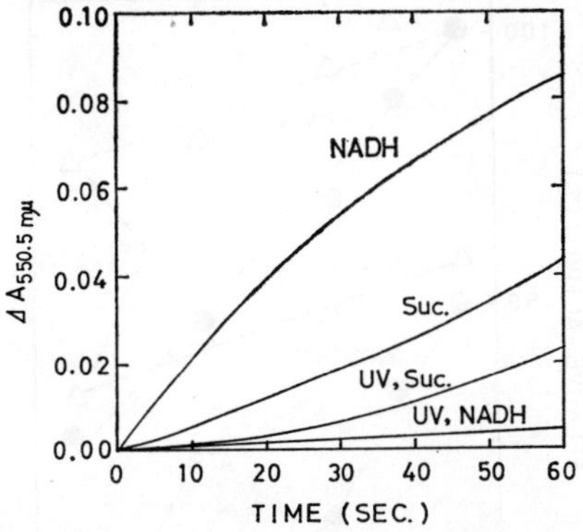

Fig. 3. Effect of ultraviolet light treatment of chromatophores on reduction of added cytochrome c_2 by NADH and succinate.

Chromatophores prepared from a blue-green mutant (G-9) were used. The standard components of the reaction mixture were as follows: 1.0 ml 0.1 *M* 3,3 dimethyl glutaric acid-Tris-amediol (pH 7.0) containing 50% sucrose, 0.1 ml $\times 10^{-4}M$ oxidized cytochrome c_2, 0.3 ml water and 0.05 ml chromatophores ($A_{873\ m\mu}$/ml=50). Reactions were started by adding 0.1 ml $10^{-2}M$ NADH or 1 *M* succinate to the sample cuvette and 0.1 ml water to the control cuvette, and carried out in the dark at room temperature. Reduction of cytochrome c_2 was followed by increase of absorption at 550.5 mμ with a Cary 14 R spectrophotometer. A chromatophore suspension was illuminated at 0°C for 3 hours with an ultraviolet lamp (Manasulu light, 2536 Å, Manasulu Chemical Industry Co. Ltd., Tokyo).[3]

hardly changed in rate after the one hour illumination. It seems likely, therefore, that ATP formation and at least most of the PPi formation differ in their coupling mechanisms and that only the former is damaged by the one hour illumination. Nishikawa[8] has reported that photosynthetic PPi formation is only partially inhibited by a simultaneous occurrence of photosynthetic ATP formation, the maximal inhibition ranging from 25 to 50%. Photosynthetic ATP (+PMS, ATP) and PPi (+PMS, PPi) formation are markedly less sensitive to ultraviolet light illumination when induced with phenazine methosulfate (PMS) than when induced with ascorbate. However, with PMS, photosynthetic ATP formation is again much more sensitive to it than photosynthetic PPi formation. These findings indicate that besides the coupling mechanism of ATP formation, the electron transport system is also damaged by ultraviolet light illumination for a prolonged period, and that PMS can bypass the damaged site of the electron transport system.

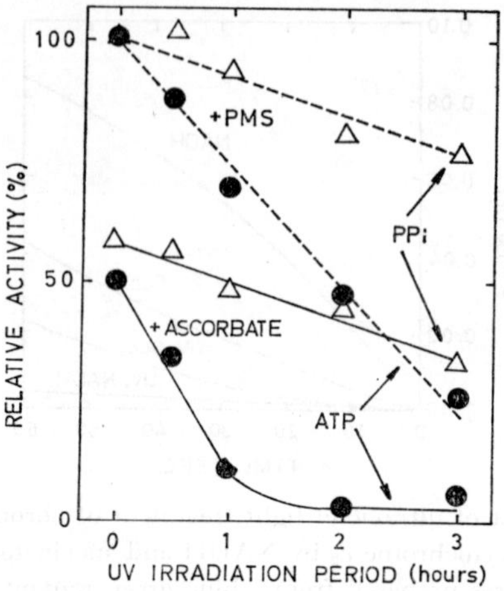

Fig. 4. Effect of ultraviolet light irradiation of chromatophores on photosynthetic ATP and PPi formation.

Chromatophores prepared from a blue-green mutant (G-9) were used. Reactions for ascorbate-induced (+ascorbate) and PMS-induced (+PMS) photosynthetic ATP and PPi formation were carried out in the presence of $6.7 \times 10^{-2} M$ ascorbate and $4 \times 10^{-4} M$ PMS, respectively. See Fig. 3 and ref. 8 for details.

The rates of ATPase and ATP-Pi exchange in the light and in the dark decrease in parallel with those of ATP formation.

Effect of oxidation-reduction potential on various activities associated with photosynthetic ATP formation

The chromatophores prepared by the grinding method still contain sufficient amounts of reductants to permit, without addition of redox-buffers, photosynthetic ATP formation occurring at a rate approximately 20% of the maximal value (Fig. 1). When DCPI is added to these chromatophores, with increasing concentrations of DCPI, the rate of photosynthetic ATP formation decreases together with those for ATP-Pi exchanges in the light and in the dark, while that for PPi-ase in the dark increases (Fig. 5). On the other hand, the rate of ATPase in the dark increases and then decreases, the maximum rate being obtained with approximately $6.7 \times 10^{-5} M$ DCPI. The uncoupler, 2,4-dinitrophenol (DNP) ($1 \times 10^{-3} M$), stimulates the rate of ATPase at DCPI concentrations lower than $6.7 \times 10^{-5} M$, but is without effect at higher concentrations. The rate of PPi-ase is also stimulated by DNP at DCPI concentrations lower than $6.7 \times 10^{-4} M$; the rates attained at all concentrations of DCPI tested in the presence of DNP are almost the same as that attained at $6.7 \times 10^{-4} M$

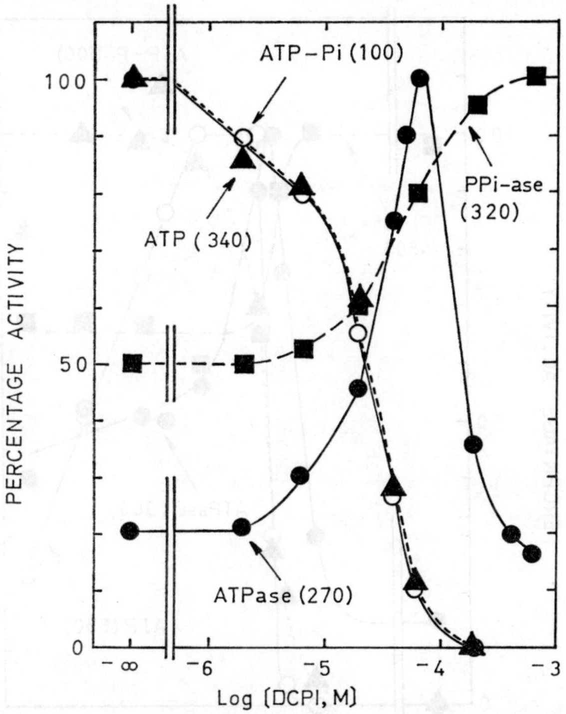

Fig. 5. Relation among various activities associated with photosynthetic ATP formation brought about by chromatophores.

Symbols used: ATP, photosynthetic ATP formation; ATP-Pi, ATP-Pi exchange. ATP-Pi, ATPase and PPi-ase represent the activities in darkness. The numerals in parentheses represent the maximum activities (mμ-mole/A_{880} mμ/hour) under the experimental conditions. See ref. 7, 8 for other experimental conditions.

DCPI in the absence of DNP. At DCPI concentrations lower than 6.7×10^{-5} *M*, the rate of ATPase is inhibited in the light, and at higher concentrations, it is stimulated, the maximum rate being almost the same for the reactions in the light and in the dark. Possibly, the inhibition and stimulation of ATPase activity in the light is caused by a photosynthetic shift of electron distribution in the cyclic electron transport system, thus producing, under particular conditions, "light-dependent" ATPase.[14] The experiment shown in Fig. 6 was carried out in the presence of 6.7×10^{-4} *M* DCPI, at which concentration photosynthetic ATP formation, ATP-Pi exchange and ATPase are almost completely inhibited (Fig. 5). When ascorbate is added, the activities depressed by the concentration of DCPI

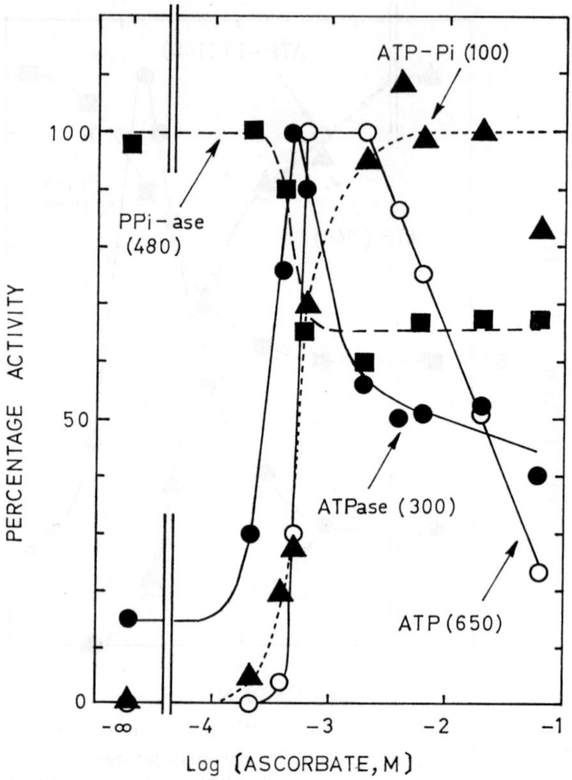

Fig. 6. Relation among various activities associated with photosynthetic ATP formation brought about by chromatophores.

All the reaction media for activity assay contained $6.7 \times 10^{-4} M$ DCPI. See ref. 4, 7, 8 for details. Symbols used, the same as for Fig. 5.

are restored (Fig. 6). The maximum rate of photosynthetic ATP formation is attained with approximately 1×10^{-3} *M* ascorbate in the presence of 6.7×10^{-4} *M* DCPI, which is equivalent to an E_h value of +0.13 volt. At higher concentrations of ascorbate, the rate of photosynthetic ATP formation decreases, while that of ATP-Pi exchange in darkness is scarcely influenced. The maximum rate of ATPase is attained with approximately 6.7×10^{-4} *M* ascorbate. DNP stimulates the rate of ATPase only when the activity has been inhibited by concentrations higher than 6.7×10^{-4} *M*.[7] It is conceivable that both ATPase and ATP-Pi exchange are exhibited by the same system, and that the system involves two adjacent oxidation-reduction components of the electron transport system, the oxidation-reduction reaction between which couples ATP formation. These hypothetical components are represented by "A" and "B" in Fig. 8 The system exhibits ATPase activity when one ("A") of the two components is in the oxidized form and the other ("B") is in the reduced form. DNP may substitute for "A" either in the oxidized or in the reduced form, reacting with "B". It may be noted that DNP is an excellent electron acceptor.[15] The system exhibits the ATP-Pi exchange activity when both "A" and "B" are in the reduced forms, and the PPi-ase activity when they are in the oxidized forms. In this consideration, the E_h-independent activities are ignored. Thus, DNP inhibits the ATP-Pi exchange which has been stimulated by high concentrations of ascorbate. The relationships of these activities as a function of E_h value of the reaction mixture are summarized in Fig. 8.

Oligomycin inhibits ATP-Pi exchange, ATPase and photosynthetic ATP formation in parallel: 3.3 μg/ml of oligomycin causes more than 90% inhibition of these activities. However, in the presence and absence of ADP, oligomycin inhibits neither the activity for photosynthetic PPi formation nor that for PPi-ase[8] (Table II).

TABLE II. Effect of several conditions on photosynthetic ATP and PPi formation and partial reactions caused by chromatophores.

Conditions	ATP formation (light)	ATP-Pi exchange	ATPase	ADP-ATP exchange	PPi formation (light)	PPi-ase
Light		↑	↓ ↑ +PMS	↑		↓
DNP	↓	L ↓ D ↓	L ↓ D ↓	L ↓ D →	↓	L ↑ D ↑
Oligomycin	↓	L ↓ D ↓	L ↓ D ↓	D →	→	L → D →
Pi			L ↑ D ↑			D ↓
As_i	↓	L → D ↓	L ↑ D ↑	L ↓ D →	↓	D ↓
Sonication	↓	L ↓ D ↓	L ↓ D ↓	L ↓ D ↑	↓	L ↑ D ↑

The arrows pointing up, down and horizontally represent stimulation, inhibition and noi nfluence, respectively. L and D show activities in the light and dark, respectively. (+PMS) shows the activity measured in the presence of PMS. See ref. 7, 8 for details.

It may be speculated that the E_h-dependent activities for ATP-Pi exchange and ATPase result from partial reactions of the energy conversion system leading to ATP formation, and that the E_h-dependent activity

TABLE III. Relations among various reactions related to photophosphorylation brought about by *R. rubrum* chromatophores.

E_h (volt)	ATP formation	ATP-Pi exchange	ATPase	PPi-ase
+0.18	0	7	31	(100)
+0.15	30	28	(100)	82
+0.12	(100)	93	55	60
+0.075	50	(100)	50	68

The numerals represent the percentage activities. All the reactions other than ATP formation were carried out in darkness. See ref. 7, 8 for details.

for PPi-ase results from a partial reaction of the PPi-forming system. The optimum E_h values are +0.08 volt or more negative for the ATP-Pi exchange activity, approximately +0.15 volt for the ATPase activity and +0.18 volt or more positive for the PPi-ase activity[7] (Table III). Thus, it seems likely that the two hypothetical components "A" and "B," carrying out an oxidation-reduction reaction leading to ATP formation, have the mid-potential (E_m) values ranging from 0 to +0.2 volt.

Oxidation-reduction components present in cells

The components shown in Table IV have been extracted and purified[9,16] from light-grown cells. All these components are extracted in water-soluble states by disrupting cells with a French pressure cell or by simple grinding procedures, and are separated from photophosphorylating chromatophores by means of centrifugation. With the chromatophores thus prepared, the maximum rate of photosynthetic ATP formation is influenced neither by further washing with the buffer solutions nor by addition of the isolated soluble components to the reaction mixture.[2] Together with the observations on the components present in chromatophores (Table V), this indicates that the components shown in Table IV are not constituents of the "internal" (associated with chromatophores) electron transport system coupling ATP formation. These components may constitute an "external" electron transport system(s), perhaps being located in the cytoplasmic fluid. It is probable that the internal and external electron transport systems are linked in function at sites such as cytochrome c_2 or cytochrome cc'; thus, the former is protected from attacks by molecular oxygen and maintains the level of electron density for maximum coupling.

Besides bacteriochlorophyll, the chromatophores prepared from the blue-green mutant (G-9) contain two different types of cytochromes,

TABLE IV. Components purified from *R. rubrum*.

	E_{mid} (volt)
Cytochrome c_2	+0.301
Cytochrome *cc'*	+0.02
Cytochrome *b*-557.5	−0.2
FMN-protein	−0.2
Ferredoxins:	
Non-PPNR	+0.2
PPNR	−0.3
Oxidoreductase:	
NADH: heme protein (FAD)[a]	
NADH: dye (FMN)	

a. The purified enzyme does not contain flavin. When the enzyme activity is measured in the presence of a mixture of FAD and FMN (7 : 2), the enzyme mostly binds FAD. It may be preferable to call this enzyme "NADH : FAD oxidoreductase." When the purified enzyme is added to chromatophores showing no NADH dehydrogenase activity, their activity for the reduction of cytochrome *C*-560 ($E_{m,7}$= −0.2 volt) is restored. When a flavin is also added, all of the cytochrome *C*-560 ($E_{m,7}$=+0.02 volt, also) is reduced. This indicates that the added enzyme catalyzes the reduction by NADH of only the flavin located close to cytochrome *C*-560 ($E_{m,7}$=−0.2 volt).

TABLE V. Components present in one chromatophore 600 Å in diameter.

Components	Number of molecules		$E_{m,7}$ (volt)
Bacteriochlorophyll	340		
Cytochrome *c*-552	10		+0.30
Cytochrome *C*-560	18		
Reducible by succinate		7[b]	+0.02
Reducible by NADH		7[b]	−0.2
Reducible in the light[a]		3[b]	
Ubiquinone-10	140		
Rhodoquinone	30		
Flavins, acid-extractable	9		
FAD		7	
FMN		2	

a. The photoreduction was measured in the presence of 3.3μg/ml of antimycin A and $6.7\times10^{-2}M$ ascorbate under illumination of light at 880 mμ. The extent of reduction is almost the same in the presence or absence of NADH and succinate, added to chromatophores sufficiently long before the addition of antimycin A that the reductions of cytochrome *C*-560 by succinate and NADH are at steady states.

b. These values vary from batch to batch of chromatophore preparations. The highest values so far obtained are shown. Ultraviolet light illumination is inhibitory on the photoreduction.

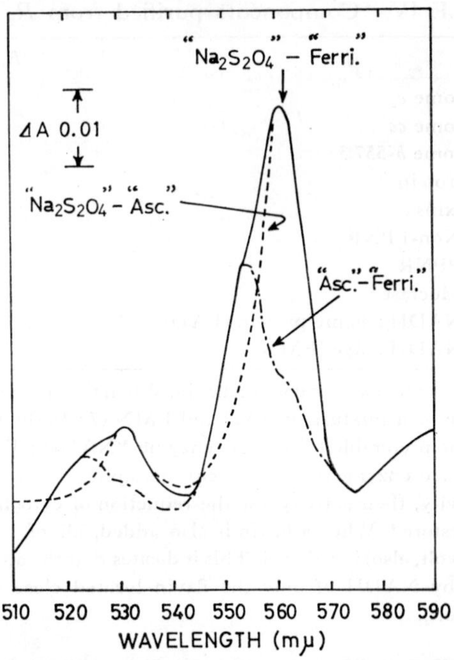

Fig. 7. Difference absorption spectrum of chromatophores.

Chromatophores prepared from a blue-green mutant (G-9) were used. The standard components of the mixture were as follows: 1.0 ml 0.1 *M* glycylglycine buffer (pH 8.0) containing 50% sucrose, 0.25 ml chromatophores (A_{873} mμ/ml=50) and water to make the total volume of 1.5 ml. "Ferri",+solid ferricyanide; "Asc," +0.1 ml 1 *M* ascorbate; "$Na_2S_2O_4$": + solid dithionite.

one showing a reduced α-peak at 552 mμ and the other at 560 mμ (Fig. 7). These cytochromes are firmly associated with chromatophores, so that their concentrations are not lowered by repeated washings with the buffer solutions. In chromatophores, the cytochrome with a reduced α-peak at 552 mμ (cytochrome *c*-552) has the $E_{m,7}$ value of +0.30 volt (n=1), a value that is almost the same as that for cytochrome c_2 (with a reduced α-peak at 550 mμ) (Table V). The other cytochrome with a reduced α-peak at 560 mμ shows a reduced γ-peak at 429–432 mμ, suggesting that this cytochrome may be of a *b*-type; however, its alkaline pyridine ferrohemochrome shows a reduced α-peak at 551 mμ and therefore we may tentatively call it "cytochrome *C*-560."

The cytochrome *C*-560 can be separated into two different kinds: With the use of proper preparations of chromatophores that are active with respect to the activities for NADH dehydrogenase and succinic dehydrogenase, approximately half of the cytochrome *C*-560 present can be reduced by the addition of succinate. The cytochrome *C*-560 reduced by succinate shows the $E_{m,7}$ value of $+0.02$ volt ($n=1$). Most of the rest of the cytochrome *C*-560 can be reduced by NADH, with $E_{m,7}=-0.2$ volt ($n=2$). Thus, most of the total cytochrome *C*-560 can be reduced by the addition of both NADH and succinate. Recently, Yamamoto (unpublished) in our laboratory extracted all these chromatophore-associated cytochromes with the aid of Triton X-100, cholate or Tween 20. Actively photophosphorylating chromatophores, which show no activity for photoreduction of NAD[3] and which are lacking in the activities for succinic dehydrogenase and NADH dehydrogenase, can be prepared. In addition, one of the ferredoxins (non-heme iron proteins), which shows the activity for photosynthetic pyridine nucleotide reduction (PPNR), has been extracted and purified only from fresh batches of light-grown cells but not from chromatophores. It is probable, therefore, that these activities are not primarily related to the reactions of photosynthetic ATP and PPi formation.

Ubiquinone-10 and rhodoquinone are extractable from lyophilized chromatophores with iso-octane. Okayama, Yamamoto *et al.*[17] have demonstrated that the resulting quinone-free chromatophores do not show any ascorbate-induced photosynthetic ATP formation, but when the same amount of ubiquinone-10 as that originally present (4 mμmoles/$A_{873\ m\mu}$), but not rhodoquinone, is added back, photosynthetic ATP formation is completely restored. The activity thus restored is sensitive to antimycin A to the same extent as the activity with non-treated chromatophores. These findings indicate that ubiquinone-10 is an essential component for the photosynthetic electron transport system coupling ATP formation, but not rhodoquinone. Kakuno *et al.* (unpublished) have shown that the reduction of chromatophore-associated cytochrome *C*-560 by NADH is inactivated when chromatophores are freed of the quinones, and that it is reactivated when ubiquinone-10 or rhodoquinone is added back. They have found also that antimycin A inhibits the reduction by NADH of cytochrome *c*-552 but not that of cytochrome *C*-560. These findings suggest an electron transport system from NADH to cytochrome *c*-552 as follows: NADH→(NADH dehydrognease)[16]→FAD[16]→quinone →cytochrome *C*-560 ($E_{m,7}=-0.2$ volt)→"antimycin A-sensitive site ($E_{m,7}$ $=+0.2$ volt)"[10]→cytochrome *c*-552.

If each chromatophore is a sphere of 600 Å in diameter with a density of 1.25 (ref. 18, 19), the number of molecules of these components in one chromatophore may be calculated (Table V).

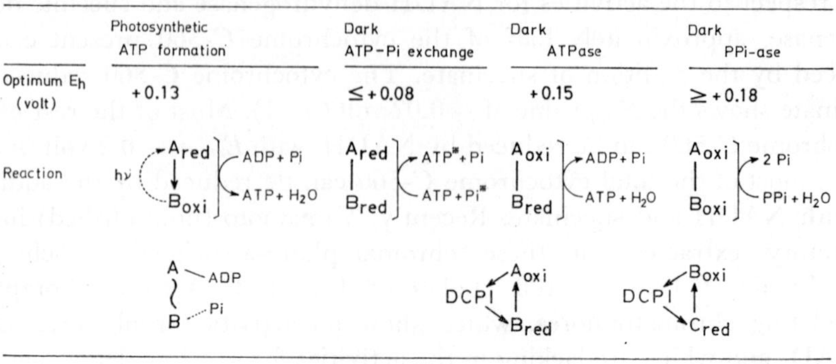

Fig. 8. Possible mechanism of coupling of ATP formation and partial reactions with electron transport system.

A, B and C represent successively adjacent oxidation-reduction components in the electron transport system coupling with the ATP-forming system; their E_m values are more positive in this order.

Possible relations among oxidation-reduction components and various partial reactions of the energy-conversion system

The possible relations among oxidation-reduction components associated with chromatophores and various reactions catalyzed by the energy-conversion system may be summarized as shown in Fig. 8. When 6.7×10^{-4} *M* DCPI and various concentrations of ascorbate are added to the reaction medium as a redox-buffer, the maximum rate of photosynthetic ATP formation is obtained at an E_h value around $+0.13$ volt[7] (Fig. 6). Apparently, this value is more closely related to the oxidation-reduction level of the internal electron transport system than the value previously reported with the use of ascorbate alone, approximately 0 volt.[2] With the redox-buffer, the rates of ATP-Pi exchange, ATPase and PPi-ase in darkness are maximal at $\leqq+0.08$ volt, $+0.15$ volt and $\geqq+0.18$ volt, respectively. Fig. 8 presents a possibility that the E_h-dependent activities for ATP-Pi exchange, ATPase and PPi-ase may be invoked by both "A" and "B" in the reduced forms, by "A" in the oxidized form and "B" in the reduced form, and by both "A" and "B" in the oxidized forms, respectively. It seems likely that DCPI produces a by-pass of electron flow from "A" and "B" when the E_h value is around $+0.15$ volt; thus, the uphill electron flow from "B" to "A," which has been driven at the expense of the hydrolysis of ADP$\sim$P, links in a cyclic fashion with the downhill by-pass of electron flow which can liberate the free energy in an uncoupled manner. The same type of speculation may be applied to the

E_h-dependent activity for PPi-ase, except that the uphill electron flow from "C" to "B" may take place at the expense of the hydrolysis of PPi.

Yamamoto *et al.* (unpublished) have demonstrated that the E_h-dependent activity for ATPase and its activity stimulated by DNP are depressed when chromatophores are freed of the quinones, and that these depressed activities are significantly restored when ubiquinone-10 or rhodoquinone is added back. Together with their findings that these ATPase activities are inactivated by ultraviolet light illumination but not by the addition of antimycin A, it seems likely that one of the quinones, ubiquinone-10, is either "A" or "B". On the other hand, Kakutani *et al.* (unpublished) found that the E_h-dependent ATP-Pi exchange activity is not influenced by the extraction of the quinones.

Arsenate (As_i) as well as Pi stimulates ATPase activity, while both reagents inhibit PPi-ase activity, when these activities are assayed based on the liberation of [^{32}P] Pi from γ-[^{32}P] ATP and from [^{32}P] PPi[8] (Table II). PPi also stimulates ATPase activity.[8] These findings suggest that either As_i or Pi is bound in a non-energized state on a specific site present in chromatophores, where PPi is also bound, and that the addition of ATP to chromatophores produces an energized state which allows the exchange of the terminal phosphate of the ATP molecule with As_i, or Pi. In addition, it was previously suggested that a high-energy intermediate was formed in photosynthetic ATP formation, and that its formation was dependent on ADP but not Pi.[5] It may be speculated that in ATPase activity, the adenine moiety of the ATP molecule is responsible for the binding of the molecule but the pyrophosphate bonds are not. Hypothetical binding sites of ADP and Pi leading to ATP formation are also shown in Fig. 8; the free energy primarily conserved as a high-energy intermediate complex between "A" and "B" is transferred to "A"-ADP, and then ATP is formed from the resulting "A"$\sim$ADP and the bound Pi.

Acknowledgements

The authors would like to express their thanks to Dr. Hiroshi Tamiya (Tokugawa Institute for Biological Research, Tokyo) and Dr. Robert G. Bartsch (Department of Chemistry, University of California, San Diego, La Jolla, Calif.) for their great help in the preparation of this paper.

These studies have been supported by U. S. Public Health Service Grant GM-09246 from the National Institutes of Health, Bethesda, Md., U.S.A.

REFERENCES

1. Frenkel, A. W., *J. Am. Chem. Soc.*, **76**, 5568 (1954).
2. Horio, T., and Kamen, M. D., *Biochemistry*, **1**, 144 (1962).
3. Horio, T., and Yamashita, J., in *Bacterial Photosynthesis* (H. Gest, A. San Pietro, and L. P. Vernon, eds.), p. 275, Antioch Press, Ohio, 1963.
4. Horio, T., Nishikawa, K., Katsumata, M., and Yamashita, J., *Biochim. Biophys. Acta*, **94**, 371 (1965).
5. Horio, T., Nishikawa, K., and Yamashita, J., *Biochem. J.*, **98**, 321 (1966).
6. Horio, T., Nishikawa, K., and Yamashita, J., *J. Biochem.*, **55**, 327 (1964).
7. Horio, T., Horiuchi, Y., and Nishikawa, K., *Biochim. Biophys. Acta,* in press.
8. Nishikawa, K., and Horio, T., *J. Biochem.*, in press.
9. Horio, T., and Kamen, M. D., *Biochim. Biophys. Acta*, **48**, 266 (1961).
10. Yamashita, J., Yoshimura, S., Matsuo, Y., and Horio, T., *Biochim. Biophys. Acta*, **143**, 154 (1967).
11. Horio, T., Stedingk, L. V., and Baltscheffsky, H., *Acta Chem. Scand.*, **20**, 1 (1966).
12. Lehninger, A. L., *The Mitochondrion*, W. A. Benjamin, Inc., New York, 1964.
13. Lehninger, A. L., *Bioenergetics*, W. A. Benjamin, Inc., New York, 1965.
14. Avron, M., *J. Biol. Chem.*, **237**, 2011 (1962).
15. Szent-Györgyi, A., *Introduction to a Submolecular Biology*, Academic Press, 1960.
16. Horio, T., Bartsch, R. G., and Kamen, M. D., in preparation.
17. Okayama, S., Yamamoto, N., Nishikawa, K., and Horio, T., submitted to *J. Biol. Chem.*
18. Oda, T., and Horio, T., *Exptl. Cell Res.*, **34**, 414 (1964).
19. Bergaron, J. A., in *The Photochemical Apparatus; Its Structure and Function*, p. 118, Brookhaven National Laboratory, 1959.

Chairman's Remarks

As has been the theme in the previous sections of this volume, this particular section on the regulatory aspects continues from the comparative point of view, with papers on bacteria, algae, higher plants, and higher plant chloroplasts. Both biochemical and biophysical approaches have been examined, and again, the striking impression is that here, in these well worked out photosynthetic systems such as the chromatophore, the defined and fully functional isolated chloroplast, and the well worked out metabolic pathways in all these organisms, are systems that lend themselves beautifully to examination into the very basic aspects of modern molecular and cellular problems. The problem of regulation of metabolism has all too often been examined in microbial and animal systems, with the primary emphasis on microbial systems. The use of photosynthetic structure and function to study the basic genetics and physical chemical aspects of regulation is, indeed, an exciting new avenue, not only to understand the mechanism of these basic activities of cells, but to further clarify at the molecular and cellular level the activities in photosynthesis.

This session at the end of the symposium was a relatively short one, and the papers speak for themselves. Dr. Horio gave a picture of some of the more basic aspects of the compounds concerned with photosynthetic electron transport in bacterial chromatophore systems. Also described are the remarks by Drs. Fuller and Anderson on the potential effects by the products of light on regulation of photosynthetic carbon metabolism in bacteria. Papers by Drs. Miyachi, Bassham, and Gibbs further elucidate the comparative aspects in higher systems. At the more molecular level, the sites of regulation in higher plant chloroplast systems in the Calvin cycle, and the various kinds of regulation that may affect the cycle, are described. Over-all cellular regulation of photosynthesis, including some further biochemical explanations of the Warburg Effect, is further elucidated by Dr. Gibbs and by Dr. Miyachi.

In the context of the over-all framework of this symposium, we have talked about molecular and structural aspects, physical chemical aspects, developmental aspects, and finally in this section, regulatory aspects. Our coverage of the regulatory aspects is by no means comprehensive or complete; rather, we have presented small selected examples. However, these papers indicate that these photosynthetic systems, in which all of us have been working so hard for so many years, offer opportunities for studying the basic molecular and cellular mechanisms of the cell that may, indeed, prove valuable additions to many other such efforts in microbial and animal systems.

R.C. Fuller

CONCLUDING REMARKS

Hans Gaffron

I have been given the honorable task of saying the proper words to conclude our lively four-day meeting. The following remarks cannot possibly be a studied evaluation of the proceedings. The last session, for instance, ended a few minutes ago. It introduced two contributions which happened to be quite new and of special interest to me. How should I balance the recent impressions with those of the first day, which I would have to reconstruct for the better part from notes? We are all too near to our own words to judge what we so enjoyed discussing with any hope of objectivity in the context of the entire research in the field. As often before on the occasion of such small special conferences, I for one regret that there is not time enough to go over the entire material again, from the first to the last paper, while the participants are still around.

There is a difference in purpose between a colossal Congress, like the one we shall attend next week, and this special conference. While the Congress provides a broad survey for all who wish to form a reasonable opinion of what is now the fashion in this or that branch of biochemistry, the small conference is a workshop, a clearing house for new and unfashionable ideas.

Only the truly good things arouse the desire for more. The sponsors of our meeting should take it as a compliment, therefore, and as a sign of success, when I wish there could be a second conference as soon as possible under the title "Hakone Revisited—Second Thoughts and Second Questions." Our sponsors expect the papers read at this meeting to be collected in a book. This is traditional, others ought to be able to know what was originally said here. But imagine the unusually stimulating kind of a book it would be if it contained not merely what has been said but mainly what thinking has been stimulated on account of what was said. Unfortunately the entire pattern of international science reporting as it has now developed will have to be changed until the dream of "Hakone Revisited," preferably in spring, will become reality.

Our little meeting has covered, of course, most of the accustomed ways to investigate photosynthesis. But we have also been shown aspects of the problem which only a few years ago were hardly thought of as being relevant to the theme. During the years past we, who have come from abroad, have tried faithfully to stay informed of the contents of the cream colored collection of papers from the Tokugawa Institute and the black-green copies of the Japanese Journal of Plant and Cell Physiology. Nevertheless we have been surprised and impressed by the variety and quality of research going on at so many places in Japan. And the little map which shows the location of the various laboratories is a nice reminder.

At the opening session Dr. Tamiya, speaking on the development of research on photosynthesis in Japan, said that 1933 was the year which saw the beginning of systematic investigations of photosynthesis in Japan. Japan was really not so far behind. At that time there were not quite two dozen scientists in the world who devoted their entire time to this problem. 1933 happened to be also the date of my first publication concerning living organisms, namely the purple bacteria, to which Otto Warburg had directed my attention in 1928. For two years or so I was under the illusion that I alone was investigating carbon dioxide assimilation in purple bacteria. This lasted until I met van Niel in 1931.

Just thirty-six years later each of the many partial problems of photosynthesis is being pursued in at least half a dozen laboratories. As our little meeting has also shown, the overlapping and sometimes the exact duplication of certain experiments is going so far that discussion questions often do not concern the understanding or interpretation of results, but just the technique or trick of how to obtain the same result more sharply, more elegantly, perhaps more convincingly—yet without bringing the underlying problem nearer to its solution. For this situation the very agencies, to whom we owe so much for vital and wonderful support, are partly to blame. I wish to plead with some of our government agencies to break the routine of expecting significant results within a year. Support of original work should be granted for several years with the stipulation that nothing be published as long as the grant is in force. Two thirds of that literature or even more, which consists only of progress reports, would thus automatically be condensed into a few publications of high quality.

The way governments understand scientific productivity is puzzling to those who actually produce science, and not only in my country. At the present time the Government of the United States supports a number of projects for growing microbes, algae, or some small higher plants in large quantities under controlled conditions to use them eventually as food for humans in outer space. The very first project of this kind, preceding these others by many years, was that of Professor Tamiya at the Tokugawa Institute. I came with the expectation to learn more about it and to see it in action. Now I have heard the Government in Japan saw no use in continuing Dr. Tamiya's mass culture experiments because they would not pay in terms of cheap food for humans for a foreseeable time. If one of these days the newspapers announce the feeding of astronauts by microbes and algae, we certainly shall remember the man who started all this and who in struggling to obtain reproducible yields discovered that even algae feel fresh and young in the morning and tired and saturated after a day's work in photosynthesis.

Fifteen years ago Professor Keilin asked me impatiently when, after spending so much time with that problem, I would finally come out with the solution. I felt very apologetic and found no better answer than to say that photosynthesis was a synthetic metabolic process and that the connection between energy source and living product in respiration, Keilin's specialty, was just as much unsolved.

Why, indeed, is photosynthesis such an intractable problem which keeps us busy year after year, particularly when what we want to know boils down to the solution of two simple questions? One is the detailed mechanism of phosphorylation, the other that of the oxidation of water. Photosynthesis shares the first of these two essential problems with all major synthetic processes in living cells. The second is unique to it. Only chlorophyll-containing chloroplasts photoxidize water. The tricks underlying these reactions are perhaps so hard to solve because they appear to be truly related to the evolution of life itself. The energetically less demanding one, the energy-trapping trick phosphorylation, may have preceded the oxidation of water by a billion years. From the point of view of thermodynamics, polymerization by heat or repetitions of the same photochemical charge separation step certainly suffice to accomplish the reaction. The true multi-quantum problem appears with the evolution of a mechanism to oxidize water. And it is for this reason that the two-step - two-quanta process in form of a coupled oxido-reduction system with overlapping potentials, has been in the center of our theoretical discussions for a number of years.

One of the surprises at this meeting was therefore the revival and reiteration of the obsolete idea that the absorption of one single quantum of visible light should be able to furnish all the free energy required to produce 1/4 free O_2 and one equivalent of reduced ferredoxin. What the present two-step hypothesis may need is perhaps an expansion, some additional assumptions such as an independent auxiliary phosphorylation system, to take care of a greater number of new observations, but certainly not a regression to a thermodynamically untenable and experimentally unsupported position.

With more time for discussion it might have been somewhat easier to answer what our friends are now bound to ask us: Can you say in a few words which problems have been settled at your meeting and what is the new trend in our research? With more time we might have been able to issue a clearer health bulletin for the two system dogma, and some interesting but perhaps not profound discrepancies between experimental observations of the same kind might have been either solved or clarified. That these matters have been left dangling may, perhaps, enhance the interest in the coming publication. The evidence will all be there, bound together. We shall need only to turn the pages back and forth instead of hunting for individual publications in separate journals. The question of the future trend of our research is relatively easy to answer. The trend is marked by attempts to isolate structural chloroplast components which not only retain some residual metabolic capacities, but which retain them unchanged and within the natural set of reaction partners.

Only a few years ago the greatest efforts were made to break down heavy particles and to solubilize the structure bound enzymes in the hope of having the principal reactions proceed in homogeneous solutions. Then came the time of the "double beam differential flash analyzing recording spectrograph" which permits us to amass a steadily increasing amount of data on the kinetics of intermediate reactions in extracts and to compare them with spectral changes in whole cells.

Recently the ever more refined techniques of electron microscopy showed that the method of making extracts by breaking down the structures very likely falsified whatever answers one might be getting, for the enzymes themselves were obviously not merely attached to, but truly parts of, the structure and could be expected to function naturally only when not dislodged from their original positions. Characteristic for our meeting have been the descriptions of how to isolate particles which preserve the right arrangement of their enzymatic components. The familiar larger chloroplast fragments, which could mainly do either all or nothing, are being gingerly cleaved into smaller structural units which hopefully embody a recognizable set of partial reaction sequences as required by our theoretical picture. The coming task is to combine different types of particles until they exhibit again the properties of the entire chloroplast. Any success of this kind ought to become visible also in the electron microscope as a kind of reconstruction of formerly divided packages of thylakoids.

It seems therefore that despite the fact that the heart of our problem remains wedded to photochemistry, we shall not be able to break away altogether from the current progress in general enzyme research. Photosynthesis is perhaps more structure bound than any other synthetic reaction.

We are still far away from the study of entirely artificial functional membranes. For the time being we must be content to alter living structures and to observe the concomitant changes in reaction patterns. The favorite objects are now plants either in various stages of development or in various mutated forms. In this way the already long subject index pertinent to studies in photosynthesis will acquire legitimately some terms which until yesterday referred only to problems in genetics. Just as the originally isolated problem of malignant growth has spread to cover virtually all cell physiology, the study of photosynthesis has emerged as an important aspect of research on the origin and evolution of life.

The international, or rather supranational, character that science brings with it is that within the confines of a conference room and during the discussion it is difficult to say in which particular part of the world the meeting is being held. But each time these past days when we stepped out onto the adjacent balconies, we who were never before in Japan realized what a great educational and cultural bonus we were receiving by having been able to attend this meeting. There is hardly an educated man on earth who has not heard of Mount Fuji, the landscapes and the refined gardens, and the architectural monuments of old Japan. To this we must add the unexpected mastery of tall modern buildings in a country shaken often by earthquakes. All these things, superficially familiar from calendars and travel posters, have impressed us beyond the power of our eager imagination. But thanks to the colleagues who are our hosts and who selected this quiet beautiful spot, we have been privileged to discover for ourselves small but very characteristic traits of Japanese living. Each one of us will probably take home some impression nobody told us about before and which we shall think of as specific and symbolic for the charming facets of life in Japan.

Here are two examples of what I mean. The porter at Odawara Station insisted most graciously that I board the wrong bus. Apparently I had put more emphasis on the word "Hakone" than "Kanko". But it turned out to be the best bus I could possibly have chosen. I was even sorry to leave it when the time came. The conductor was a pretty girl who between stops sang to herself a most charming little song. She repeated it with variations while the bus wound its way from village to village up a deep ravine. Outside Japan young and pretty bus conductors are rare anywhere in the world, but imagine one in the United States who would sing while on duty.

The other example is also musical. On television or in a music school you may have listened to lectures on the many types of percussion instruments which are used in Japan. One quiet afternoon this week I was therefore not too surprised to hear through the open window a sequence of peculiar ringing sounds. Obviously some people in the garden were practising modern music. To see which kind of instrument they used I looked out and saw a group of five or six girls coming down the stone-paved path along the rear of the garden. I am still wondering how they managed to have their wooden shoes properly tuned.

Indelible impressions of this kind we also owe to our hosts because they arranged so skillfully to have this meeting here and not in a big city. We, your international guests—and though this was specifically a Japanese-American undertaking we have two delegates from Australia and one from Israel among us, each counting intellectually for ten; thus we are a strongly international group—we wish to express our gratitude to you for this successful meeting two times over—first for having surprised us by the unexpected variety and intensity of your scientific presentations, second, for the opportunity to appreciate a very old but to us so new culture in the country where it developed during fifteen hundred and more years. One final word: together we did not solve the problem of photosynthesis—quite. We must therefore look forward cheerfully to our next meeting. May it be as fruitful as this one has been.

Participants in the Conference

1) Arnon, D. I.
Department of Cell Physiology,
University of California,
Berkeley, California 94720.

2) Avron, M.
Biochemistry Department,
Weizmann Institute of Science,
Rehovoth, Israel.

3) Bartsch, R. G.
Department of Chemistry,
University of California San Diego,
La Jolla, California 92037.

4) Bassham, J. A.
Lawrence Radiation Laboratory,
University of California,
Berkeley, California 94720.

5) Benson, A. A.
Scripps Institution
of Oceanography,
University of California San Diego,
La Jolla, California 92037.

6) Boardman, N. K.
C.S.I.R.O.,
Division of Plant Industry,
P.O. Box 109, Canberra, A.C.T.,
Australia.

7) Bogorad, L.
The Biological Laboratories,
Harvard University,
Cambridge, Massachusetts 02138.

8) Butler, W. L.
Department of Biology,
Revelle College,
University of California San Diego,
La Jolla, California 92037.

9) Chiba, Y.
Department of Biology,
Faculty of Science,
Kyushu University,
Fukuoka.

10) Cost, K.
Department of Botany,
College of Biological Sciences,
University of Minnesota,
Minneapolis, Minnesota 55455.

11) French, C. S.
Carnegie Institution of Washington,
Department of Plant Biology,
Stanford, California.

12) Frenkel, A. W.
Department of Botany,
College of Biological Sciences,
University of Minnesota,
Minneapolis, Minnesota 55455.

13) Fujita, Y.
Institute of Applied Microbiology,
University of Tokyo,
Hongo, Bunkyoku,
Tokyo.

14) Fuller, R. C.
Oak Ridge Graduate School
of Biomedical Sciences,
Oak Ridge National Laboratory,
P.O. Box Y, Oak Ridge,
Tennessee 37830.

15) Gaffron, H.
Institute of Molecular Biophysics,
Florida State University,
Tallahassee, Florida 32306.

16) Gibbs, M.
Department of Biology,
Brandeis University,
Waltham, Massachusetts 02154.

17) Gibson, J.
Microbiology Section,
Division of Biological Sciences,
Cornell University,
Itaca, New York 14850.

18) Hase, E.
Institute of Applied Microbiology,
University of Tokyo,
Hongo, Bunkyoku,
Tokyo (Postal, Address), and
Tokugawa Institute
for Biological Research,
Mejiromachi, Toshimaku, Tokyo.

19) Hattori, A.
Institute of Applied Microbiology,
University of Tokyo,
Hongo, Bunkyoku, Tokyo.

20) Hess, E. L.
Molecular Biology Section,
Division of Biological and Medical Sciences,
National Science Foundation,
Washington, D.C. 20550

21) Hildreth, W. W.
Johnson Research Foundation,
University of Pennsylvania,
Philadelphia, Pennsylvania 19104.

22) Hodge, W. H.
National Science Foundation,
American Embassy,
Akasaka, Minatoku, Tokyo.

23) Horio, T.
Division of Enzymology,
Institute for Protein Research,
Osaka University,
Kitaku, Osaka.

24) Huzisige, H.
Department of Biology,
Faculty of Science,
Okayama University,
Okayama.

25) Ishida, J.
Department of Zoology,
University of Tokyo,
Hongo, Bunkyoku, Tokyo.

26) Itoh, M.
The Tokugawa Institute for Biological Research,
Meijiromachi, Toshimaku,
Tokyo (Postal Address), and
Tokyo Institute of Technology,
Ookayama, Meguroku, Tokyo.

27) Iwamura, T.
The Tokugawa Institute for Biological Research,
Meijiromachi, Toshimaku,
Tokyo (Postal Address), and
The Institute of Applied Microbiology,
University of Tokyo, Tokyo.

28) Izawa, S.
Department of Biology,
Queen's University,
Kingston, Ontario, Canada.

29) Jagendorf, A. T.
Section of Genetics, Development and Physiology,
Division of Biological Sciences,
Cornell University,
Ithaca, New York 14850.

30) Kanai, R.
Department of Biochemistry,
Saitama University,
Urawa, Saitamaken.

31) Katoh, S.
Department of Biophysics and Biochemistry,
Faculty of Science,
University of Tokyo,
Hongo, Bunkyoku, Tokyo.

32) Kikuchi, G.
Department of Biochemistry,
Tohoku University,
School of Medicine,
Sendai.

33) Mauzerall, D.
Rockefeller University,
New York, New York 10021.

34) Mitsui, A.
Department of Biochemistry,
Medical School,
Yokohama City University,
Urahunecho, Yokohama.

35) Miyachi, S.
Institute of Applied Microbiology,
University of Tokyo,
Hongo, Bunkyoku,
Tokyo (Postal Address), and
The Tokugawa Institute for Biological Research,
Mejiromachi, Toshimaku, Tokyo.

36) Miyayama, H.
Ministry of Education,
Kasumigaseki, Chiyodaku, Tokyo.

37) Morita, S.
Department of Biophysics and Biochemistry,
Faculty of Science,
University of Tokyo,
Hongo, Bunkyoku, Tokyo.

38) Moudrianakis, E. N.
Biology Department,
The Johns Hopkins University,
Baltimore, Maryland 21218.

39) Mukohata, Y.
Department of Biology,
Faculty of Science,
Osaka University,
Toyonaka, Osaka.

40) Murakami, S.
Department of Biology,
University of Tokyo,
Komaba, Meguroku, Tokyo.

41) Nakayama, H.
Japanese Society
for the Promotion of Science,
Kanda-Hitotsubashi, Chiyodaku,
Tokyo.

42) Nishida, K.
Botanical Institute,
Faculty of Science,
Kanazawa University,
Kanazawa.

43) Nishimura, M.
Johnson Research Foundation,
University of Pennsylvania,
Philadelphia, Pennsylvania 19104.

44) Ogawa, T.
The Tokugawa Institute
for Biological Research,
Meijiromachi, Toshimaku,
Tokyo (Postal Address), and
The Tokyo Institute of Technology,
Ookayama, Meguroku, Tokyo.

45) Oh-hama, T.
Institute of Applied Microbiology,
University of Tokyo,
Hongo, Bunkyoku,
Tokyo (Postal Address), and
The Tokugawa Institute
for Biological Research,
Mejiromachi, Toshimaku, Tokyo.

46) Park, R. B.
Botany Department,
University of California,
Berkeley, California 94720.

47) Pon, N. G.
Department of Biochemistry,
College of Agriculture,
University of California,
Reverside, California 92502.

48) Ries, R. R.
National Science Foundation,
American Embassy,
Akasaka, Minatoku, Tokyo.

49) Sakurai, H.
Department of Biology,
School of Education,
Waseda University,
Totsuka, Shinjukuku, Tokyo.

50) San Pietro, A.
Charles F. Kettering
Research Laboratory,
Yellow Springs, Ohio 45387.

51) Sasa, T.
Department of Biology,
Faculty of Science,
Kyushu University,
Fukuoka.

52) Shibata, K.
The Tokugawa Institute
for Biological Research,
Mejiromachi, Toshimaku,
Tokyo (Postal Address), and
Tokyo Institute of Technology,
Ookayama, Meguroku, Tokyo.

53) Smillie, R. M.
Plant Physiology Unit,
Botany Department,
University of Sydney,
Sydney, Australia.

54) Takamiya, A.
Department of Biophysics
and Biochemistry,
Faculty of Science,
University of Tokyo,
Hongo, Bunkyoku, Tokyo.

55) Tamiya, H.
The Tokugawa Institute
for Biological Research,
Meijiromachi, Toshimaku, Tokyo.

56) Vernon, L. P.
Charles F. Kettering
Research Laboratory,
Yellow Springs, Ohio 45387.

57) Weaver, E. C.
Exobiology Division,
Ames Research Center,
National Aeronautics and
Space Administration,
Moffett Field, California 94035.

58) Yakushiji, E.
Faculty of Science,
Toho University,
Narashino, Chibaken.

AUTHOR INDEX

A

B

C

D

H

I

J

K

L

M

N

O

T

U